Combinatorial Inference in Geometric Data Analysis

Chapman & Hall/CRC
Computer Science and Data Analysis Series

SERIES EDITORS
David Blei, Princeton University
David Madigan, Rutgers University
Marina Meila, University of Washington
Fionn Murtagh, Royal Holloway, University of London

Recently Published Titles:

Combinatorial Inference in Geometric Data Analysis
Brigitte Le Roux, Solène Bienaise, Jean-Luc Durand

Time Series Clustering and Classification
Elizabeth Ann Maharaj, Pierpaolo D'Urso, Jorge Caiado

Bayesian Regression Modeling with INLA
Xiaofeng Wang, Yu Yue Ryan, Julian J. Faraway

Chain Event Graphs
Rodrigo A. Collazo, Christiane Goergen, Jim Q. Smith

Data Science Foundations
Geometry and Topology of Complex Hierarchic Systems and Big Data Analytics
Fionn Murtagh

Exploratory Data Analysis with MATLAB
Wendy L. Martinez, Angel R. Martinez, Jeffrey Solka

Microarray Image Analysis: An Algorithmic Approach
Karl Fraser, Zidong Wang, Xiaohui Liu

Introduction to Data Technologies
Paul Murrell

Exploratory Multivariate Analysis by Example Using R
Francois Husson, Sebastien Le, Jérôme Pagès

Music Data Analysis: Foundations and Applications
Claus Weihs, Dietmar Jannach, Igor Vatolkin, Guenter Rudolph

Computational Statistics Handbook with MATLAB
Wendy L. Martinez, Angel R. Martinez

Statistics in MATLAB: A Primer
MoonJung Cho, Wendy L. Martinez

For more information about this series, please visit:
https://www.crcpress.com/Chapman--HallCRC-Computer-Science--Data-Analysis/
book-series/CHCOSCDAANA

Combinatorial Inference in Geometric Data Analysis

Brigitte Le Roux

Solène Bienaise

Jean-Luc Durand

CRC Press
Taylor & Francis Group
Boca Raton London New York

CRC Press is an imprint of the
Taylor & Francis Group, an **informa** business

CRC Press
Taylor & Francis Group
6000 Broken Sound Parkway NW, Suite 300
Boca Raton, FL 33487-2742

First issued in paperback 2021

Version Date: 20190201

ISBN-13: 978-1-03-209373-4 (pbk)
ISBN-13: 978-1-4987-8161-9 (hbk)

Library of Congress Cataloging-in-Publication Data

Names: Le Roux, Brigitte, author. | Bienaise, Soláene, 1986- author. |
Durand, Jean-Luc (Mathematician), author.
Title: Combinatorial inference in geometric data analysis / Brigitte Le Roux,
Soláene Bienaise, Jean-Luc Durand.
Description: Boca Raton, Florida : CRC Press, [2019] | Series: Chapman &
Hall/CRC computer science and data analysis series | Includes
bibliographical references and index.
Identifiers: LCCN 2018052973| ISBN 9781498781619 (hardback : alk. paper) |
ISBN 9781315155289 (ebook)
Subjects: LCSH: Geometric analysis. | Combinatorial analysis. | Statistics.
Classification: LCC QA360 .L37 2019 | DDC 515/.9--dc23
LC record available at https://lccn.loc.gov/2018052973

Visit the Taylor & Francis Web site at
http://www.taylorandfrancis.com

and the CRC Press Web site at
http://www.crcpress.com

Contents

Preface

The main purpose of the book is to face the issue of statistical inference in Geometric Data Analysis (GDA) and, more generally, in Exploratory Data Analysis (EDA), which is usually not related to hypothesis testing, and for which classical assumptions on the distribution of observations—such as normal distributions, random sampling, etc.—are not valid.

The real question about statistical inference is not whether or not it can be used in GDA—it certainly can, as reflected, e.g., in Principal Component Analysis (PCA)—but when and how to use it in a fruitful way. Our point is that the mechanical application of the "fit-&-test" technique, so widely spread today, is not especially fruitful. In our opinion, statistical inference should be conducted as an *Inductive Data Analysis* (IDA), which is a natural extension of descriptive data analysis (Rouanet et al., 1990, 1998)[1]. In fact, there is a tradition of inference, in the line of Fisherian thinking, that is still very much alive in laboratory practice—even though, it must be said, it is rather belittled in statistical teaching (in spite of Tukey's valiant efforts). In this researcher's tradition, significance testing is the natural tool for *extending the descriptive conclusion of existence of effects*. The first thing to do, we submit, should be to enforce this tradition of inference in research.

Now, a casual look at research studies involving GDA shows that statistical inference is scarcely used beyond the routine χ^2-test. But, in many studies, researchers would like to substantiate the descriptive conclusions obtained by GDA methods. For instance, suppose that it is found that the mean point of a group is far from the centre of the cloud; this finding triggers the following natural query from the researcher (informally stated): "Is the observed deviation a genuine one, or might it be due to chance?". To cope with this query, the usual sampling models, with their drastic assumptions, are simply not appropriate. *Combinatorial methods*, based on permutation tests that are distribution-free and not restricted by any assumptions about population, are the most in harmony with IDA.

In light of this, we began to develop methods of combinatorial inference in GDA: see the research article by Le Roux (1998), then Chapter 8 of the book by Le Roux and Rouanet (2004). But at that time, the practice of permutation tests was hindered by computational obstacles (computers lacked both the speed and the storage for calculating exact permutation tests), therefore,

---

[1] In fact, IDA has been a long-standing (even although non-dominant) part of statistical inference, from Fisher to Tukey, in strong opposition to the Neyman–Pearson school.

we used approximate methods. Nowadays, the exact tests, using exhaustive method whenever possible or Monte Carlo method if not, can be performed, hence approximate solutions only serve to get an order of magnitude of results. Thus, it was clear to us that an in-depth study of permutation tests applied to GDA ought to be done. It was the subject of the PhD dissertation of Solène Bienaise (2013). In this book, we outline and develop the ideas she presented in her PhD dissertation.

It should be stressed that these methods apply not only to clouds produced by a GDA method but more generally to any *Euclidean cloud*. As a result, in the line of pure geometry, we prove general properties, which orient the *choice of test statistic*, lead us to define *compatibility regions* and, last but not least, permit to *optimize the computational time*.

To support our work, we developed computer programs using the R statistical language and the menu-driven Coheris Analytics SPAD software that has an internal method to execute R programs in a user-friendly way[2].

Organisation of the book (see Table of Contents)

— *Chapter 1* provides an *introduction* to the issue of statistical inference in GDA.

— *Chapter 2* is devoted to *clouds of points in a geometric space*. Basic descriptive statistics are outlined. The covariance structure of a cloud and the theory of principal axes are introduced, and then the Mahalanobis distance and the principal ellipsoids of a cloud are defined. Lastly, the partition of a cloud with the between- and the within-classes decomposition is presented.

— *Chapter 3* introduces the principle of permutation testing. Then it deals with the *combinatorial typicality test* for mean point and for variance of a cloud with respect to a reference cloud, complemented by compatibility regions. Then, the combinatorial methods applied to GDA methods are studied.

— *Chapter 4* is devoted to the *geometric typicality test* answering the question of comparison of the mean point of a cloud to a reference point. Then, the privileged case of a design with two repeated measures is studied. Finally a brief overview of some other methods is outlined.

— *Chapter 5* develops *homogeneity permutation tests*, that is, the comparison of the mean points of several subclouds; two structures of data are studied: independent groups design and repeated measures design.

— *Chapter 6* (*Research Case Studies*) is nearly a book in itself. It presents the methodology in action with four extensive applications, one from medicine,

---

[2]The R scripts, together with the data sets used in this work, are available from the authors and on the first author's website.

one from political science, one from sociology and one from cognitive science. In each case, detailed information on how the analyses using our R scripts interfaced with SPAD can be carried out is provided.

— *Chapter 7* gathers, in a compact way, the *mathematical bases*, that is, the mathematical background necessary for reading this book.

We have tried to allow for a "non-linear reading" of the book, at the expense of some redundancy. Indeed, after reading the introductory chapter, the reader will perhaps wish to have a look at Chapter 6 (Research Case Studies), before turning to technical chapters.

The *mathematical background* mainly consists in *abstract linear algebra* (the short overview given in Chapter 7 should make the book self-contained).

The *statistical background* consists in elementary multivariate descriptive statistics: means, variances, covariances, correlation and regression coefficients[3]. The properties of a *cloud of points* (the central object of the book) are recalled in Chapter 2, making the book self-contained in this respect too. However, some knowledge of GDA *methods*[4], especially Principal Component Analysis and Multiple Correspondence Analysis, are used in Chapter 6.

About the authors

BRIGITTE LE ROUX is associate researcher at Laboratoire de Mathématiques Appliquées (MAP5/CNRS) of Paris Descartes University and at the political research centre of Sciences Po Paris (CEVIPOF/CNRS). She completed her doctoral dissertation in applied mathematics at the Faculté des Sciences de Paris in 1970; this work was supervised by Jean-Paul Benzécri.

e-mail: Brigitte.LeRoux@mi.parisdescartes.fr

website homepage: www.mi.parisdescartes.fr/∼lerb/

SOLÈNE BIENAISE is data scientist. She completed her doctoral dissertation in applied mathematics in 2013 at the Paris-Dauphine University, under the direction of Pierre Cazes and Brigitte Le Roux.

e-mail: bienaise.solene@hotmail.fr

JEAN-LUC DURAND is associate professor at the Psychology department and researcher at LEEC (Laboratoire d'Éthologie Expérimentale et Comparée) of Paris 13 University. He completed his doctoral dissertation in psychology at Paris Descartes University in 1989, supervised by Henry Rouanet.

e-mail: jean-luc.durand@univ-paris13.fr

---

[3] As a prerequisite, we recommend the introductory book by Freedman et al. (1991).

[4] An elementary presentation of GDA methods can be found in Le Roux and Rouanet (2010) and an extensive one in Le Roux and Rouanet (2004).

Acknowledgements

We are grateful to our mathematician colleague Pierre Cazes (Paris-Dauphine University) who scrutinized the entire manuscript, and to Fionn Murtagh (University of Huddersfield) for many helpful discussions.

We also wish to thank, for their remarks, Philippe Bonnet (Paris Descartes University) and above all Frédéric Lebaron (École normale supérieure Paris-Saclay, formerly ENS Cachan) who shared his data with us.

Special thanks are due to Karine Debbasch (Language Centre of Paris Descartes University), for assistance in English.

Last but not least, we thank Rob Calver and Lara Spieker of CRC Press, for their helpful attention and friendly patience.

BRIGITTE LE ROUX
SOLÈNE BIENAISE
JEAN-LUC DURAND

Symbols

Symbol Description

$\circ$	composition of maps	M.Pcov	M-covariance endomorphism of permutation P-cloud
$\langle \cdot \vert \cdot \rangle$	scalar product		
$\Vert \cdot \Vert$	Euclidean norm	n	cardinality of the set I
$[\cdot \vert \cdot]$	M-scalar product	n_i	absolute frequency of i
$\vert \cdot \vert$	M-norm (Mahalanobis norm)	$\mathcal{N}(0,1)$	standard normal distribution
$\vert \cdot \vert_P$	generalized Mahalanobis norm with respect to point P	P^J	permutation P-cloud
		$p(D\!=\!d)$	proportion of the values of variable D that are equal to d
$<\!\cdot\!>$	nesting symbol		
$\sim$	distributed as	p	p-value
		$\widetilde{p}$	approximate p-value
α	reference level	$\mathbb{R}$	set of real numbers
χ^2_L	χ^2 with L d.f.	$\mathcal{U}$	geometric space
δ^j_i	$= 1$ if i belongs to sample j and 0 if not	$\mathcal{V}$	vector space
		V_{cloud}	variance of cloud
D^2	squared M-distance statistic	V_M	M-variance statistic
η^2	eta-squared coefficient	Var_C	between-C variance
f_i	relative frequency of i	$\text{Var}_{I(C)}$	within-C variance
G	mean point of a cloud	$\mathcal{W}_{r,n-1}(\boldsymbol{\Sigma})$	$\boldsymbol{\Sigma}$-scaled r-dimensional Wishart with r and $n-1$ d.f.
Gcov	covariance endomorphism of permutation G-cloud		
		y^i_ℓ	principal coordinate of point M^i on principal axis ℓ
H^J	sample cloud		
Hcov	covariance endomorphism of sample cloud H^J	z^i_ℓ	standardized coordinate of point M^i on principal axis ℓ
I	set-theoretic support of cloud		
$I\!<\!C\!>$	partition of I into C classes		

Abbreviations

CA	Correspondence Analysis
d.f.	degrees of freedom
EDA	Exploratory Data Analysis
GDA	Geometric Data Analysis
IDA	Inductive Data Analysis
i.i.d.	independent and identically distributed
MANOVA	Multivariate Analysis of Variance
MC	Monte Carlo
MCA	Multiple Correspondence Analysis
MDS	Multidimensional Scaling
PCA	Principal Component Analysis

J	sample set	
$\mathcal{J}$	sample space	
$\mathcal{L}$	direction vector space of affine support of cloud M^I	
L	dimension of cloud M^I	
λ_ℓ	eigenvalue *or* variance of axis ℓ	
M^I	cloud of points $(M^i)_{i \in I}$	
$\mathcal{M}$	affine support of cloud M^I	
Mcov	covariance endomorphism of cloud M^I	
Mcov^P	generalized covariance endomorphism of M^I with respect to P	

1

Introduction

Today, exploratory and confirmatory
can—and should—proceed side by side[1].
John W. Tukey

In the present book, we outline statistical inference procedures for Geometric Data Analysis (GDA) that are based on a combinatorial framework and that highlight the role of *permutation tests*. The methods we develop relate mainly to studying a *Euclidean cloud*, that is, a family of statistical observations conceptualized as points in a multidimensional space.

1.1 On Combinatorial Inference

Permutation tests belong to the set of resampling methods, so called because the data are resampled and reexamined repeatedly in order to obtain results. They are data-dependent, in that all the information requested for analysis is contained within the observed data. No assumptions regarding the distribution of the data are required. Because permutation tests are computationally intensive, it took the advent of powerful computers to make them practical and, thus, they have been really used only recently.

Permutation tests were initiated by Fisher (1935) and Pitman (1937), then further developed by Romano (1989); Edgington (2007); Pesarin and Salmaso (2010); Good (2011) and others[2].

Performing a permutation test

The steps for performing a permutation test are the following.

1. Outline the *effect of interest*, then define the *group of permutations* of observations that is consistent with the absence of effect ("null hypothesis").

   > For instance, let us consider the issue of the homogeneity of two groups of observations g_1 and g_2 with respectively n_1 and n_2 observations. Saying that the two groups are homogeneous amounts to

---

[1] Tukey (1977, p. VII).

[2] For the historical development of statistical permutation tests, see Berry et al. (2014).

saying that the subdivision into groups may be ignored, then we pool the two groups of observations. The permutation group consists in reallocating n_1 observations of the pooled group to g_1 and the others to g_2 in all possible ways. Thus, it comprises $\frac{(n_1+n_2)!}{n_1! \times n_2!}$ possible rearrangements of the $n_1 + n_2$ observations.

2. Choose a suitable *test statistic* and compute it for the observed data.

   In general, the test statistics we choose will be based on *mean points* and will be dependent on the *covariance structure* of the reference cloud. Their properties will be studied in detail, and thus calculation algorithms become fairly fast.

3. Determine the *permutation distribution* of the test statistic by calculating the values of the test statistic for all possible rearrangements or for a large sample thereof.

4. Determine the *p-value* by locating the observed value of the test statistic on the permutation distribution, that is, calculate the proportion of rearrangements for which test statistic values are more extreme than or as extreme as the observed one[3].

Permutation tests generally consist of two types: exact and approximate tests.

In an *exact test*, the permutation distribution of the test statistic is computed for all possible rearrangements of observations (*exhaustive method*), or, when the number of permutations is too large, for a random subset of all possible rearrangements (*Monte Carlo method*)[4].

For large data sets, whenever possible, an *approximate test*, which consists in replacing the discrete distribution of the test statistic by a *classical distribution* (normal distribution, χ^2-distribution, etc.), may be performed; the result of the approximate method is obtained with very few calculations and thus it is easy to get an order of magnitude of the p-value.

It is often useful to give a *compatibility region* along with a test. To create it, we test different "null hypotheses". For instance, let us consider the comparison of the mean point of a cloud to a reference point (the "null hypothesis" is that the deviation between the mean point and the reference point is null). From a pure geometry standpoint, every point of the space may be chosen as a reference point; the compatibility region at level α is defined as the set of points for which the test is nonsignificant at level α.

---

[3]This combinatorial conception of significance level exactly corresponds to the "non-stochastic interpretation" presented by Freedman and Lane (1983).

[4]It should be noted that it is equivalent to carry out simulations by means of algorithms in which possible rearrangements are done *without replacement* from the observed data. Consequently, permutation procedures are substantially different from the well-known bootstrap techniques in which possible rearrangements are done *with replacement* from the observed data.

Permutation modelling

As is well-known, traditional parametric methods, such as the Student t-test and the F-test, depend on assumptions on the distribution of the observations (such as homoscedasticity, normality and random sampling), which are rarely satisfied in practice. By contrast, permutation tests try to keep assumptions at a lower level, avoiding those that are difficult to justify or to interpret; in particular, they do not depend on assumptions on the distribution of observations, which is why they are termed *distribution-free*.

Permutation tests are distribution-free and nonparametric.

Several inductive problems may be correctly and effectively solved within the permutation framework, and others are very difficult or even impossible to solve outside it. Combinatorial inference may be used in all situations where the *exchangeability* of observations with respect to groups in the "null hypothesis" can be assumed. However, when the exchangeability of observations is not satisfied or cannot be assumed in the "null hypothesis", it may be useful to use bootstrap techniques, which are less demanding in terms of assumptions and may be effective at least for exploratory purposes.

Chance formulations and tests for randomness

Chance formulations in statistical inference have a long-standing history, significance tests being interpreted as *tests for randomness*. When the conclusion of "no effect" is compatible with observed data (nonsignificant result), it is commonly said that the result "might have occurred by chance" suggesting, by implication, that attempting to interpret the effect any further would be fruitless. Contrariwise, a significant effect is regarded as an event that "is not due to chance", and then attempting to search an interpretation is in order.

Randomization and permutation

In most experimental studies, *units* are randomly assigned to treatment levels (groups or samples). In these cases, many authors prefer the name randomization tests (Pitman, 1937; Edgington, 2007) or even re-randomization tests (Gabriel and Hall, 1983; Lunneborg, 2000) in place of permutation tests[5]. We prefer the latter name because it is more general. In many situations—for instance, in a "before–after" design, or when comparing two "natural groups" (e.g., boys and girls, etc.)— randomization is either nonexistent or impossible, but under the hypothesis of null effect the *observations* may be indifferently assigned to the two groups. In fact, a sufficient condition for properly applying permutation tests is that, if the effect of interest is null (null hypothesis), observed data are exchangeable with respect to groups.

---

[5]In the frequentist framework, permutation tests are confined to the situations where there is a random allocation of units to treatments: the so-called "physical act" of *randomization*, see Edgington (2007, esp. pp. 17–21), Cox and Hinkley (1974, pp. 179–204).

1.2   On Geometric Data Analysis

Geometric Data Analysis[6] (GDA) may be defined broadly as a set of multivariate models and methods for representing objects as points in a multidimensional Euclidean space. The core of GDA is the Euclidean cloud with its principal axes and Euclidean clusterings. In GDA, Euclidean clouds are constructed
from contingency tables by Correspondence Analysis (CA), from dissimilarity
tables by Multidimensional Scaling (MDS), from Individuals×Variables tables
by Principal Component Analysis (PCA) in the case of numerical variables
and by Multiple Correspondence Analysis (MCA) in the case of categorical
variables, etc.

The *three key ideas of* GDA are the following.

1. *Geometric modelling.* Data are represented by clouds of points in a multidimensional geometric space.

2. *Formal approach.* The mathematical foundations of GDA are the structures
   of linear algebra and multidimensional Euclidean geometry.

3. *Inductive philosophy.* Descriptive analysis comes prior to inductive analysis
   and probabilistic modelling.

Throughout the book, we will study a Euclidean cloud in the line of pure
geometry, that is, independently of any specific choice of a coordinate system
(*coordinate-free approach*). Of course, a coordinate system has to be chosen
to carry out computations, but we will not resort to them until it is really
necessary.

Frame model

Any empirical study involves the statement of a frame model, which guides
the collection of data and the interpretation of results.

In GDA, two principles should be followed (see Benzécri, 1992, pp. 382–383):

(1) *Homogeneity*: the topic of a study determines the fields wherefrom data
    are collected, but at times one has to take into account heterogeneous data
    collected at different levels, hence the preliminary phase of data coding[7];

(2) *Exhaustiveness*: data should constitute an exhaustive or at least a repre-
    sentative inventory of the domain under study.

---

[6]This name was suggested by Patrick Suppes in 1996 (see the foreword of the book by
Le Roux and Rouanet, 2004).

[7]See, e.g., Murtagh (2005).

Structured Data Analysis

A GDA of an Individuals×Variables table brings out the structure of the data table, that is, the relations between the set of individuals and the set of variables. However, they do not take into account the structures the two sets themselves may be equipped with, and which are defined by what we call *structuring factors*, that is, in most cases, variables that describe the basic sets but do not serve to construct the geometric space.

In *Structured Data Analysis*[8], we start from a geometric model of data, and we "graft" descriptive procedures (decomposition of variances) and inductive procedures (permutation tests), in the line of Multivariate Analysis of Variance (MANOVA). In a geometric setup, the effects of structuring factors on individuals become effect-vectors in the geometric space.

More often than not, the questions of interest, which are connected with structuring factors, may not only be relevant, but central to the study that has led to the geometric construction. Clearly, structured data, that is, data tables whose basic sets are equipped with structuring factors, constitute the rule rather than the exception (see the case studies presented in Chapter 6).

1.3 On Inductive Data Analysis

In the data analysis philosophy, statistical procedures should dig out "what the data have to say", and depend as little as possible on gratuitous hypotheses, unverifiable assumptions, etc. It is why the combinatorial framework, which is data-driven and entirely free from assumptions, is the most in harmony with geometric data analysis methods.

In this book, we develop permutations tests for clouds of points that stem from a GDA method or not. We present two *typicality tests* for a mean point, one of which is the combinatorial procedure for comparing the mean point of a subcloud to the mean point of a reference cloud, and the other is the permutation geometric procedure for comparing the mean point of a subcloud to a reference point. Then, in the same line, combinatorial inference is extended to various *homogeneity tests*. The rationale of homogeneity tests is apparent for structured data where the set of rearrangements is generated by a permutation group associated with the data structure. We study two cases: independent groups design and repeated measures design.

In accordance with the data analysis philosophy, the descriptive analysis of data should be conducted first, as a fully-fledged analysis. Therefore, an analysis consists in two steps:

---

[8]For a detailed presentation of Structured Data Analysis, see Le Roux and Rouanet (1984, 2004); Le Roux (2014b).

(1) *descriptive analyses*, that is, looking at the importance of effects and stating the descriptive conclusions;

(2) as a follow-up of descriptive analysis, *inductive analyses* the main objective of which is to corroborate (whenever possible) descriptive conclusions.

Description comes first and inference later.

A lot of statistical inference work, generally using bootstrap method and without emphasis on geometric clouds, has been done in GDA[9]; to name a few references: Lebart (1976, 2007); Gilula and Haberman (1986); Saporta and Hatabian (1986); de Leeuw and van der Burg (1986); Daudin et al. (1988); Gifi (1990); Le Roux (1998); Le Roux and Rouanet (2004, 2010); Beh and Lombardo (2014); Greenacre (2016), etc.

1.4   Computational Aspects

One might well make use of both types of software: menu-driven programs because of their convenience and a programming language because sooner or later we will encounter a new application beyond a menu-driven program's capabilities. We can combine these two aspects by using SPAD software and R scripts interfaced with SPAD. Thus, using SPAD, we offer a user-friendly approach not only to data management, statistical analyses but also to combinatorial inference through our R scripts interfaced with SPAD. In addition, R scripts may be amended at the request of the user.

Coheris SPAD software

SPAD[10] software started being developed in the 1970s by a group of researchers at the Crédoc (Research centre for the study and observation of living conditions in Paris); it is a pioneering software for the implementation of the data analysis methods developed in France.

Since 2007, SPAD software has been developed within the Coheris company—a provider of analytics and CRM (Customer Relationship Management) solutions, especially with the introduction of a user-friendly workflow-based graphical interface and an enlargement of statistical methods. It allows a process of *chain of analyses*—data import, data mining, data exploration and data modelling, predictive models, not to mention a powerful graphical data visualisation method—, in various environments (e.g., Hadoop). The possibility of seeing the data evolve in real time with each treatment directly in the tool is very significant.

---

[9] J.-P. Benzécri has constantly insisted on the *inductive logic* embodied in Correspondence Analysis (CA). In Benzécri et al. (1973, Vol. 2), there is an entire chapter (pp. 210–230) on inference in CA.

[10] Système Portable d'Analyse des Données.

Here are some highlights of the tool.

- The *data management* methods offer a wide range of codings.
- For most *statistical methods* (geometric data analysis, neural networks, time series, etc.), the SPAD Research and Development team works in partnership with academic experts who make their recent work available and help to set up new algorithms.
- SPAD internal method permits to *run an external R script*[11] and in order to make this feature easier to use, a parameter interface calling from R code is available. As we know, the methods developed in R are often difficult to access at the level of data management, thus the interfaced scripts aim to alleviate the user of these issues. Moreover, the data resulting from the R scripts can be transmitted to SPAD and thus become accessible for possible subsequent analysis (graphics, etc.).
- In recent years, the programs, which were initially written in Fortran, have been rewritten in Java. Thus, the code optimization and its parallelization would reduce calculation time considerably.

The computer programs of the methods we develop in Chapters 3, 4 and 5 of the present book come in two forms. On the one hand, *basic R scripts* with data management and graphics reduced to the minimum are outlined; on the other, guides for performing analyses with SPAD and *interfaced R scripts*[12] are described.

Basic R scripts

The R scripts, which are presented at the end of Chapters 3 (p. 55), 4 (p. 97) and 5 (p. 141), are mere applications of the main methods outlined in the chapters. They do not develop all the possibilities of the methods but they allow one to grasp the core aspects of the combinatorial methods. They are based on matrix calculus; they are written in a simple way and fully commented in order to allow the reader to grasp algorithms and/or to perform methods at least on the examples in the book.

R scripts interfaced with SPAD

The execution of R scripts within SPAD has been very useful in order to implement the combinatorial methods outlined in Chapters 3, 4 and 5. For each method, configuration windows of standard SPAD type are devised, which allows the user to set the R script parameters very easily.

The global process is the following (see Figure 1.1).

- SPAD and R must be installed on the computer.
- Data and R programs of combinatorial methods must be stored on the computer.

---

[11] It is also possible to use Java scripts.
[12] Programs and data are available from the authors.

- SPAD then allows importing the data, possibly preparing the data, implementing the tests and performing interactive graphs.

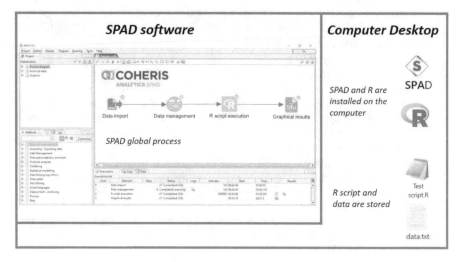

Figure 1.1
Presentation of SPAD software.

The parameter settings of the methods are described in each chapter (see §3.6.2, §4.6.2 and §5.7.2). SPAD with our interfaced R scripts has been used to perform the analyses of the real case studies presented in Chapter 6.

2

Cloud of Points in a Geometric Space

A cloud without metric is a cloud without shape[1].
J.-P. Benzécri

The methods of statistical inference developed in this book relate mainly to studying a *Euclidean cloud*, that is, a family of statistical observations conceptualized as points in a multidimensional Euclidean space.

Ready-made Euclidean clouds occur whenever observations are points recorded in the ambient physical space. For instance, the impacts of bullets on a target or the positions of bees in a swarm define Euclidean clouds.

In general, Euclidean clouds are built using numerical data sets. In Geometric Data Analysis (GDA), clouds are constructed from contingency tables by Correspondence Analysis (CA), from dissimilarity tables by Multidimensional Scaling (MDS), from Individuals×Variables tables by Principal Component Analysis (PCA) in the case of numerical variables and by Multiple Correspondence Analysis (MCA) in the case of categorical variables, etc.

Throughout the book, we will study a Euclidean cloud in the line of pure geometry, that is, independently of any specific choice of a coordinate system (*coordinate-free approach*). Of course, a coordinate system has to be chosen to carry out computations, but we will not resort to them until it is really necessary.

Without loss of generality, the study will be conducted on the *Target example*. Let us consider a plane target area with centre O, then the data set consists of a cloud of ten impact points, as depicted in Figure 2.1. When it comes to numerical computations, we will take a unit length u such that the frame of the picture is a $18u \times 18u$ square.

Figure 2.1
Target example.

The chapter is organized as follows. Firstly, we define basic statistics for a cloud (§2.1). Secondly, we introduce the covariance structure of a cloud and briefly present the search of principal directions[2] (§2.2). Then, we define the Mahalanobis distance and principal (or inertia) ellipsoids of a cloud (§2.3.1).

---

[1]Benzécri (1973, Volume 2, p. 32).

[2]For an elementary presentation of Euclidean clouds of points and the determination of principal directions, see Le Roux and Rouanet (2010, Chapter 2).

Finally we study the partition of a cloud into subclouds with the between-class cloud and the within-class cloud (§2.4).

The numerical expressions and matrix formulas are given at the end of each section.

2.1   Basic Statistics

Let $\mathcal{U}$ be a multidimensional geometric space[3] and $\mathcal{V}$ be the Euclidean vector space over $\mathbb{R}$ associated with $\mathcal{U}$ (see Chapter 7, §7.3, p. 229 and §7.4, p. 236).

Notations

Points, elements of the geometric space $\mathcal{U}$, are denoted by capital letters: M, P, A, etc.

Geometric vectors, elements of the vector space $\mathcal{V}$, are "arrowed": $\overrightarrow{\varepsilon}$, $\overrightarrow{u}$,... The vector associated with the bipoint (M,P) is denoted $\overrightarrow{MP}$, or P − M ("terminus minus initial") as the *deviation* from M to P.

The *scalar product* on $\mathcal{V}$ is denoted $\langle\cdot|\cdot\rangle$ and the *Euclidean norm* $\|\cdot\|$.

The *geometric distance* between two points M and P is denoted MP (or PM) with MP $= \|\overrightarrow{MP}\|$.

Definition 2.1 (Euclidean cloud). *Given a (nonempty) finite set I, a Euclidean cloud (in short, a cloud) is defined by the map*

$$
\begin{array}{rccl}
\mathrm{M}^I : & I & \longrightarrow & \mathcal{U} \quad \text{(geometric space)} \\
& i & \longmapsto & \mathrm{M}^i \quad \text{(point)}
\end{array}
$$

In general, the points of a cloud are weighted by positive masses that are usually derived from an absolute frequency measure $n_I = (n_i)_{i \in I}$ (*weighted cloud*), possibly elementary $n_I = (1)_{i \in I}$ (*elementary cloud*). The associated relative frequency measure is denoted $f_I = (f_i)_{i \in I}$ with $f_i = n_i/n$, letting[4] $n = \sum n_i$.

The set I is the *set-theoretic support* of the cloud. The smallest affine subspace of $\mathcal{U}$ containing the points of the cloud is the *affine support* (in short, support) of the cloud. It is denoted $\mathcal{M}$ and its direction vector space is denoted $\mathcal{L}$. By definition, the *dimension of the cloud*, denoted L, is the dimension of $\mathcal{L}$. By *plane cloud* we mean a two-dimensional cloud ($L = 2$).

---

[3]By space (without qualification), we always mean a *multidimensional geometric space*, whose elements are *points*, as opposed to a *vector space*, whose elements are *vectors*.

[4]*Subscripts and summing notation.* When there is no ambiguity, we write $\sum n_i$ rather than $\sum_{i \in I} n_i$, etc. For $\sum n_i$, we simply write n (rather than $n.$): omitting a subscript indicates summing over this subscript. Contrary to the dot notation, subscripts are position-independent.

The basic statistics of a Euclidean cloud (mean point, variance) readily extend those pertaining to numerical variables.

2.1.1 Mean Point

Definition 2.2. *Let* $(\mathrm{M}^i, n_i)_{i \in I}$ *be a weighted cloud in* $\mathcal{U}$ *(with* $n = \sum n_i \neq 0$ *and* $f_i = n_i/n$*); for any point* $\mathrm{P} \in \mathcal{U}$*, the point* G *such that* $\overrightarrow{\mathrm{PG}} = \sum f_i \overrightarrow{\mathrm{PM}^i}$ *is called the* **mean point** *of the cloud.*

Point G does not depend on point P, it is written $\mathrm{G} = \sum f_i \mathrm{M}^i$ (see Chapter 7, p. 238). The equations $\sum f_i \overrightarrow{\mathrm{GM}^i} = \overrightarrow{0}$ or $\sum n_i \overrightarrow{\mathrm{GM}^i} = \overrightarrow{0}$ express the *barycentric property* of the mean point: the weighted mean (or sum) of the deviations from the points of the cloud to the mean point is the null vector.

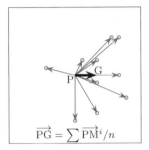 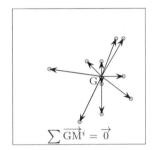

$$\overrightarrow{\mathrm{PG}} = \sum \overrightarrow{\mathrm{PM}^i}/n \qquad \sum \overrightarrow{\mathrm{GM}^i} = \overrightarrow{0}$$

Figure 2.2
Target example. Mean point and barycentric property.

The mean point G of the *cloud of two points* (M^i, n_i) and $(\mathrm{M}^{i'}, n_{i'})$ belongs to the line segment $[\mathrm{M}^i, \mathrm{M}^{i'}]$ and verifies $n_i \overrightarrow{\mathrm{GM}^i} + n_{i'} \overrightarrow{\mathrm{GM}^{i'}} = \overrightarrow{0}$. When $n_i = n_{i'}$, the mean point is the middle of the segment $[\mathrm{M}^i, \mathrm{M}^{i'}]$. For $n_i = 3$ and $n_{i'} = 2$, one has $\mathrm{GM}^i/\mathrm{GM}^{i'} = n_{i'}/n_i = 2/3$ and $\overrightarrow{\mathrm{GM}^i} = -\frac{n_{i'}}{n_i + n_{i'}} \overrightarrow{\mathrm{M}^i \mathrm{M}^{i'}}$ (see Figure 2.3).

Figure 2.3
Construction of the mean point of a cloud of two weighted points.

2.1.2 Variance

Definition 2.3. *The* **variance of a cloud,** *denoted* V_{cloud}*, is the weighted mean of the squared distances from the points of the cloud to the mean point.*

$$V_{\text{cloud}} = \sum n_i (\text{GM}^i)^2 / n = \sum f_i (\text{GM}^i)^2$$

Remark. We say "variance" for the weighted *mean* of squared distances and "inertia" for the weighted *sum* of squared distances.

Proposition 2.1 (Huygens's formula). *The weighted mean of the squared distances from the points of a cloud to a reference point* P *is equal to the variance of the cloud plus the squared distance between the mean point* G *of the cloud and the reference point* P.

$$\sum f_i (\text{PM}^i)^2 = V_{\text{cloud}} + \text{PG}^2$$

Proof. From $\overrightarrow{\text{PM}^i} = \overrightarrow{\text{PG}} + \overrightarrow{\text{GM}^i}$ (Chasles's identity, p. 237), one obtains $(\text{PM}^i)^2 = \text{PG}^2 + (\text{GM}^i)^2 + 2\langle\overrightarrow{\text{PG}}|\overrightarrow{\text{GM}^i}\rangle$, hence $\sum f_i (\text{PM}^i)^2 = \text{PG}^2 + \sum f_i (\text{GM}^i)^2 + 2\langle\overrightarrow{\text{PG}}|\sum f_i \overrightarrow{\text{GM}^i}\rangle$ (linearity of scalar product). One has $\sum f_i \overrightarrow{\text{GM}^i} = \overrightarrow{0}$ (barycentric property), hence the proposition. ☐

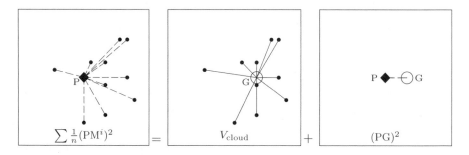

Figure 2.4
Target example. Illustration of Huygens's formula.

The following proposition shows that the variance of a cloud can also be expressed as a function depending on the distances between pairs of points.

Proposition 2.2.
$$V_{\text{cloud}} = \tfrac{1}{2} \sum \sum f_i f_{i'} (\text{M}^i \text{M}^{i'})^2$$

Proof. By taking point M^i as a reference point, the Huygens's formula can be written $\sum_{i' \in I} f_{i'} (\text{M}^i \text{M}^{i'})^2 = V_{\text{cloud}} + (\text{GM}^i)^2$, hence $\sum_{i \in I} f_i \sum_{i' \in I} f_{i'} (\text{M}^i \text{M}^{i'})^2 = \sum_{i \in I} f_i V_{\text{cloud}} + \sum_{i \in I} f_i (\text{GM}^i)^2 = 2 V_{\text{cloud}}$. ☐

In particular, the variance of a *cloud of two points* (M^i, n_i) and $(\text{M}^{i'}, n_{i'})$ can be written

$$\frac{n_i n_{i'}}{(n_i + n_{i'})^2} (\text{M}^i \text{M}^{i'})^2 \tag{2.1}$$

2.1.3 Numerical Expressions and Matrix Formulas

By choosing a Cartesian frame of the geometric space (see Chapter 7, p. 240) and working on the coordinates of points, we will go "from points to numbers", and get "numerical formulas" with their matrix expressions.

Let $\left(O, (\overrightarrow{\varepsilon_k})_{k\in K}\right)$ be a Cartesian frame of the geometric space $\mathcal{U}$ of dimension K. Let $q_{kk'}$ denote the scalar product between the basis vectors $\overrightarrow{\varepsilon_k}$ and $\overrightarrow{\varepsilon_{k'}}$: $q_{kk'} = \langle \overrightarrow{\varepsilon_k} | \overrightarrow{\varepsilon_{k'}} \rangle$.

Numerical expressions

The *Cartesian coordinates* of point M^i with respect to the Cartesian frame $\left(O, (\overrightarrow{\varepsilon_k})_{k\in K}\right)$ are $(x_k^i)_{k\in K}$, that is, $\overrightarrow{OM^i} = \sum_{k\in K} x_k^i \overrightarrow{\varepsilon_k}$ (see Chapter 7, p. 240).

- The *profile* of point M^i (in short profile of i) is $x_K^i = (x_k^i)_{k\in K}$.

- The *coordinate variable* associated with vector $\overrightarrow{\varepsilon_k}$ (in short *variable k*) is $x_k^I = (x_k^i)_{i\in I}$. Its mean is denoted $\overline{x}_k$, with $\overline{x}_k = \sum_i f_i x_k^i$, and its variance is denoted $\mathrm{Var}\, x_k^I$, with $\mathrm{Var}\, x_k^I = \sum_i f_i (x_k^i - \overline{x}_k)^2$.

  The covariance between variables k and k' is denoted $v_{kk'}$ with $v_{kk'} = \sum_i f_i (x_k^i - \overline{x}_k)(x_{k'}^i - \overline{x}_{k'})$ (in particular $v_{kk} = \mathrm{Var}\, x_k^I$).

It can be shown that the *Cartesian coordinates of the mean point* of the cloud are equal to the means $(\overline{x}_k)_{k\in K}$ of the coordinate variables, that is, $\overrightarrow{OG} = \sum \overline{x}_k \overrightarrow{\varepsilon_k}$ and the *variance of the cloud* is equal to $\sum \sum v_{kk'} q_{kk'}$.

Matrix formulas

The following auxiliary matrices[5] will be used: $\mathbf{I}$ (or $\mathbf{I}_I$) denotes the identity $I \times I$-matrix and $\mathbf{e}$ the I-column with all terms equal to 1.

We denote $\mathbf{f}_I$ the I-column of relative frequencies $(f_i)_{i\in I}$, and $\mathbf{F}_I$ the diagonal matrix whose diagonal terms are the relative frequencies $(f_i)_{i\in I}$.

- $\mathbf{X} = [x_k^i]$ is the $(I \times K)$-matrix of coordinates of points;

- $\mathbf{x}^i$ is the K-column of the profile of i;

- $\mathbf{x}_k$ is the I-column of variable k.

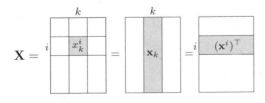

---

[5]Elements of matrix calculus and matrix notations can be found in Chapter 7, §7.1.

- $\mathbf{X}_0 = [x_k^i - \overline{x}_k]$, with $\mathbf{X}_0 = (\mathbf{I} - \mathbf{e}\,\mathbf{e}^\top \mathbf{F}_I)\mathbf{X}$, is the centred matrix of coordinates;

- $\mathbf{V} = [v_{kk'}]$ is the $K{\times}K$-covariance matrix between the K coordinate variables with $\mathbf{V} = \mathbf{X}_0^\top \mathbf{F}_I \mathbf{X}_0$;

- $\mathbf{Q} = [q_{kk'}]$ is the $K{\times}K$-matrix of scalar products between the basis vectors $(\overrightarrow{\varepsilon_k})_{k\in K}$.

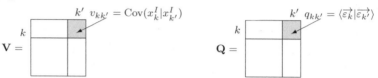

The expression of the squared distance from point M^i to point O is:

$$(GM^i)^2 = \sum_{k\in K} \sum_{k'\in K} q_{kk'}\,(x_k^i - \overline{x}_k)(x_{k'}^i - \overline{x}_{k'})^2 = (\mathbf{x}_0^i)^\top \mathbf{Q}\,\mathbf{x}_0^i$$

The variance of the cloud is $V_{\mathrm{cloud}} = \mathrm{tr}(\mathbf{VQ})$.

Target example

If the plane of the target (p. 9) is referred to two rectangular axes (horizontal and vertical) passing through point O (centre of the target), the graduations of both axes corresponding to the distance unit u (Figure 2.5), then one obtains abscissas and ordinates of points (Table 2.1).

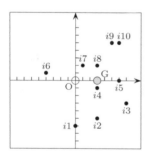

Figure 2.5
Cloud referred to axes.

Table 2.1
Coordinates of points (matrix $\mathbf{X}$) and basic statistics.

	x_{k1}^I	x_{k2}^I
$i1$	0	-6
$i2$	3	-5
$i3$	7	-3
$i4$	3	-1
$i5$	6	0
$i6$	-4	1
$i7$	1	2
$i8$	3	2
$i9$	5	5
$i10$	6	5

Coordinates of the mean point: $\begin{bmatrix} 3 \\ 0 \end{bmatrix}$

Coordinate variables x_{k1}^I and x_{k2}^I

Var $x_{k1}^I = 10$, Var $x_{k2}^I = 13$

Cov$(x_{k1}^I | x_{k2}^I) = +2$.

Covariance matrix: $\mathbf{V} = \begin{bmatrix} 10 & 2 \\ 2 & 13 \end{bmatrix}$

$V_{\mathrm{cloud}} = 10 + 13 = 23$.

2.2   Covariance Structure of a Cloud

In this section, we study the covariance structure of a Euclidean cloud and we introduce the problem of determining the principal axes of a cloud.

2.2.1 Covariance Endomorphism

The key element of the covariance structure of a cloud is the endomorphism—which we call the *covariance endomorphism* of the cloud—denoted Mcov; it is defined as follows.

Definition 2.4 (Endomorphism Mcov). *Given a weighted cloud* (M^I, n_I) *together with its mean point* G *in a geometric space* $\mathcal{U}$ *with associated vector space* $\mathcal{V}$, *the covariance endomorphism of cloud* M^I *is such that*

$$\begin{aligned} \mathrm{Mcov}: \quad \mathcal{V} &\longrightarrow \mathcal{V} \\ \overrightarrow{u} &\longmapsto \sum f_i \langle \overrightarrow{\mathrm{GM}^i} | \overrightarrow{u} \rangle \overrightarrow{\mathrm{GM}^i} \end{aligned} \qquad (2.2)$$

Proposition 2.3. *The endomorphism* Mcov *is symmetric and positive.*

Proof. $\forall \overrightarrow{u}, \overrightarrow{v} \in \mathcal{V}$, $\langle \mathrm{Mcov}(\overrightarrow{u}) | \overrightarrow{v} \rangle = \langle \sum f_i \langle \overrightarrow{\mathrm{GM}^i} | \overrightarrow{u} \rangle \overrightarrow{\mathrm{GM}^i} | \overrightarrow{v} \rangle$. By linearity of the scalar product, $\langle \sum f_i \langle \overrightarrow{\mathrm{GM}^i} | \overrightarrow{u} \rangle \overrightarrow{\mathrm{GM}^i} | \overrightarrow{v} \rangle = \sum f_i \langle \overrightarrow{\mathrm{GM}^i} | \overrightarrow{u} \rangle \langle \overrightarrow{\mathrm{GM}^i} | \overrightarrow{v} \rangle$, that is, $\langle \mathrm{Mcov}(\overrightarrow{u}) | \overrightarrow{v} \rangle = \langle \overrightarrow{u} | \mathrm{Mcov}(\overrightarrow{v}) \rangle$, hence the symmetry (see Definition 7.5, p. 231). $\langle \mathrm{Mcov}(\overrightarrow{u}) | \overrightarrow{u} \rangle = \sum f_i \langle \overrightarrow{\mathrm{GM}^i} | \overrightarrow{u} \rangle^2 \geq 0$, hence the positivity. $\square$

One has $\mathrm{Mcov}(\mathcal{V}) = \mathcal{L}$ (direction of the affine support of the cloud M^I, see p. 10) and $\forall \overrightarrow{u} \in \mathcal{L}^\perp$, $\mathrm{Mcov}(\overrightarrow{u}) = \overrightarrow{0}$. In particular, the covariance endomorphism of a *cloud of two points* has rank one.

Proposition 2.4. *The variance of cloud* (M^I, n_I) *in the direction defined by the unit vector* $\overrightarrow{u}$ *is equal to* $\langle \mathrm{Mcov}(\overrightarrow{u}) | \overrightarrow{u} \rangle$.

Proof. The variance of cloud M^I in the direction $\overrightarrow{u}$ is, by definition, the variance of the projected cloud A^I onto the line $(\mathrm{G}, \overrightarrow{u})$. The projected point A^i of point M^i onto this line is such that $\overrightarrow{\mathrm{GA}^i} = \langle \overrightarrow{\mathrm{GM}^i} | \overrightarrow{u} \rangle \overrightarrow{u}$ (see Proposition 7.6, p. 242) and $(\mathrm{GA}^i)^2 = \| \langle \overrightarrow{\mathrm{GM}^i} | \overrightarrow{u} \rangle \overrightarrow{u} \|^2 = \langle \overrightarrow{\mathrm{GM}^i} | \overrightarrow{u} \rangle^2$. Hence $\mathrm{Var}\,\mathrm{A}^I = \sum f_i \langle \overrightarrow{\mathrm{GM}^i} | \overrightarrow{u} \rangle^2 = \sum f_i \langle \overrightarrow{\mathrm{GM}^i} | \overrightarrow{u} \rangle \langle \overrightarrow{\mathrm{GM}^i} | \overrightarrow{u} \rangle = \langle \sum f_i \langle \overrightarrow{\mathrm{GM}^i} | \overrightarrow{u} \rangle \overrightarrow{\mathrm{GM}^i} | \overrightarrow{u} \rangle$ (linearity of the scalar product), that is, $\mathrm{Var}\,\mathrm{A}^I = \langle \mathrm{Mcov}(\overrightarrow{u}) | \overrightarrow{u} \rangle$. $\square$

If $\| \overrightarrow{u} \| \neq 1$, the variance in the direction defined by the vector $\overrightarrow{u}$ is equal to $\langle \mathrm{Mcov}(\overrightarrow{u}) | \overrightarrow{u} \rangle / \| \overrightarrow{u} \|^2$.

Let $\overrightarrow{u}$ and $\overrightarrow{v}$ be two unit vectors, the coordinates of M^i over axes $(\mathrm{G}, \overrightarrow{u})$ and $(\mathrm{G}, \overrightarrow{v})$ are $u^i = \langle \overrightarrow{\mathrm{GM}^i} | \overrightarrow{u} \rangle$ and $v^i = \langle \overrightarrow{\mathrm{GM}^i} | \overrightarrow{v} \rangle$ with $\sum f_i u^i = 0 = \sum f_i v^i$. The quantity $\sum f_i \langle \overrightarrow{\mathrm{GM}^i} | \overrightarrow{u} \rangle \langle \overrightarrow{\mathrm{GM}^i} | \overrightarrow{v} \rangle$ is the mean of the products of these coordinates $(\sum f_i u^i v^i)$, that is, the *covariance between the coordinate variables* of cloud M^I associated with $\overrightarrow{u}$ and $\overrightarrow{v}$: $\mathrm{Cov}(u^I | v^I) = \langle \mathrm{Mcov}(\overrightarrow{u}) | \overrightarrow{v} \rangle$.

It is with these formulas in mind that the endomorphism Mcov *was defined*[6] *and called "covariance endomorphism" of the cloud.*

---

[6]See Benzécri (1984, pp. 148–150), Benzécri (1992, pp. 208–210).

Endomorphism Mcovᴾ

The endomorphism Mcovᴾ defined hereafter is the *generalization of the covariance endomorphism* of a cloud when the reference point is a point P ≠ G. It is especially used in the geometric typicality test, where a point in the geometric space is chosen as a *reference point* (Chapter 4).

Definition 2.5 (Endomorphism Mcovᴾ). *Let* P *be a point in* $\mathcal{U}$, *the generalized covariance endomorphism of cloud* M^I *with respect to point* P, *denoted* Mcovᴾ, *is such that*

$$\mathrm{Mcov^P}: \quad \begin{array}{ccc} \mathcal{V} & \rightarrow & \mathcal{V} \\ \overrightarrow{u} & \longmapsto & \sum f_i \langle \overrightarrow{\mathrm{PM}^i} | \overrightarrow{u} \rangle \overrightarrow{\mathrm{PM}^i} \end{array}$$

Proposition 2.5. $\forall \overrightarrow{u} \in \mathcal{V}$, $\mathrm{Mcov^P}(\overrightarrow{u}) = \mathrm{Mcov}(\overrightarrow{u}) + \langle \overrightarrow{\mathrm{PG}} | \overrightarrow{u} \rangle \overrightarrow{\mathrm{PG}}$

Proof. $\mathrm{Mcov^P}(\overrightarrow{u}) = \sum f_i \langle \overrightarrow{\mathrm{PM}^i} | \overrightarrow{u} \rangle \overrightarrow{\mathrm{PM}^i} = \sum f_i \langle \overrightarrow{\mathrm{PG}} | \overrightarrow{u} \rangle \overrightarrow{\mathrm{PM}^i} + \sum f_i \langle \overrightarrow{\mathrm{GM}^i} | \overrightarrow{u} \rangle \overrightarrow{\mathrm{PM}^i}$ since $\overrightarrow{\mathrm{PM}^i} = \overrightarrow{\mathrm{PG}} + \overrightarrow{\mathrm{GM}^i}$ (Chasles's identity, p. 237). We have $\sum f_i \overrightarrow{\mathrm{PM}^i} = \overrightarrow{\mathrm{PG}}$, then $\mathrm{Mcov^P}(\overrightarrow{u}) = \langle \overrightarrow{\mathrm{PG}} | \overrightarrow{u} \rangle \overrightarrow{\mathrm{PG}} + \sum f_i \langle \overrightarrow{\mathrm{GM}^i} | \overrightarrow{u} \rangle (\overrightarrow{\mathrm{PG}} + \overrightarrow{\mathrm{GM}^i})$. By barycentric property $\sum f_i \overrightarrow{\mathrm{GM}^i} = \overrightarrow{0}$, hence $\mathrm{Mcov^P}(\overrightarrow{u}) = \langle \overrightarrow{\mathrm{PG}} | \overrightarrow{u} \rangle \overrightarrow{\mathrm{PG}} + \sum f_i \langle \overrightarrow{\mathrm{GM}^i} | \overrightarrow{u} \rangle \overrightarrow{\mathrm{GM}^i}$. □

Corollary 2.1. *The endomorphism* Mcovᴾ *is symmetric and positive.*

Proof. $\forall \overrightarrow{u}, \overrightarrow{v} \in \mathcal{V}$, $\langle \mathrm{Mcov^P}(\overrightarrow{u}) | \overrightarrow{v} \rangle = \langle \mathrm{Mcov}(\overrightarrow{u}) | \overrightarrow{v} \rangle + \langle \overrightarrow{\mathrm{PG}} | \overrightarrow{u} \rangle \langle \overrightarrow{\mathrm{PG}} | \overrightarrow{v} \rangle$ hence the symmetry; $\langle \mathrm{Mcov^P}(\overrightarrow{u}) | \overrightarrow{u} \rangle = \langle \mathrm{Mcov}(\overrightarrow{u}) | \overrightarrow{u} \rangle + \langle \overrightarrow{\mathrm{PG}} | \overrightarrow{u} \rangle^2 \geq 0$. □

Proposition 2.6. *The covariance endomorphism* Mcov *of cloud* M^I *depends only on distances between the points of the cloud:*

$$\forall \overrightarrow{u} \in \mathcal{L}, \ \mathrm{Mcov}(\overrightarrow{u}) = \sum \sum f_i f_{i'} \langle \overrightarrow{\mathrm{M}^i \mathrm{M}^{i'}} | \overrightarrow{u} \rangle \overrightarrow{\mathrm{M}^i \mathrm{M}^{i'}}$$

Proof. After Proposition 2.5 with M^i as a reference point, one has $\mathrm{Mcov^{M^i}}(\overrightarrow{u}) = \mathrm{Mcov}(\overrightarrow{u}) + \langle \overrightarrow{\mathrm{GM}^i} | \overrightarrow{u} \rangle \overrightarrow{\mathrm{GM}^i} = \mathrm{Mcov}(\overrightarrow{u}) + \sum f_i \langle \overrightarrow{\mathrm{GM}^i} | \overrightarrow{u} \rangle \overrightarrow{\mathrm{GM}^i} = 2\mathrm{Mcov}(\overrightarrow{u})$. Since $\mathrm{Mcov^{M^i}}(\overrightarrow{u}) = \sum_{i' \in I} f_{i'} \langle \overrightarrow{\mathrm{M}^i \mathrm{M}^{i'}} | \overrightarrow{u} \rangle \overrightarrow{\mathrm{M}^i \mathrm{M}^{i'}}$, we deduce the formula. □

As a consequence, the covariance endomorphism of a *cloud of two points* (M^i, n_i) and $(\mathrm{M}^{i'}, n_{i'})$ is such that

$$\overrightarrow{u} \mapsto \frac{n_i n_{i'}}{(n_i + n_{i'})^2} \langle \overrightarrow{\mathrm{M}^i \mathrm{M}^{i'}} | \overrightarrow{u} \rangle \overrightarrow{\mathrm{M}^i \mathrm{M}^{i'}} \quad (\overrightarrow{u} \in \mathcal{V}) \tag{2.3}$$

2.2.2   Principal Directions

The principal directions of a cloud play a crucial role in studying the statistical inference problems that we tackle in the following chapters. Therefore we will briefly address the issue of searching the principal directions of a cloud[7].

---

[7]For a full study of principal axes of a cloud in a geometric space, see Benzécri (1992, Chapter 11), Le Roux and Rouanet (2004, Chapter 3, §3.3), Le Roux (2014a, Chapter 5, §5.4). For presentations based on matrix calculus, see Greenacre (1984); Gower and Hand (1996); Lebart et al. (2006); Pagès (2014), etc.

The *problem of principal directions* is to best approximate a high dimensional cloud (whose dimension equals L) by a cloud in a lower dimensional space—a line (one-dimensional cloud), a plane (two-dimensional cloud), etc.—such that the approximate cloud is intuitively "of greatest elongation". When operationalized in terms of *variance*, this requirement leads to search, among all the L'-dimensional subspaces $(L' < L)$, the one such that the cloud projected onto it has maximum variance.

For any given L', the first principal direction always exists and is generally unique. Thus there is in general a unique first principal line $(L' = 1)$, then a unique first principal plane $(L' = 2)$, etc., hence a *hierarchy of principal directions*. Furthermore, this hierarchy possesses the following *heredity property*: the first principal plane contains the first principal line, etc.

Theorem 2.1. *The direction vectors of the principal lines of the cloud are eigenvectors of the covariance endomorphism of the cloud.*

Let us denote L the set of natural numbers from 1 to L. There exists an orthonormal basis $(\overrightarrow{\alpha_\ell})_{\ell \in L}$ of the direction $\mathcal{L}$ of the affine support $\mathcal{M}$ of the cloud constituted of L eigenvectors of endomorphism Mcov associated with the non-null eigenvalues; it is called *orthonormal principal basis* of $\mathcal{L}$.

$$\text{M}cov(\overrightarrow{\alpha_\ell}) = \lambda_\ell \overrightarrow{\alpha_\ell} \quad (\ell \in L)$$

From now on, we assume that $\lambda_1 \geq \dots \lambda_\ell \geq \dots \lambda_L > 0$.

The first L'-dimensional principal direction $(L' \leq L)$ is generated by the first L' eigenvectors $(\overrightarrow{\alpha_\ell})_{\ell=1,\dots L'}$.

Principal axis. The oriented principal line passing through point G and with direction vector $\overrightarrow{\alpha_\ell}$ is called principal axis ℓ or simply *Axis ℓ*.

The following property is a direct consequence of Proposition 2.4 (p. 15).

Corollary 2.2. *The variance of the cloud projected on Axis ℓ is equal to the eigenvalue λ_ℓ; it is called* variance of Axis ℓ.

With respect to the orthonormal principal basis $(\overrightarrow{\alpha_\ell})_{\ell \in L}$ of $\mathcal{L}$, the coordinates of point M^i, called *principal coordinates*, are denoted $(y_\ell^i)_{\ell \in L}$, with

$$y_\ell^i = \langle \overrightarrow{\text{GM}^i} | \overrightarrow{\alpha_\ell} \rangle \quad (\ell \in L) \quad \text{and} \quad (\text{GM}^i)^2 = \sum_{\ell \in L} (y_\ell^i)^2$$

Let M$_\ell^i$ denote the orthogonal projection of point M on Axis ℓ, the *quality of representation* of point M^i on Axis ℓ is equal to the squared cosinus of the angle θ_ℓ between $\overrightarrow{\text{GM}^i}$ and Axis ℓ, that is:

$$\cos^2 \theta_\ell = \left(\frac{\text{GM}_\ell^i}{\text{GM}^i} \right)^2 \qquad (2.4)$$

The quality of representation of point M^i in the principal plane associated with axes ℓ and ℓ' is equal to $\cos^2 \theta_\ell + \cos^2 \theta_{\ell'}$.

Principal variables. The set of principal coordinates of points $(M^i)_{i \in I}$ on axis $(G, \overrightarrow{\alpha_\ell})$ defines the ℓ-th *principal variable* that is denoted y_ℓ^I.

We have the following properties: Mean $y_\ell^I = 0$, Var $y_\ell^I = \lambda_\ell$ and, for $\ell \neq \ell'$, ℓ-th and ℓ'-th principal variables are *uncorrelated* since $\mathrm{Cov}(y_\ell^I | y_{\ell'}^I) = \langle \mathrm{Mcov}(\overrightarrow{\alpha_\ell}) | \overrightarrow{\alpha_{\ell'}} \rangle = \lambda_\ell \langle \overrightarrow{\alpha_\ell} | \overrightarrow{\alpha_{\ell'}} \rangle = 0$ $(\overrightarrow{\alpha_\ell} \perp \overrightarrow{\alpha_{\ell'}})$.

Proposition 2.7. *The variance of the cloud is equal to the sum of the eigenvalues:*

$$V_{\mathrm{cloud}} = \sum \lambda_\ell$$

Proof. The orthogonal projection onto Axis ℓ of point M^i is the point M_ℓ^i such that $\overrightarrow{GM_\ell^i} = \langle \overrightarrow{GM^i} | \overrightarrow{\alpha_\ell} \rangle \overrightarrow{\alpha_\ell}$ (Proposition 7.6, p. 242). $\forall i \in I$, $\overrightarrow{GM^i} = \sum_{\ell \in L} \overrightarrow{GM_\ell^i}$, with $\overrightarrow{GM_\ell^i} \perp \overrightarrow{GM_{\ell'}^i}$ for $\ell \neq \ell'$, therefore $(GM^i)^2 = \sum_{\ell \in L} (\overrightarrow{GM_\ell^i})^2$ and $\sum_{i \in I} f_i (GM^i)^2 = \sum_{\ell \in L} \sum_{i \in I} f_i (\overrightarrow{GM_\ell^i})^2 = \sum_{\ell \in L} \mathrm{Var}\, M_\ell^I = \sum_{\ell \in L} \lambda_\ell$ (Proposition 2.2, p. 17). $\quad\square$

Proposition 2.8. *If* P *is a point in* $\mathcal{M}$ *such that* $\overrightarrow{PG} = x \overrightarrow{\alpha_\ell}$ *($x \in \mathbb{R}$), then the endomorphism* Mcov^P *(defined p. 16) has the same eigenvectors with the same eigenvalues as* Mcov, *except that* $\overrightarrow{\alpha_\ell}$ *is associated with* $\lambda_\ell + x^2$.

Proof. $\mathrm{Mcov}^P(\overrightarrow{\alpha_{\ell'}}) = \mathrm{Mcov}(\overrightarrow{\alpha_{\ell'}}) + \langle \overrightarrow{PG} | \overrightarrow{\alpha_{\ell'}} \rangle \overrightarrow{PG} = \lambda_{\ell'} \overrightarrow{\alpha_{\ell'}} + x^2 \langle \overrightarrow{\alpha_\ell} | \overrightarrow{\alpha_{\ell'}} \rangle \overrightarrow{\alpha_\ell}$.
For $\ell' \neq \ell$, we have $\langle \overrightarrow{\alpha_{\ell'}} | \overrightarrow{\alpha_\ell} \rangle = 0$, hence $\mathrm{Mcov}^P(\overrightarrow{\alpha_{\ell'}}) = \lambda_{\ell'} \overrightarrow{\alpha_{\ell'}}$.
For $\ell' = \ell$, we have $\langle \overrightarrow{\alpha_{\ell'}} | \overrightarrow{\alpha_\ell} \rangle = \|\overrightarrow{\alpha_\ell}\|^2 = 1$, hence $\mathrm{Mcov}^P(\overrightarrow{\alpha_\ell}) = \lambda_\ell \overrightarrow{\alpha_\ell} + x^2 \overrightarrow{\alpha_\ell}$. $\quad\square$

2.2.3   Numerical Expressions and Matrix Formulas

The notations of this section are the same as those used in §2.1.3 (p.13).

Proposition 2.9. *With respect to the basis* $(\overrightarrow{\varepsilon_k})_{k \in K}$ *of* $\mathcal{V}$ *the matrix representation of* Mcov *is* $\mathbf{VQ}$, *and that of* Mcov_P *is* $(\mathbf{V} + \mathbf{dd}^\top)\mathbf{Q}$, *where* $\mathbf{d} = [d_k]$ *is the* K*-column of coordinates of* $\overrightarrow{PG}$.

Proof. One has $\langle \overrightarrow{GM^i} | \overrightarrow{\varepsilon_k} \rangle = \left\langle \sum_{k'} (x_{k'}^i - \overline{x}_{k'}) \overrightarrow{\varepsilon_{k'}} \big| \overrightarrow{\varepsilon_k} \right\rangle = \sum_{k'} (x_{k'}^i - \overline{x}_{k'}) q_{kk'}$.
$\mathrm{Mcov}(\overrightarrow{\varepsilon_k}) = \sum_i f_i \langle \overrightarrow{GM^i} | \overrightarrow{\varepsilon_k} \rangle \overrightarrow{GM^i} = \sum_i f_i \sum_{k'} (x_{k'}^i - \overline{x}_{k'}) q_{kk'} \sum_{k''} (x_{k''}^i - \overline{x}_{k''}) \overrightarrow{\varepsilon_{k''}} = \sum_{k'} \sum_{k''} \left(\sum_i f_i (x_{k'}^i - \overline{x}_{k'})(x_{k''}^i - \overline{x}_{k''}) \right) q_{kk'} \overrightarrow{\varepsilon_{k''}} = \sum_{k''} \left(\sum_{k'} v_{k'k''} q_{kk'} \right) \overrightarrow{\varepsilon_{k''}}$ with $\mathbf{VQ} = [\sum_{k'} v_{k'k''} q_{kk'}]$.
$\langle \overrightarrow{PG} | \overrightarrow{\varepsilon_k} \rangle = \sum_{k'} d_{k'} q_{kk'}$, hence $\langle \overrightarrow{PG} | \overrightarrow{\varepsilon_k} \rangle \overrightarrow{PG} = \sum_{k''} d_{k''} \left(\sum_{k'} d_{k'} q_{kk'} \right) \overrightarrow{\varepsilon_{k''}}$. $\quad\square$

Remark. When the basis $(\overrightarrow{\varepsilon_k})_{k \in K}$ is orthonormal, the matrix representation of Mcov is the covariance matrix $\mathbf{V}$.

Let $\overrightarrow{\alpha_\ell}$ be the eigenvector of Mcov associated with λ_ℓ and $\mathbf{a}_\ell = [a_{k\ell}]$ be the K-column of coordinates of $\overrightarrow{\alpha_\ell}$ with respect to the basis $(\overrightarrow{\varepsilon_k})_{k \in K}$.

Proposition 2.10. *The principal direction equation writes:*

$$\mathbf{VQa}_\ell = \lambda_\ell \mathbf{a}_\ell, \quad \text{with } \mathbf{a}_\ell^\top \mathbf{Qa}_{\ell'} = 0 \text{ for } \ell \neq \ell'$$

Let $\mathbf{A} = [a_{k\ell}]$ be the $K \times L$-matrix of the K-columns $(\mathbf{a}_\ell)_{\ell \in L}$, $\boldsymbol{\Lambda}$ the diagonal $L \times L$-matrix of eigenvalues $(\lambda_\ell)_{\ell \in L}$ and $\mathbf{I}_L$ the identity $L \times L$-matrix, looking for *unit vectors* one has:

$$\mathbf{VQA} = \mathbf{A\Lambda} \quad \text{with } \mathbf{A}^\top \mathbf{QA} = \mathbf{I}_L$$

The principal coordinates of point M^i are $y_\ell^i = \langle \overrightarrow{GM^i} | \overrightarrow{\alpha_\ell} \rangle = (\mathbf{x}_0^i)^\top \mathbf{Qa}_\ell$ $(\ell \in L)$. Let $\mathbf{Y} = [y_\ell^i]$ be the $I \times L$-matrix of *principal coordinates* of points:

$$\mathbf{Y} = \mathbf{X_0 QA} \text{ with } \mathbf{Y}^\top \mathbf{F}_I \mathbf{Y} = \boldsymbol{\Lambda} \tag{2.5}$$

The $I \times L$-matrix of *standardized principal coordinates* is

$$\mathbf{Z} = \mathbf{X_0 QA\Lambda}^{-1/2} \tag{2.6}$$

Proposition 2.11. *With respect to the* orthonormal principal basis $(\overrightarrow{\alpha_\ell})_{\ell \in L}$ *of $\mathcal{L}$, the matrix of* Mcov $: \mathcal{L} \to \mathcal{L}$ *is $\boldsymbol{\Lambda}$ and, denoting $\mathbf{u} = [u_\ell]$ the L-column of the coordinates of $\overrightarrow{PG}$, that of* McovP $: \mathcal{L} \to \mathcal{L}$ *is $\boldsymbol{\Lambda} + \mathbf{uu}^\top$.*

Proof. Mcov$(\overrightarrow{\alpha_\ell}) = \lambda_\ell \overrightarrow{\alpha_\ell}$; McovP$(\overrightarrow{\alpha_\ell}) = \lambda_\ell \overrightarrow{\alpha_\ell} + u_\ell \sum_{\ell' \in L} u_{\ell'} \overrightarrow{\alpha_{\ell'}}$. $\qquad \square$

Target example

Recall (see page 14) that $\mathbf{V} = \begin{bmatrix} 10 & 2 \\ 2 & 13 \end{bmatrix}$, and $\overline{\mathbf{x}} = \mathbf{X}^\top \mathbf{e}/10 = \begin{bmatrix} 3 \\ 0 \end{bmatrix}$.

Let $\overrightarrow{\varepsilon_1}$ and $\overrightarrow{\varepsilon_2}$ be the unit vectors of the horizontal and vertical axes. With respect to this orthonormal basis $(\overrightarrow{\varepsilon_1}, \overrightarrow{\varepsilon_2})$, the matrix of scalar products is $\mathbf{Q} = \begin{bmatrix} 1 & 0 \\ 0 & 1 \end{bmatrix}$, the matrix of the covariance endomorphism Mcov is equal to $\mathbf{V}$ and that of McovO is $\mathbf{V_O} = \begin{bmatrix} 10 & 2 \\ 2 & 13 \end{bmatrix} + \begin{bmatrix} 3 \\ 0 \end{bmatrix} \begin{bmatrix} 3 & 0 \end{bmatrix} = \begin{bmatrix} 19 & 2 \\ 2 & 13 \end{bmatrix}$.

The unit principal vectors (eigenvectors of $\mathbf{V}$) are $\overrightarrow{\alpha_1} = \frac{1}{\sqrt{5}}(\overrightarrow{\varepsilon_1} + 2\overrightarrow{\varepsilon_2})$ and $\overrightarrow{\alpha_2} = \frac{1}{\sqrt{5}}(-2\overrightarrow{\varepsilon_1} + \overrightarrow{\varepsilon_2})$, respectively associated with $\lambda_1 = 14$ and $\lambda_2 = 9$, hence the change of basis matrix (cf. §7.2.4, p.228) is $\mathbf{A} = \frac{1}{\sqrt{5}} \times \begin{bmatrix} 1 & -2 \\ 2 & 1 \end{bmatrix}$.

The matrix of principal coordinates is $\mathbf{Y} = (\mathbf{X} - \mathbf{e}\overline{\mathbf{x}}^\top)\mathbf{A}$.

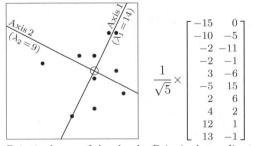

$$\frac{1}{\sqrt{5}} \times \begin{bmatrix} -15 & 0 \\ -10 & -5 \\ -2 & -11 \\ -2 & -1 \\ 3 & -6 \\ -5 & 15 \\ 2 & 6 \\ 4 & 2 \\ 12 & 1 \\ 13 & -1 \end{bmatrix}$$

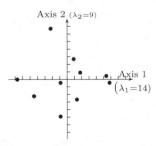

Principal axes of the cloud. Principal coordinates. Cloud referred to principal axes.

2.3 Mahalanobis Distance and Principal Ellipsoids

In this section, we present two notions that are both linked to the covariance structure and extensively used in the following chapters: firstly the Mahalanobis distance, which will be used as a test statistic of the combinatorial test for the mean point of a cloud (Chapter 3), and secondly the principal (or inertia) ellipsoids of a cloud, which provide fundamental elements to define compatibility regions.

Let us consider the L-dimensional affine support of the cloud $\mathcal{M} \subseteq \mathcal{U}$ (see p. 10) and its direction vector subspace $\mathcal{L} \subseteq \mathcal{V}$ with its two orthogonal principal bases: the *orthonormal principal basis* $(\overrightarrow{\alpha_\ell})_{\ell \in L}$, and the M-orthonormal one $(\overrightarrow{\beta_\ell})_{\ell \in L}$ with $\overrightarrow{\beta_\ell} = \sqrt{\lambda_\ell}\, \overrightarrow{\alpha_\ell}$, which is called *"orthocalibrated" principal basis*.

From now on, we will consider the restrictions of endomorphisms to the subspace $\mathcal{L} \subseteq \mathcal{V}$ (see p. 233).

2.3.1 Mahalanobis Distance

The restriction of the covariance endomorphism of cloud M^I to $\mathcal{L}$, which, for simplicity of notation, is still denoted Mcov, is non-singular then invertible and $\forall \overrightarrow{u} \in \mathcal{L}$, $\langle \mathrm{Mcov}^{-1}(\overrightarrow{u}) | \overrightarrow{u} \rangle > 0$, hence the Mahalanobis norm on $\mathcal{L}$ is defined as follows.

Definition 2.6. *The* Mahalanobis norm *on $\mathcal{L}$, denoted $|\cdot|$, is such that*

$$\forall \overrightarrow{u} \in \mathcal{L}, \; |\overrightarrow{u}|^2 = \langle \mathrm{Mcov}^{-1}(\overrightarrow{u}) | \overrightarrow{u} \rangle$$

Proposition 2.12. $\forall \overrightarrow{u} = \sum u_\ell \overrightarrow{\alpha_\ell}$, $|\overrightarrow{u}|^2 = \sum u_\ell^2 / \lambda_\ell$ *and, letting* $z_\ell = u_\ell / \sqrt{\lambda_\ell}$, $|\overrightarrow{u}|^2 = \sum z_\ell^2$.

We now study the covariance structure of the cloud when the support of the cloud is equipped with the scalar product induced by $\mathrm{Mcov} : \mathcal{L} \to \mathcal{L}$. More precisely, this scalar product on $\mathcal{L}$, denoted $[\cdot|\cdot]$, is defined by

$$\forall \overrightarrow{u}, \overrightarrow{v} \in \mathcal{L}, \; [\overrightarrow{u}|\overrightarrow{v}] = \langle \overrightarrow{u} | \mathrm{Mcov}^{-1}(\overrightarrow{v}) \rangle = \langle \mathrm{Mcov}^{-1}(\overrightarrow{u}) | \overrightarrow{v} \rangle \qquad (2.7)$$

Since the restriction of Mcov to $\mathcal{L}$ is symmetric and positive definite, the bilinear form $[\cdot|\cdot]$ defines a scalar product on $\mathcal{L}$ (see Definition 7.3, p. 229).

Notations

. The scalar product related to Mcov is called the M-*scalar product*. The *Mahalanobis distance* between two points A and B in $\mathcal{M}$ is called the M-*distance* and simply denoted $|AB|$.

Recall that the geometric norm of vector $\overrightarrow{u}$ is denoted $\|\overrightarrow{u}\|$, the *geometric distance* between two points A and B is denoted AB and the scalar product is denoted $\langle \cdot | \cdot \rangle$.

The covariance endomorphism of cloud M^I defined from the M-scalar product, called M-*covariance endomorphism* and denoted M.Mcov, is such that

$$\forall \vec{u} \in \mathcal{L}, \ \mathrm{M.Mcov}(\vec{u}) = \sum f_i [\overrightarrow{\mathrm{GM}^i} | \vec{u}] \overrightarrow{\mathrm{GM}^i} \tag{2.8}$$

The M-*variance of the cloud* is equal to $\sum f_i |\mathrm{GM}^i|^2$; the M-variance in a direction is the M-variance of the M-projected cloud onto this direction, etc.

Proposition 2.13. *The* M-*covariance endomorphism* $\mathcal{L} \to \mathcal{L}$ *of cloud* M^I *is the identity endomorphism of* $\mathcal{L}$ *(denoted* id_L*):* M.Mcov $= id_L$.

Proof. $\forall \vec{u} \in \mathcal{L}, \ \sum f_i [\overrightarrow{\mathrm{GM}^i} | \vec{u}] \, \overrightarrow{\mathrm{GM}^i}$ $= \sum f_i \langle \overrightarrow{\mathrm{GM}^i} | \mathrm{Mcov}^{-1}(\vec{u}) \rangle \overrightarrow{\mathrm{GM}^i}$
$= \mathrm{Mcov}(\mathrm{Mcov}^{-1}(\vec{u})) = \vec{u}.$ □

Corollary 2.3. *The* M-*variance of the cloud* M^I *in any direction of* $\mathcal{M}$ *is equal to 1 and the total* M-*variance is equal to L (dimension of the support).*

Proof. By Proposition 2.4 (p.15), the M-variance of the cloud in the direction defined by the M-unit vector $\vec{u}$ is equal to $[id_L(\vec{u}) | \vec{u}] = |\vec{u}|^2 = 1$. Therefore, the total variance is equal to $\sum_{\ell \in L} [id_L(\vec{\beta_\ell}) | \vec{\beta_\ell}] = \sum_{\ell \in L} |\vec{\beta_\ell}|^2 = L$. □

Definition 2.7. *Given* $\mathrm{P} \in \mathcal{M}$, *the generalized Mahalanobis norm with respect to point* P, *denoted* $|\cdot|_\mathrm{P}$, *is defined by*

$$\forall \vec{u} \in \mathcal{L}, \ |\vec{u}|_\mathrm{P}^2 = [\mathrm{Mcov}^{-1}_\mathrm{P}(\vec{u}) | \vec{u}]$$

Theorem 2.2. $\forall \vec{u} \in \mathcal{L}, \ |\vec{u}|_\mathrm{P}^2 = |\vec{u}|^2 - \dfrac{[\overrightarrow{\mathrm{PG}} | \vec{u}]^2}{1 + |\mathrm{PG}|^2}.$

Proof. The inverse of Mcov^P is such that $\vec{u} \mapsto \mathrm{Mcov}^{-1}(\vec{u}) - \frac{[\overrightarrow{\mathrm{PG}} | \vec{u}]}{1 + |\mathrm{PG}|^2} \mathrm{Mcov}^{-1}(\overrightarrow{\mathrm{PG}})$
(see Proposition 7.3, p. 233); hence $|\vec{u}|_\mathrm{P}^2 = |\vec{u}|^2 - \frac{[\overrightarrow{\mathrm{PG}} | \vec{u}]^2}{1 + |\mathrm{PG}|^2}.$ □

The following corollary is a direct consequence of the preceding theorem.

Corollary 2.4 (Reciprocity formula). $\forall \mathrm{P} \in \mathcal{M}, \ |\mathrm{PG}|_\mathrm{P}^2 = \dfrac{|\mathrm{PG}|^2}{1 + |\mathrm{PG}|^2}$

2.3.2 Principal (or Inertia) Ellipsoids

The principal ellipsoid of a cloud gives geometric characterizations of the statistical properties regarding location and scattering.

Definition 2.8. *Given* $\kappa > 0$, *the* principal *or* inertia κ-*ellipsoid* $\mathcal{E}_\kappa$ *of the cloud is defined as the set of points* $\mathrm{P} \in \mathcal{M}$ *such that* $|\mathrm{GP}| = \kappa$.

$$\mathcal{E}_\kappa = \{\mathrm{P} \in \mathcal{M} \text{ such that } |\mathrm{GP}| = \kappa"\}$$

κ is called the *scale parameter* of the ellipsoid.

The *indicator ellipsoid* is the principal κ-ellipsoid with $\kappa = 1$, and the *concentration ellipsoid* is that with $\kappa = \sqrt{L+2}$ (see Cramér, 1946, p. 300).

Proposition 2.14. *The principal κ-ellipsoid of the cloud is centred at point* G *and the half-lengths of its principal axes are equal to* $(\kappa \sqrt{\lambda_\ell})_{\ell \in L}$.

Proposition 2.15 (Projection of an ellipsoid). *If a cloud is orthogonally projected onto a subspace, the κ-ellipsoid of the projected cloud is the projection of the κ-ellipsoid of the cloud onto this subspace.*

For a proof, see Le Roux and Rouanet (2004, p. 97).

Magnitude of deviation

Recall that a numerical variable is *standardized* by subtracting its mean and dividing the difference by its standard deviation. It is *scaled* with respect to a reference population by subtracting the mean of the reference population and dividing the difference by the standard deviation of the reference population.

The (absolute) *scaled deviation* between means is commonly taken as a descriptive indicator of the magnitude of deviation[8]. It will be regarded as "notable" if it is greater than 0.4 and small if not. This convention is a rule of thumb; the value of notable limit should be reconsidered in each practical situation (see, e.g., Cohen, 1977; Rouanet, 1996).

For any point $P \in \mathcal{M}$ with principal coordinates $(y_\ell)_{\ell \in L}$, the M-distance between points P and G, which is equal to $(\sum y_\ell^2 / \lambda_\ell)^{1/2}$, is an *index of magnitude* of the deviation between points P and G that takes into account the shape of the cloud. This index is the multidimensional extension of the scaled deviation for numerical variables. It can be used as an *extremality index* of point P among the points of the cloud: the higher the index, the more extreme the point (as will be seen on page 25 for the *Target Example*).

In combinatorial inference, this index applied to mean points of subclouds obtained by sampling will be fundamental (see Chapter 3).

Concentration ellipsoids and continuous distributions

The concept of principal ellipsoids applies to *continuous distributions*. For the multidimensional normal and related distributions (such as the t-distribution), principal ellipsoids are equal density contours, and the density at every point inside an ellipsoid is higher than at every point outside.

Given an L-dimensional cloud together with its κ-ellipsoid, one may consider a distribution of uniform density over the domain bounded by the ellipsoid; it can be shown that this distribution has the same principal axes

---

[8]Let n be the size of the reference population; the scaled deviation is equal to $\sqrt{n/(n-1)}$ times the "effect size index" defined by Cohen (1977) which seems commonly used in behavioral research.

as the cloud and that its eigenvalues are equal to $\left(\lambda_\ell \, \kappa^2/(L+2)\right)_{\ell \in L}$. Consequently, a uniform distribution over the domain bounded by the L-dimensional κ-ellipsoid with $\kappa^2 = L + 2$ has the same eigenvalues as the cloud; such an ellipsoid is called the *concentration ellipsoid* of the cloud; see Cramér (1946, p. 300); Anderson (2003, p. 44); Malinvaud (1981).

For the normal L-dimensional distribution, the proportion of the distribution inside the κ-ellipsoid is given by $p(\chi_L^2 \leq \kappa^2)$, where χ_L^2 denotes the classical chi-squared variable with L degrees of freedom. For a given value of κ, the higher the value of L, the smaller the corresponding proportion.

For $L = 1$, the "indicator interval" ($\kappa = 1$) whose half-length is equal to one standard deviation contains 68.27% of the distribution; the "concentration interval" ($\kappa = 2$) whose half-length is equal to two standard deviations contains 95.45% of the distribution.

For $L = 2$, the *indicator ellipse* ($\kappa = 1$) contains 39.35% of the distribution; the *concentration ellipse* ($\kappa = 2$) contains 86.47% of the distribution.

2.3.3 Numerical Expressions and Matrix Formulas

For any point $\mathrm{P} \in \mathcal{M}$ whose coordinates with respect to the *orthonormal principal frame* $(\mathrm{G}, (\overrightarrow{\alpha_\ell})_{\ell \in L})$ of $\mathcal{M}$ are $(y_\ell)_{\ell \in L}$ (with $\overrightarrow{\mathrm{GP}} = \sum y_\ell \overrightarrow{\alpha_\ell}$), then $|\mathrm{GP}|^2 = \sum y_\ell^2/\lambda_\ell$ and the equation of the κ-ellipsoid is

$$\sum_{\ell \in L} \frac{y_\ell^2}{\lambda_\ell} = \kappa^2$$

Letting $\mathbf{y} = [y_\ell]$, one has $|\mathrm{GP}|^2 = \mathbf{y}^\top \mathbf{\Lambda}^{-1} \mathbf{y}$. The Cartesian equation of the κ-ellipsoid is written $\mathbf{y}^\top \mathbf{\Lambda}^{-1} \mathbf{y} = \kappa^2$.

If $(z_\ell)_{\ell \in L}$ are the coordinates of point P with respect to the *orthocalibrated principal frame* $(\mathrm{G}, (\overrightarrow{\beta_\ell})_{\ell \in L})$, we have $\overrightarrow{\mathrm{GP}} = \sum z_\ell \overrightarrow{\beta_\ell}$, then $|\mathrm{GP}|^2 = \sum z_\ell^2$ and the equation of the κ-ellipsoid is

$$\sum_{\ell \in L} z_\ell^2 = \kappa^2$$

Letting $\mathbf{z} = [z_\ell]$, the Cartesian equation of the κ-ellipsoid writes $\mathbf{z}^\top \mathbf{z} = \kappa^2$.

Plane Cloud

With respect to any orthonormal basis $(\overrightarrow{e}, \overrightarrow{e}')$ of the support of the plane cloud, where $v = \langle \mathrm{Mcov}(\overrightarrow{e}) | \overrightarrow{e} \rangle$ and $v' = \langle \mathrm{Mcov}(\overrightarrow{e}') | \overrightarrow{e}' \rangle$ are the variances in the directions $\overrightarrow{e}$ and $\overrightarrow{e}'$ and $c = \langle \mathrm{Mcov}(\overrightarrow{e}) | \overrightarrow{e}' \rangle$ the covariance, the equation of the principal κ-ellipse is:

$$\frac{1}{vv'-c^2}(v'x^2 - 2cxx' + vx'^2) = \kappa^2$$

With respect to the principal orthonormal basis $(\overrightarrow{\alpha_1}, \overrightarrow{\alpha_2})$, the equation of the principal κ-ellipse is $y_1^2/\lambda_1 + y_2^2/\lambda_2 = \kappa^2$

The half-length of a principal axis of the concentration ellipse is equal to two times the standard deviation of the cloud in the direction of the axis[9]. This ellipse often constitutes a good geometric summary of the plane cloud.

Target example

Now, we will represent the indicator and concentration ellipses of the *"Target"*, firstly in the geometric space (Figure 2.6), secondly in the Mahalanobis space, that is, in the subspace $\mathcal{M}$ of $\mathcal{U}$ associated with the direction vector subspace $\mathcal{L} \in \mathcal{V}$ equipped with the M-scalar product (Figure 2.7).

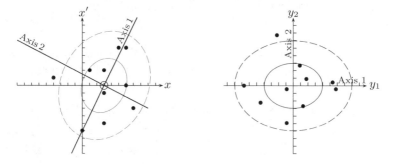

Figure 2.6
Target example. Indicator ellipse ($\kappa = 1$, solid line) and concentration ellipse ($\kappa = 2$, dashed line) referred to orthonormal basis (left figure) and to principal basis (right figure), with $v = 10$, $v' = 13$, $c = 2$ and $\lambda_1 = 14$, $\lambda_2 = 9$.

The locus of points P whose M-distance to point G is equal to 1 is the indicator ellipse in the geometric space (Figure 2.6) and a circle with centre G and radius 1 in the Mahalanobis space (Figure 2.7).

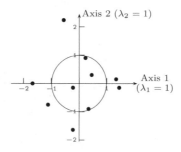

Figure 2.7
Target example. Representation of the cloud in the Mahalanobis space (the graphic unit is that of the M-distance) and circle with centre G and radius 1.

---

[9]In the seminal paper by Pearson (1901), a geometric definition of PCA with the use of principal ellipses is given.

Extremality index

From Figure 2.8, we see that points M and M′ are both at the same geometric distance (here 5) from the mean point (circle whose radius is equal to 5, in dashed line).

Now for M, one has $|GM| = 1.41$ (black ellipse) and for M′, one has $|GM'| = 1.66$ (grey ellipse); therefore point M′ (which is closer to Axis 2) is less extreme than point M (which is closer to Axis 1).

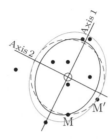

Figure 2.8
Extremality index for points M and M′.

2.4   Partition of a Cloud

The concepts presented in this section are especially useful for determining compatibility regions in the case of homogeneity situations, as we will see in Chapter 5 (p. 120).

Subcloud

Consider a subset of I termed "class c" and denoted $I{<}c{>}$ (read "I in c") or I_c. The associated subcloud is denoted $M^{I{<}c{>}}$.

- Its *weight* is $n_c = \sum\limits_{i \in I{<}c{>}} n_i$;

- its *mean point*, denoted G^c, is such that $G^c = \sum\limits_{i \in I{<}c{>}} n_i M^i / n_c$,

- its *variance* is equal to $\sum\limits_{i \in I{<}c{>}} n_i (G^c M^i)^2 / n_c$

- its *covariance endomorphism* is such that

$$\vec{u} \mapsto \sum\limits_{i \in I{<}c{>}} n_i \langle \overrightarrow{G^c M^i} | \vec{u} \rangle \overrightarrow{G^c M^i} / n_c \quad (\vec{u} \in \mathcal{U})$$

Between-C cloud

Consider a partition of I into C classes[10], denoted $I{<}C{>}$, hence the partition of cloud M^I into C subclouds $(M^{I<c>})_{c \in C}$ (see, e.g., Figure 2.9)[11].

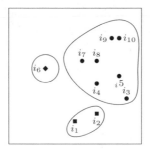

Figure 2.9

Target example. Three-class partition of cloud M^I: $I{<}c_1{>}= \{i_1, i_2\}$ (■), $I{<}c_2{>}= \{i_6\}$ (♦) and $I{<}c_3{>}= \{i_3, i_4, i_5, i_7, i_8, i_9, i_{10}\}$ (●).

The *between-C cloud* consists of the C mean points $(G^c)_{c \in C}$ of subclouds; it is denoted G^C. Each point G^c is weighted by n_c.

- Its mean point is the mean point G of the overall cloud M^I;

- its variance, denoted Var_C, is called *between-C variance*. One has (see Definition 2.3, p. 11 and Proposition 2.2, p. 12):

$$\mathrm{Var}_C = \sum n_c (GG^c)^2 / n = \tfrac{1}{2} \sum \sum n_c n_{c'} (G^c G^{c'})^2 / n^2 \qquad (2.9)$$

- its *covariance endomorphism*, denoted Mcov_C, is such that

$$\forall \overrightarrow{u} \in \mathcal{U}, \ \mathrm{Mcov}_C(\overrightarrow{u}) = \sum_{c \in C} \tfrac{n_c}{n} \langle \overrightarrow{GG^c} | \overrightarrow{u} \rangle \overrightarrow{GG^c} \qquad (2.10)$$

In particular, if the between-C cloud only consists of two points (G^c, n_c) and $(G^{c'}, n_{c'})$, and denoting the relative frequencies by $f_c = n_c/n$ and $f_{c'} = n_{c'}/n$, one has (see Corollary 2.3, p. 16)

$$\forall \overrightarrow{u} \in \mathcal{U}, \ \mathrm{Mcov}_C(\overrightarrow{u}) = f_c f_{c'} \langle \overrightarrow{G^c G^{c'}} | \overrightarrow{u} \rangle \overrightarrow{G^c G^{c'}}$$

---

[10]In general we denote the cardinality (number of elements) of a finite set as the set itself; the cardinality of I, C, etc., often denoted $|I|, |C|$, etc., will be denoted I, C, etc.

[11]Owing to the definition of a cloud (see Definition 2.1, p. 10), two points can belong to different classes and have the same location in the geometric space.

Within-C cloud

For each class c of the partition, we define the *within-c cloud* as the translation of the subcloud $M^{I<c>}$ by vector $\overrightarrow{G^cG}$ (see p. 240): for $i \in I<c>$, the point $M^{i(c)}$ (weighted by n_i) is the point $M^i + \overrightarrow{G^cG}$ (see opposite figure).

The weight of the within-c cloud is equal to n_c; its mean point is G; its variance, denoted $\text{Var}_{I(c)}$, is equal to the variance of the subcloud $M^{I<c>}$:

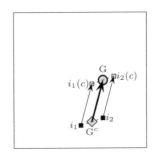

Figure 2.10
Target example. Within-c_2 cloud.

$$\text{Var}_{I(c)} = \sum_{i \in I<c>} n_i (G^cM^i)^2 / n_c$$

Within notation. In $I(c)$ (read "*I* within *c*"), "*c*" is between parentheses in order to mark the difference between the point M^i of the overall cloud (black point) and the point $M^{i(c)}$ of the within-cloud (grey point).

The *within-C cloud*, denoted $M^{I(C)}$, is the union of the C within-c clouds.

$$\forall c \in C, \forall i \in I<c>, \; M^{i(c)} = M^i + \overrightarrow{G^cG} \tag{2.11}$$

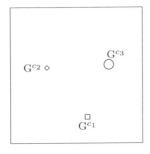

Figure 2.11
Target example. Between-C cloud G^C and within-C cloud $M^{I(C)}$.

- Its *mean point* is G;

- its variance, denoted $\text{Var}_{I(C)}$ and called *within-C variance*, is the weighted mean of the variances of the subclouds:

$$\text{Var}_{I(C)} = \sum_{c \in C} f_c \, \text{Var}_{I(c)}$$

- its *covariance endomorphism*, denoted $\text{Mcov}_{I(C)}$, is the weighted mean of the covariance endomorphisms of the subclouds.

$$\text{Mcov}_{I(C)} = \sum_{c \in C} f_c \text{Mcov}_{I(c)} \tag{2.12}$$

Theorem 2.3. *The covariance endomorphism of the cloud is the sum of that of the between-C cloud and of that of the within-C cloud:*

$$\text{Mcov} = \text{Mcov}_C + \text{Mcov}_{I(C)}$$

Proof. By applying Proposition 2.5 (p.16) to subcloud $\text{M}^{I<c>}$, one has: $\forall \vec{u} \in \mathcal{V}$,
$\sum_{i \in I<c>} \frac{n_i}{n_c} \langle \overrightarrow{\text{GM}^i} | \vec{u} \rangle \overrightarrow{\text{GM}^i} = \langle \overrightarrow{\text{G}\text{G}^c} | \vec{u} \rangle \overrightarrow{\text{G}\text{G}^c} + \sum_{i \in I<c>} \frac{n_i}{n_c} \langle \overrightarrow{\text{G}^c\text{M}^i} | \vec{u} \rangle \overrightarrow{\text{G}^c\text{M}^i}$ and thus, by
multiplying by n_c/n and summing over c we get the theorem. $\square$

The following formula is an immediate consequence of Theorem 2.3.

Corollary 2.5. *The total variance is the sum of the between-C variance and of the within-C variance:*

$$V_{\text{cloud}} \quad = \quad \text{Var}_C \quad + \quad \text{Var}_{I(C)}$$
Total variance = between-C variance + within-C variance

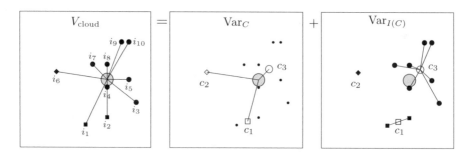

Figure 2.12
Target example. Illustration of variance components.

The ratio of the between-C variance to the total variance defines the eta-squared coefficient (η^2) of the partition.

$$\eta^2 = \text{Var}_C / V_{\text{cloud}} \tag{2.13}$$

The coefficient η is called *correlation ratio*, and it takes values between 0 and 1. The limit $\eta = 0$ represents the special case of no dispersion among the means of the classes, while $\eta = 1$ refers to no dispersion within the classes.

The η^2 coefficient is often taken as a *descriptive index of the magnitude* of the discrepancy between the C classes. For C' classes ($C' < C$), we will take n'/n times the ratio of the variance of the cloud of the C' mean points to the total variance (partial η^2):

$$\eta'^2 = \tfrac{n'}{n} \text{Var}_{C'} / V_{\text{cloud}} \tag{2.14}$$

These indexes will be regarded as "notable"[12] if they are greater than 0.04.

---

[12] As said previously in this chapter (p. 22), this notable limit is a rule of thumb that may be reconsidered in each practical situation.

3

Combinatorial Typicality Tests

In this chapter we will present the *combinatorial* (or set-theoretic) *test of typicality*, in short, *typicality test*. It consists in comparing the observations of a group with the ones of a reference population of which the group may or may not be a subset. Intuitively, the test answers the *question*:

> *Can the group be assimilated to the reference population?*
>
> *Is it typical of it?*

or, more specifically,

> *"How can a typicality level be assessed for the group with respect to the population, according to some statistic of interest?*

The test will be applied to case studies in Chapter 6.

3.1 The Typicality Problem

In this section, we begin by describing three situations for which the typicality problem can be raised, then we outline briefly the principle of the combinatorial typicality test.

3.1.1 Typicality Situations

Let us consider the following situations.

Gifted children

In a follow-up study on five gifted children, a psychologist found that for a certain task the mean grade of the group is 20, whereas, for a reference population of children of the same age, the mean is known to be 15 and the standard deviation 6.

> *Is the psychologist entitled to claim that the group of gifted children is, on average, superior to the reference population of children?*

Robespierre's speeches

In *textual analysis*, many situations are relevant to combinatorial inference. For instance, let us consider the ten speeches delivered by Robespierre at the "*Convention*" between November 1793 and July 1794[1]. The word "*je*"— translation: "*I*", subject pronoun in the first person singular—occurs 194 times in the ten speeches (61,449 words) and 121 times in the tenth one (13,933 words) (see Lafon, 1984; Rouanet et al., 1990, p. 165).

The question is:

Does the last speech differ from all ten as far as the frequency of occurrence of the word "je" is concerned?

In this case, putting a probabilistic model on the data would be incongruous, as Robespierre surely did not pick his words randomly from an urn!

Target Example

To exemplify the typicality test in a geometric setup, we will take again the *Target Example* (see Chapter 2, p. 9). The data set consists of a target with 10 impact points and, in addition, a target with 4 impact points (Figure 3.1).

The question is:

Is the 4-impact target shooter the same as the 10-impact one?

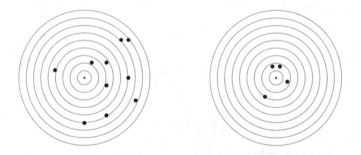

Figure 3.1
Target example. Target with 10 impacts and target with 4 impacts.

In order to offer a solution to the typicality problem, the basic idea is to compare the *group* (5 gifted children, last speech, 4-impact target) to the samples of the *reference set* or *reference population*[2] (children of the same age, the ten speeches, the 10-impact target), where samples are simply defined

---

[1]Robespierre delivered his ultimate speech at the "*Convention*" two days before he was guillotined in July 28, 1794.

[2]In the context of *finite sampling*, the term *population* will always refer to a *finite set*. In the sequel we will speak indifferently of reference population or reference set.

as subsets of the reference set. For this purpose, we construct the set of all *possible samples* that have the same number of elements as the group, and we locate the observations of the group among the ones of the possible samples according to some *statistic of interest*.

3.1.2 Principle of Combinatorial Typicality Test

The reference population consisting of n individuals[3] is denoted by I and the group consisting of n_c $(1 < n_c < n)$ individuals by C; the group may or may not be a subset of I.

To answer the question, we construct a combinatorial typicality test[4] according to the following general principle.

1. *Sample set*

   The first step is the definition of a set of *samples* of the reference population. In combinatorial inference, a "sample of the population" is defined as an n_c-element *subset* of the reference set I ("samples" in a plain set-theoretic sense).

   The set of all $\binom{n}{n_c}$ samples defines the *sample set*.

2. *Test statistic*

   The next step is to choose a test statistic that measures the *effect of interest*. In this chapter, we will focus on two types of statistic; one is based on mean points (location parameters), the other on variances (scale parameters).

3. *Combinatorial p-value*

   Then, the proportion of samples, whose statistic value is greater than or equal to the observed one, is taken as defining the degree of typicality of the group with respect to the reference population and called *combinatorial p-value*. The smaller the p-value, the lower the typicality.

   A low p-value will be declared "statistically significant" in a combinatorial sense[5] and will point to a genuine effect, presumably "not due to chance".

4. *Conclusion*

   Taking a reference level α, called α-level[6], we state the conclusion in the following words:

---

[3] In statistics, the elements of the population are often called members or *individuals* (see Cramér, 1946, p. 324). Individuals mean *statistical individuals* (also called cases), that is to say, people, or states, or firms, etc.

[4] See Le Roux (1998); Le Roux and Rouanet (2004); Bienaise (2013, Chapter 3); Bienaise and Le Roux (2017).

[5] This combinatorial conception of significance level exactly corresponds to the "non-stochastic interpretation" presented by Freedman and Lane (1983).

[6] To assess the more or less extreme character of p-values it is useful to use landmarks, that is, *conventional reference levels* (between 0 and 1) traditionally called α-levels. In the sequel, we will take $\alpha = .05$.

- if the p-value is less than or equal to α, the group is said to be *atypical* of the reference population at level α for the property of interest;

- if the p-value is greater than α, the group cannot be declared atypical; reference population and group are said to be *compatible at level α*.

Throughout the book, we will deal with procedures for multidimensional data conceptualized as clouds in a geometric space.

Firstly, we will focus on the inference on the mean point of a cloud using a test statistic based on the Mahalanobis distance between mean points (§3.2). Secondly, we will study the particular case of one-dimensional cloud (§3.3). Thirdly, we will introduce the typicality test for studying the variance of a cloud (§3.4). Then, we will apply the procedures to a cloud stemming from a GDA method (§3.5). Finally, we will give the R script used in connection with the SPAD software (§3.6).

3.2   Combinatorial Typicality Test for Mean Point

As already said, we deal with procedures for multidimensional data conceptualized as clouds in a geometric space, that is, Euclidean clouds. The study is done in the line of *pure geometry*, namely independently of any specific choice of a coordinate system.

The set I of individuals is represented by a cloud of n points in a geometric space $\mathcal{U}$, called *reference cloud* and denoted M^I; its mean point is denoted G and called *reference mean point*.

The set C of the individuals of the group is represented[7] by a cloud of n_c points in $\mathcal{U}$, called *group cloud* and denoted M^C; its mean point is denoted C and called *observed mean point*.

> Of course, a coordinate system has to be chosen to carry out computations. In the R scripts (see §3.6, p. 55), for the sake of simplicity, we choose an orthonormal basis, and we work on the database consisting of coordinates of points in this basis. In most cases, we choose the *orthonormal principal basis* associated with the reference cloud and work on the *principal coordinates* of points (see Chapter 2, p. 17).

The combinatorial typicality test presented in this section is based on mean points, that is, the property of interest brings into play *mean points* of subclouds. In a certain sense, we can say it is a procedure for comparing the mean point of a cloud to the mean point of a reference cloud.

---

[7] As already noted, the cloud M^C may or may not be a subcloud of the reference cloud. If it is not a subcloud of M^I, and if it does not lie in the affine support of the reference cloud, we will study its orthogonal projection on it.

3.2.1 Sample Set

The *sample set* is the set of all n_c-element subsets of I, it is denoted J and its cardinality is equal to $\binom{n}{n_c}$. The subset of the n_c individuals of I belonging to sample j is denoted I_j.

In the sequel, it will often be convenient to use the table of correspondence between sets J and I. This table, called *sample table*, is a $J \times I$ array of numbers 0 and 1, with in cell (j, i) the value $\delta_i^j = 1$ if individual i belongs to sample j and $\delta_i^j = 0$ if not.

$$\delta_i^j = \left\{ \begin{array}{ll} 1 & \Longleftrightarrow \quad i \in I_j \\ 0 & \Longleftrightarrow \quad i \notin I_j \end{array} \right.$$

The following properties hold:

- $\forall j \in J, \; \sum_{i \in I} \delta_i^j = n_c$;

- $\forall i \in I, \; \sum_{j \in J} \delta_i^j = \binom{n-1}{n_c-1}$: number of combinations of $n_c - 1$ elements among the $n - 1$ elements of I other than i;

- $\forall i, i' \in I, i \neq i', \; \sum_{j \in J} \delta_i^j \delta_{i'}^j = \binom{n-2}{n_c-2}$: number of combinations of $n_c - 2$ elements among the $n - 2$ elements of I other than i and i'.

3.2.2 Sample Space

The subset I_j is represented by the subcloud of cloud M^I consisting of the n_c points $(\mathrm{M}^i)_{i \in I_j}$; it is denoted by M^{I_j}. The family $(\mathrm{M}^{I_j})_{j \in J}$ of these $\binom{n}{n_c}$ subclouds is called *sample space* and denoted by $\mathcal{J}$.

$$\mathcal{J} = (\mathrm{M}^{I_j})_{j \in J}$$

3.2.3 Sample Cloud

Let $\mathrm{H} : \mathcal{J} \to \mathcal{U}$ be the map that associates with each subcloud of the sample space its mean point.

$$\begin{array}{rcl} \mathrm{H} : & \mathcal{J} & \to \quad \mathcal{U} \\ & \mathrm{M}^{I_j} & \mapsto \quad \mathrm{H}^j = \sum_{i \in I_j} \mathrm{M}^i / n_c \end{array}$$

The map H defines the cloud H^J which is called *sample cloud*; its number of points is equal to $\binom{n}{n_c}$.

Using the sample table, the mean point H^j can be defined by the relation

$$\forall \mathrm{P} \in \mathcal{U}, \; \overrightarrow{\mathrm{PH}^j} = \sum_{i \in I} \delta_i^j \overrightarrow{\mathrm{PM}^i} / n_c$$

The two following propositions are, in the geometric setup, the generalization of the properties of the sampling distribution of the mean of a numerical variable, when the sampling is without replacement in a finite population (see Cramér, 1946, p. 523). The first one is related to the mean point of the sample cloud, the second one to its covariance endomorphism.

Proposition 3.1. *The mean point of the sample cloud is the mean point of the reference cloud.*

Proof. $\forall \mathrm{P} \in \mathcal{U}$, $\sum_{j \in J} \overrightarrow{\mathrm{PH}^j} = \sum_{i \in I} \left(\sum_{j \in J} \delta_i^j \right) \overrightarrow{\mathrm{PM}^i} / n_c = \binom{n-1}{n_c-1} \sum_{i \in I} \overrightarrow{\mathrm{PM}^i} / n_c$ since $\sum_{j \in J} \delta_i^j = \binom{n-1}{n_c-1}$. Yet $\sum_{i \in I} \overrightarrow{\mathrm{PM}^i} / n = \overrightarrow{\mathrm{PG}}$ hence $\sum_{j \in J} \overrightarrow{\mathrm{PH}^j} / \binom{n}{n_c} = \frac{n}{n_c} \binom{n-1}{n_c-1} \overrightarrow{\mathrm{PG}} / \binom{n}{n_c} = \overrightarrow{\mathrm{PG}}$, therefore the mean point of cloud H^J is G. $\square$

Let us recall the definition of the covariance endomorphism of a cloud (see Chapter 2, p. 15). If $\mathcal{V}$ denotes the Euclidean vector space associated with $\mathcal{U}$, the covariance endomorphism Mcov of the cloud M^I is defined as the map

$$\mathrm{Mcov}: \quad \begin{aligned} \mathcal{V} &\rightarrow \mathcal{V} \\ \overrightarrow{u} &\mapsto \sum_{i \in I} \frac{1}{n} \langle \overrightarrow{\mathrm{GM}^i} | \overrightarrow{u} \rangle \overrightarrow{\mathrm{GM}^i} \end{aligned}$$

Proposition 3.2. *The covariance endomorphism of the sample cloud, denoted* Hcov, *is proportional to that of the reference cloud:*

$$\mathrm{Hcov} = \frac{n - n_c}{n - 1} \times \frac{1}{n_c} \times \mathrm{Mcov}$$

Proof. By definition, the covariance endomorphism of cloud H^J is such that $\forall \overrightarrow{u} \in \mathcal{V}$, $\mathrm{Hcov}(\overrightarrow{u}) = \sum_{j \in J} \langle \overrightarrow{\mathrm{GH}^j} | \overrightarrow{u} \rangle \overrightarrow{\mathrm{GH}^j} / \binom{n}{n_c}$. Hence

$$\mathrm{Hcov}(\overrightarrow{u}) = \sum_{j \in J} \left\langle \sum_{i \in I} \delta_i^j \overrightarrow{\mathrm{GM}^i} / n_c \, \middle| \, \overrightarrow{u} \right\rangle \left(\sum_{i' \in I} \delta_{i'}^j \overrightarrow{\mathrm{GM}^{i'}} / n_c \right) / \binom{n}{n_c}$$

$$= \frac{1}{n_c^2} \sum_{i \in I} \sum_{i' \in I} \left(\sum_{j \in J} \delta_i^j \delta_{i'}^j \right) \langle \overrightarrow{\mathrm{GM}^i} | \overrightarrow{u} \rangle \overrightarrow{\mathrm{GM}^{i'}} / \binom{n}{n_c}$$

$$= \frac{1}{n_c^2} \left[\sum_{i \in I} \sum_{j \in J} (\delta_i^j)^2 \langle \overrightarrow{\mathrm{GM}^i} | \overrightarrow{u} \rangle \overrightarrow{\mathrm{GM}^i} + \sum_{i \in I} \sum_{\substack{i' \in I \\ i' \neq i}} \sum_{j \in J} \delta_i^j \delta_{i'}^j \langle \overrightarrow{\mathrm{GM}^i} | \overrightarrow{u} \rangle \overrightarrow{\mathrm{GM}^{i'}} \right] / \binom{n}{n_c}$$

One has $\sum_{j \in J} (\delta_i^j)^2 = \sum_{j \in J} (\delta_i^j) = \binom{n-1}{n_c-1}$ and for $i \neq i'$, $\sum_{j \in J} \delta_i^j \delta_{i'}^j = \binom{n-2}{n_c-2}$; therefore

$$\mathrm{Hcov}(\overrightarrow{u}) = \frac{1}{n_c^2} \left[\binom{n-1}{n_c-1} \sum_{i \in I} \langle \overrightarrow{\mathrm{GM}^i} | \overrightarrow{u} \rangle \overrightarrow{\mathrm{GM}^i} + \binom{n-2}{n_c-2} \sum_{i \in I} \sum_{\substack{i' \in I \\ i' \neq i}} \langle \overrightarrow{\mathrm{GM}^i} | \overrightarrow{u} \rangle \overrightarrow{\mathrm{GM}^{i'}} \right] / \binom{n}{n_c};$$

$$\mathrm{Hcov}(\overrightarrow{u}) = \frac{1}{n_c} \mathrm{Mcov}(\overrightarrow{u}) + \frac{(n_c-1)}{n_c \, n(n-1)} \left(\sum_{i \in I} \sum_{i' \in I} \langle \overrightarrow{\mathrm{GM}^i} | \overrightarrow{u} \rangle \overrightarrow{\mathrm{GM}^{i'}} - \sum_{i \in I} \langle \overrightarrow{\mathrm{GM}^i} | \overrightarrow{u} \rangle \overrightarrow{\mathrm{GM}^i} \right) =$$

$\frac{1}{n_c} \mathrm{Mcov}(\overrightarrow{u}) + \overrightarrow{0} - \frac{n_c-1}{n_c(n-1)} \mathrm{Mcov}(\overrightarrow{u})$. Hence one has: $\mathrm{Hcov} = \frac{1}{n_c} \mathrm{Mcov} - \frac{n_c-1}{n_c(n-1)} \mathrm{Mcov} = \frac{1}{n_c}(1 - \frac{n_c-1}{n-1}) \times \mathrm{Mcov} = \frac{1}{n_c} \times \frac{n-n_c}{n-1} \times \mathrm{Mcov}$. $\square$

Let $\mathcal{M} \subseteq \mathcal{U}$ be the affine support of the reference cloud M^I and $\mathcal{L} \subseteq \mathcal{V}$ the direction of $\mathcal{M}$ (see Chapter 2, §2.1, p. 10).

In the sequel, the clouds will be studied in the affine support $\mathcal{M}$ of the cloud M^I or in a subspace of $\mathcal{M}$, hence the covariance endomorphisms will be their restrictions to $\mathcal{L}$ or to a subspace of $\mathcal{L}$.

The next two properties are related to the Mahalanobis distance attached to the reference cloud M^I, called M-distance. Let us recall (see §2.3.1, p. 20) that the M-distance between points P and Q in $\mathcal{M}$, denoted $|PQ|$, is equal to $\langle \mathrm{Mcov}^{-1}(\overrightarrow{PQ})|\overrightarrow{PQ}\rangle^{1/2}$ and that the principal κ-ellipsoid of cloud M^I is the set of points $P \in \mathcal{M}$ such that $|GP| = \kappa$.

Proposition 3.3. *Every principal ellipsoid of the sample cloud is a principal ellipsoid of the reference cloud.*

This proposition is an immediate consequence of Proposition 3.2.

Target example

The sample cloud, whose number of points is equal to $\binom{10}{4} = 210$, is shown in Figure 3.2. From the properties of the two preceding propositions we deduce that the concentration ellipse of the sample cloud is $\sqrt{6}$ times smaller than that of the reference cloud as can be seen in the figure below.

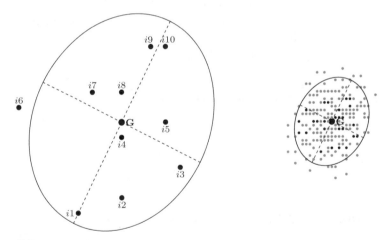

Figure 3.2
Target example. Reference cloud M^I (left) and sample cloud H^J (right) (210 points, superimposed points are in black) with concentration ellipses (the graphic scale is twice that of Figure 3.1, p. 30.).

Proposition 3.4. *The M-distance between every point H^j of the sample cloud and point G is less than or equal to $\sqrt{\frac{n-n_c}{n_c}}$ $(1 \leq n_c \leq n)$:*

$$\forall j \in J, \ |GH^j| \leq \sqrt{\frac{n-n_c}{n_c}}$$

Proof. If $n_c = n$, $I_j = I$ and M^{I_j} is concentrated in G: the formula is verified.

If $n_c < n$, let $\mathcal{D} \subset \mathcal{M}$ be the line passing through points G and H^j. Let M'^i be the M-projection of point M^i on line $\mathcal{D}$. From Corollary 2.3 (p. 21), the M-variance of cloud M^I is equal to 1 in all directions of $\mathcal{M}$, thus in the direction $\mathcal{D}$, therefore $\sum \frac{1}{n}|GM'^i|^2 = 1$.

The mean points of the subclouds associated with the partition of I into two classes I_j and $I \setminus I_j$ are $H^j = \sum_{i \in I_j} M^i/n_c$ (weighted by n_c) and $H'^j = \sum_{i \notin I_j} M^i/(n - n_c)$ (weighted by $n - n_c$). By barycentric property (see p. 11), $|GH^j| = \frac{n-n_c}{n}|H^j H'^j|$. The M-variance of the cloud of two points H^j and H'^j, or between-variance, is equal to $\frac{n_c}{n}\frac{n-n_c}{n}|H'^j H^j|^2$ (see Proposition 2.2, p. 12), that is, $\frac{n_c}{n}\frac{n-n_c}{n}\left(\frac{n}{n-n_c}\right)^2|GH^j|^2 = \frac{n_c}{n-n_c}|GH^j|^2$. After the between-within decomposition of the M-variance (see Corollary 2.5, p. 28) in projection on line $\mathcal{D}$, one has: $\sum_{i \in I}|GM'^i|^2/n = 1 = \frac{n_c}{n-n_c}|GH^j|^2 +$ within-variance, hence $\frac{n_c}{n-n_c}|GH^j|^2 \leq 1$ and we deduce the inequality. $\qquad\square$

Note that the least upper bound is attained when the two subclouds are both reduced to one point. In this case, the within-variance is null.

3.2.4   Test Statistic

In the typicality test for the mean point, the "difference" between subclouds will be evaluated through the deviation between their mean points. As already indicated in Chapter 2 (p. 22), the Mahalanobis distance between two points with respect to the covariance structure of a cloud is an index of the magnitude of the deviation between points of the cloud. Therefore, we choose this type of index as a *test statistic*[8], that is, a test statistic that depends not only on mean points but also on the *covariance structure of the reference cloud*.

More precisely, the map $D^2 : \mathcal{J} \to \mathbb{R}_{\geq 0}$ (set of non-negative real numbers) that associates which each subcloud of the sample space the squared M-distance between its mean point and the one of the reference cloud, defines D^2 as a statistic.

$$D^2 : \begin{array}{ccc} \mathcal{J} & \to & \mathbb{R}_{\geq 0} \\ M^{I_j} & \mapsto & |GH^j|^2 \end{array} \tag{3.1}$$

The value of the statistic D^2 for sample j is simply denoted d_j^2:

$$d_j^2 = D^2(M^{I_j}) = |GH^j|^2$$

Let us now consider the mean point C of the group cloud, the squared M-distance between point C and the reference point G is called *observed value* of D^2 and denoted d_{obs}^2.

$$d_{obs}^2 = |GC|^2$$

---

[8]In multivariate analysis, it is usual to consider the statistic $\mathbf{y}^\top \mathbf{S}^{-1}\mathbf{y}$, where $\mathbf{S} = \frac{n}{n-1}\mathbf{V}$ is the covariance matrix corrected by the number of degrees of freedom and $\mathbf{y}$ the vector of the components of deviation between two points.

If the sample j is such that $d_j^2 \geq d_{obs}^2$, it is said to satisfy the property on $\mathcal{J}$ denoted by $(D^2 \geq d_{obs}^2)$.

Remarks

(1) *Choice of the test statistic.* The test statistic is expressed as a function of the covariance structure of the reference cloud, hence as a function of the covariance structure of the sample cloud. This statistic is a Euclidean distance on $\mathcal{M}$ which takes into account the shape of the sample cloud. So it seems a "better" test statistic compared to the one based on the geometric distance (see §2.3.2, p. 22).

(2) *Notation.* The test statistics are denoted by italic capital letters (D^2).

Combinatorial distribution of the test statistic

Let $p(D^2 = d_j^2)$ denote the proportion of samples which satisfy the property $(D^2 = d_j^2)$. The *distribution* of statistic D^2 is defined by the map from $D^2(\mathcal{J})$ to $[0, 1]$ that associates with each value d_j^2 the proportion $p(D^2 = d_j^2)$.

$$
\begin{aligned}
D^2(\mathcal{J}) &\rightarrow [0, 1] \\
d_j^2 &\mapsto p(D^2 = d_j^2)
\end{aligned}
$$

This distribution is discrete and is most often represented by a histogram: see Figure 3.3 (p. 39) for the *Target example*.

Combinatorial p-value

We want to know how extreme the observed value d_{obs}^2 is over the distribution of D^2. To this end, we look for sample subclouds that satisfy the property $(D^2 \geq d_{obs}^2)$. The proportion of those subclouds will define the typicality level of the test, that is, the *combinatorial p-value*.

$$
p = p\big(D^2 \geq d_{obs}^2\big)
$$

Remarks:

(1) The p-value is usually defined by including the observed value d_{obs}^2. But this choice is a matter of convention, as the p-value might as well have been defined by using the exclusive convention $p(D^2 > d_{obs}^2)$. The *inclusive convention* is known to be conservative and the *exclusive convention* anti-conservative. This is why some authors have proposed an intermediate solution where the p-value is defined by $\big(p(D^2 \geq d_{obs}^2) + p(D^2 > d_{obs}^2)\big)/2$. For a sample set of small size the proportion $p(D^2 = d_{obs}^2)$ may be not negligible so that the conclusion may be affected by the choice between these two conventions; for sample sets of large size, this proportion is negligible so that using one or the other of these conventions would be of no practical consequence. Throughout this book, we will comply with the tradition by using the inclusive convention.

(2) It can easily be shown that both statistics D and D^2 give the same p-values, they are said to be *equivalent combinatorial test statistics*. We choose

D^2 due to the approximation of its distribution by a χ^2-distribution (see §3.2.6, p. 43), thus leading to the approximate test.

Geometric interpretation of the p-value

We will now give the *geometric interpretation of the p-value*. Let us recall that the principal κ-ellipsoid of cloud M^I (see Definition 2.8, p.21), denoted $\mathcal{E}_\kappa$, has centre G and scale parameter κ: $\mathcal{E}_\kappa = \{P \in \mathcal{M} \text{ such that } |GP| = \kappa\}$. The *p*-value is the proportion of points H^j such that $|GH^j|^2 \geq d^2_{obs} = |GC|^2$ or equivalently $|GH^j| \geq |GC|$. Therefore, the *p*-value can be construed as the proportion of points of the sample cloud H^J located on or outside the principal ellipsoid of the reference cloud M^I passing through point G; see Figure 3.4 (p. 40) for the *Target example*.

Properties of statistic D^2

The combinatorial typicality test can be applied to every sample of the reference population viewed as a particular group of observations. For any specified α-level, the test will separate out those samples that are atypical at level α. The *fundamental typicality property* states that the proportion of such samples is at most α (the qualification "at most" rather than "equal to" is due to the discreteness of the distribution).

The following property is the basis for the approximate test.

Proposition 3.5. *The mean of the test statistic D^2 is such that*

$$\text{Mean } D^2 = L \times \frac{1}{n-1} \times \frac{n-n_c}{n_c}$$

Proof. By Corollary 2.3 (p. 21), the H-variance of the sample cloud H^J (i.e., the variance calculated with the Mahalanobis distance associated with endomorphism Hcov) is equal to the dimension L of the affine support of the sample cloud, therefore $\sum \langle \text{Hcov}^{-1}(\overrightarrow{GH^j}) | \overrightarrow{GH^j} \rangle / \binom{n}{n_c} = L$. By Proposition 3.2 (p. 34), one has $\text{Hcov} = \frac{n-n_c}{n-1} \times \frac{1}{n_c} \times \text{Mcov}$, hence $\sum \langle n_c \frac{n-1}{n-n_c} \text{Mcov}^{-1}(\overrightarrow{GH^j}) | \overrightarrow{GH^j} \rangle / \binom{n}{n_c} = L$. As a consequence, Mean $D^2 = \sum \langle \text{Mcov}^{-1}(\overrightarrow{GH^j}) | \overrightarrow{GH^j} \rangle / \binom{n}{n_c} = L \times \frac{1}{n_c} \times \frac{n-n_c}{n-1}$. □

Target example

Let $\overrightarrow{e_1}$ and $\overrightarrow{e_2}$ be the unit vectors of the horizontal and vertical axes of the plane. With respect to the orthonormal frame $(O, \overrightarrow{e_1}, \overrightarrow{e_2})$, the coordinates of the mean point G are $(3, 0)$, the matrix associated with Mcov is the covariance matrix $\mathbf{V} = \begin{bmatrix} 10 & 2 \\ 2 & 13 \end{bmatrix}$ (see p. 14). The inverse of $\mathbf{V}$ is $\mathbf{V}^{-1} = \frac{1}{126} \begin{bmatrix} 13 & -2 \\ -2 & 10 \end{bmatrix}$.

Descriptive appraisal

The mean point C of the 4-impact points coincides with the centre of the target, therefore its coordinates are $(0, 0)$.

The components of $\overrightarrow{GC}$ are $(-3, 0)$, and the squared M-distance between points G and C is

$$|GC|^2 = \tfrac{1}{126} \times \begin{bmatrix} -3 & 0 \end{bmatrix} \begin{bmatrix} 13 & -2 \\ -2 & 10 \end{bmatrix} \begin{bmatrix} -3 \\ 0 \end{bmatrix} = 0.92857 = (0.964)^2.$$

The M-distance between two points is an indicator of the magnitude of the deviation between these points (see Chapter 2, p. 22).

Since $|GC| = 0.964 \geq 0.4$, we descriptively conclude that:

The deviation between points C and G is of notable magnitude.

Combinatorial typicality test for mean point

This finding triggers the natural query from the researcher (informally stated):

"Is the observed deviation a genuine one, or might it due to chance?"

To cope with this query, we will use the typicality test for the mean point. The $\binom{10}{4} = 210$ values $(d_j^2)_{j \in J}$ generate the *distribution of the statistic D^2* (the maximum value is equal to 1.18651) (see Table 3.1 and Figure 3.3).

Table 3.1

Target example. Distribution of D^2: the highest 15 values ($d_{\mathrm{obs}}^2 = 0.92857$).

I_j	d_j^2	count	$p(D^2 \geq d_j^2)$
$\{i1i2i3i5\}$	1.18651	1	0.005
$\{i1i2i3i4\}$	1.15228	1	0.010
$\{i1i2i6i7\}$	1.05556	1	0.014
$\{i6i7i9i10\}$	1.04464	1	0.019
$\{i1i2i4i6\}$	1.02679	1	0.024
$\{i3i5i9i10\}$	1.00496	1	0.029
$\{i6i7i8i9\}$	0.95089	1	0.033
$\{i7i8i9i10\}$	0.94692	1	0.038
$\{i5i8i9i10\}$	0.93651	1	0.043
$\{i6i8i9i10, i1i2i3i6\}$	0.91567	2	0.052
$\{i1i4i6i7\}$	0.91270	1	0.057
$\{i1i6i7i8\}$	0.90972	1	0.062
$\{i6i7i8i10\}$	0.84722	1	0.067
$\{i2i3i4i5\}$	0.84276	1	0.071
$\{i1i2i6i8, i5i7i9i10\}$	0.80357	2	0.081
...			

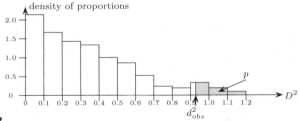

Figure 3.3

Target example. Histogram of the D^2 distribution; the surface area under the curve past the observed value $d_{\mathrm{obs}}^2 = 0.92857$ (shaded grey area) is equal to the interpolated p-value.

We will now attempt to evaluate the degree of typicality of the 4-impact target with respect to the 10-impact target. From Table 3.1 (p. 39), the number of samples whose values of statistic D^2 are greater than the observed value $d^2_{obs} = |\mathrm{GC}|^2 = 0.92857$ is equal to 9, hence the *combinatorial p-value* is equal to $9/210 < .05$. We conclude that the 4-impact target is *atypical* of the 10-impact target, at level .05. In response to the question raised on page 30, it can be said that:

the data are in favour of different shooters.

The geometric interpretation of the *p*-value is shown in Figure 3.4.

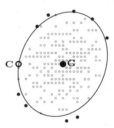

Figure 3.4

Target example. Geometric interpretation of the combinatorial *p*-value: nine points (in black) of the sample cloud are on or outside of the principal ellipse of cloud M^I passing through point C (with $\kappa = d_{obs} = 0.9636$): $p = 9/210 = .0429$.

3.2.5   Compatibility Region

By using the preceding test, the mean point C of the group cloud can be compared to every point $\mathrm{P} \in \mathcal{M}$ which is the mean point of a cloud with the same covariance structure as the reference cloud M^I. Once the "reference cloud" attached to point P has been constructed, the typicality level of the group cloud (*p*-value) with respect to this reference cloud is computed and compared to a given α-level. If the *p*-value is greater than α, the point P is said to be *compatible with point* C *at level* α; the $(1 - \alpha)$ *compatibility region*[9] for point C is the set of points $\mathrm{P} \in \mathcal{M}$ compatible with C at level α.

To construct the compatibility region, we proceed as follows. Firstly, we define the reference cloud attached to a point $\mathrm{P} \in \mathcal{M}$; secondly, we determine the sample space and the sample cloud depending on P; thirdly, we determine the associated *p*-value; finally, we characterize the compatibility region.

Step 1. Reference cloud attached to point P

The reference cloud attached to point P must verify the following two properties: its mean point is point P and its covariance structure is the same as the one of cloud M^I.

---

[9]In combinatorial inference, we say "compatibility region", "confidence region" being the term generally used in traditional inference (see, e.g., Anderson, 2003, §5.3.2, p. 178).

Such a cloud is obtained by translation of cloud M^I. Given a point $P \in \mathcal{M}$, let us consider the cloud P^I, image of cloud M^I by the translation of vector $\overrightarrow{GP}$ (see figure on the right):

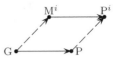

$$\forall i \in I, \ P^i = M^i + \overrightarrow{GP} \text{ or } \overrightarrow{PP^i} = \overrightarrow{GM^i}$$

One has $\overrightarrow{PP^i} = \overrightarrow{GM^i}$ $(i \in I)$, hence the mean point of cloud P^I is point P and its covariance endomorphism $Pcov : \overrightarrow{u} \mapsto \sum_{i \in I} \frac{1}{n} \langle \overrightarrow{PP^i} | \overrightarrow{u} \rangle \overrightarrow{PP^i}$ is equal to $Mcov$.

As a consequence, the Mahalanobis norm relating to cloud P^I is the M-norm: $\forall \overrightarrow{u} \in \mathcal{L}, |\overrightarrow{u}|^2_P = \langle Pcov^{-1}(v) | \overrightarrow{u} \rangle = \langle Mcov^{-1}(v) | \overrightarrow{u} \rangle = |\overrightarrow{u}|^2$.

Step 2. Sample space and sample cloud attached to point P

Taking the cloud P^I as a reference cloud, we define, as in §3.2.1 (p. 33), the set of the $\binom{n}{n_c}$ subclouds $(P^{I_j})_{j \in J}$ with $P^{I_j} = (P^i)_{i \in I_j}$, that is, the sample space that we denote $\mathcal{J}_P$. Then, we deduce the sample cloud, denoted H^J_P, constituted of the $\binom{n}{n_c}$ mean points of subclouds $(P^{i_j})_{j \in J}$, with $H^j_P = \sum_{i \in I_j} P^i / n_c$.

The sample cloud H^J_P is the translation of the sample cloud H^J by vector $\overrightarrow{GP}$ since $\sum_{i \in I_j} \overrightarrow{PP^i} / n_c = \sum_{i \in I_j} \overrightarrow{GM^i} / n_c$:

$$\overrightarrow{PH^j_P} = \overrightarrow{GH^j}$$

Thus, the mean point of cloud H^J_P is P and its covariance endomorphism $Pcov$ is equal to the covariance endomorphism $Hcov$ of cloud H^J, therefore by Proposition 3.2 (p. 34),

$$Pcov = Hcov = \tfrac{n-n_c}{n-1} \times \tfrac{1}{n_c} \times Mcov$$

Step 3. Test statistic and combinatorial p-value attached to point P

The *test statistic* is

$$\begin{aligned} D^2_P : \quad \mathcal{J}_P &\to \ \mathbb{R}_{\geq 0} \\ P^{I_j} &\mapsto \ \langle Pcov^{-1}(\overrightarrow{PH^j_P}) | \overrightarrow{PH^j_P} \rangle \end{aligned}$$

From $\overrightarrow{PH^j_P} = \overrightarrow{GH^j}$ and $Pcov = Mcov$, we deduce $D^2_P(P^{I_j}) = D^2(H^{I_j}) = d^2_j$ $(j \in J)$. Since $D^2_P(P^{I_j}) = d^2_j$ and $|PC|^2_P = |PC|^2$, the property $(D^2_P \geq |PC|^2_P)$ is equivalent to the property $(D^2 \geq |PC|^2)$, hence the *combinatorial p-value* relating to cloud P^I which is, by definition, $p(D^2_P \geq |PC|^2_P)$ is equal to

$$p(D^2 \geq |PC|^2)$$

Step 4. Characterization of compatibility region

Let us consider a point $P \in \mathcal{M}$ such that the property $p(D^2 \geq |PC|^2) > \alpha$ is verified. Such a point is said to be compatible with point C at level α.

Definition 3.1. *The* $(1 - \alpha)$ *compatibility region for point* C *is the set of points* $\mathrm{P} \in \mathcal{M}$ *compatible with point* C *at level* α.

The characterization of the compatibility region for point C will now be established from the distribution of the test statistic D^2.

Lemma 3.1. $\forall \kappa \in \mathbb{R}_{\geq 0}$, *the map* $\kappa \to p(D^2 \geq \kappa)$ *is decreasing.*

$$\kappa' \geq \kappa \Rightarrow p(D^2 \geq \kappa') \leq p(D^2 \geq \kappa)$$

Proof. This property is obvious. For $\kappa' \geq \kappa > 0$, the set $\{j \in J \,|\, d_j^2 \geq \kappa\}$ is the union of the sets $\{j \in J \,|\, d_j^2 \geq \kappa'\}$ and $\{j \in J \,|\, \kappa < d_j^2 < \kappa')\}$, the latter set being possibly empty. Hence the inequality of proportions. $\square$

Proposition 3.6. *Let* $d_\alpha^2 = \max\limits_{j \in J}\{d_j^2 \,|\, p(D^2 \geq d_j^2) > \alpha\}$. *The* $(1 - \alpha)$ *compatibility region for point* C *is the set of points* P *verifying* $|\mathrm{PC}| \leq d_\alpha$ $(d_\alpha > 0)$.

Proof. Given a point $\mathrm{P} \in \mathcal{M}$ such that $|\mathrm{PC}| \leq d_\alpha$, the p-value associated with point P is equal to $p(D^2 \geq |\mathrm{PC}|^2)$. By Lemma 3.1, $p(D^2 \geq |\mathrm{PC}|^2) \geq p(D^2 \geq d_\alpha^2)$ hence $p(D^2 \geq |\mathrm{PC}|^2) > \alpha$: point P is compatible with point C at level α. $\square$

Geometrically, the $(1 - \alpha)$ compatibility region for point C is the set of points P located inside or on the ellipsoid defined by $|\mathrm{PC}| = d_\alpha$, that is, the principal d_α-ellipsoid of cloud M^I translated by vector $\overrightarrow{\mathrm{GC}}$ in order to be centred at point C.

Target example

Given $\alpha = .05$, one has $p(D^2 \geq 0.91567) = \frac{11}{210} = .0524 > \alpha$ and $p(D^2 > 0.91567) = \frac{9}{210} = .0429 < \alpha$ (see Table 3.1, p. 39), hence $d_\alpha = \sqrt{0.91567} = 0.9569$; the compatibility region is depicted in Figure 3.5.

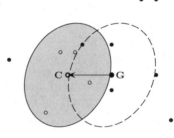

Figure 3.5
Target example. Cloud M^I (black points) with its principal d_α-ellipse (dashed line) ($d_\alpha = 0.957$) and cloud M^C (empty circles) and 95% compatibility region (in grey).

Point G is outside of the ellipse ($|\mathrm{GC}| = 0.9636 > 0.9569$): hence, as above, we conclude that point G is not compatible with point C at level .05.

3.2.6 Approximate Test

When n_c and $n - n_c$ are both large, the sample cloud of $\binom{n}{n_c}$ mean points can be fitted by the L-dimensional normal distribution with centre G and covariance structure Hcov, so that the statistic $n_c \times \frac{n-1}{n-n_c} \times D^2$ is approximately distributed as a χ^2 with L d.f. (degrees of freedom). Then, the combinatorial *p-value* can be approximated by $\widetilde{p} = p\left(\chi_L^2 \geq n_c \times \frac{n-1}{n-n_c} \times d_{obs}^2\right)$.

Let us denote $\chi_L^2[\alpha]$ the critical value of the chi-squared distribution with L d.f. at level α, that is, $p(\chi_L^2 \geq \chi_L^2[\alpha]) = \alpha$. The $1 - \alpha$ *approximate compatibility region* is the interior and boundary of the κ-ellipsoid of the initial cloud, with $\kappa = \widetilde{d}_\alpha$ with $\widetilde{d}_\alpha^2 = \frac{1}{n_c}\frac{n-n_c}{n-1}\chi_L^2[\alpha]$.

Target example

The approximate test applied to the *Target example* is a mere illustration since the data size ($n = 10$, $n_c = 4$) is too small to justify its use. We will nonetheless give the results.

$d_{obs}^2 = 0.92857$; $\chi_{obs}^2 = 4 \times \frac{10-1}{10-4} \times 0.92857$; $\widetilde{p} = p(\chi_2^2 \geq 5.5714) = .062 > .05$. One has $\chi_2^2[.05] = 5.9915$, hence $\widetilde{\kappa} = \left(\frac{1}{4} \times \frac{10-4}{10-1} \times 5.9915\right)^{1/2} = 0.999$. The approximate 95% compatibility region is larger than the exact one ($\kappa = 0.957$) (see Figure 3.7).

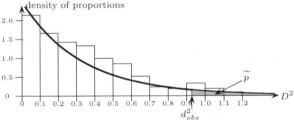

Figure 3.6
Target example. Exact and approximate distributions of D^2; the surface area under the curve past $d_{obs}^2 = 0.92857$ (shaded grey area) is equal to the approximate p-value.

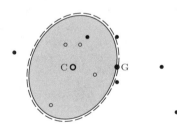

Figure 3.7
Target example. Cloud M^I; exact 95% compatibility region (in grey); 95% approximate compatibility region: inside of the dashed ellipse ($\widetilde{\kappa}_{.05} = 0.9993$).

Unlike the exact test, the approximate test produces a non-significant result. The data size is small and the p-value is close to $\alpha = .05$, therefore the approximate test is simply not appropriate to such a small sample; it is recommended to perform the exact test.

In any case, the *approximate test* provides an order of magnitude of the p-value and of the compatibility region.

Remarks.

(1) *Exhaustive vs. Monte Carlo methods.*

Nowadays it is possible to perform the *combinatorial test* using the complete *enumeration of samples* as soon as their number is not too large (say < $1,000,000$) (exhaustive method). For all cases where sample sizes are not small, the cardinality of the sample set is too large to enumerate all samples. According to many authors since Dwass (1957), we can inspect the sample set by means of *random sample without replacement* from the reference set[10], which corresponds to a random sampling from the sample set. Then we speak of MC method (Monte Carlo method).

In this book, we say *approximate method* when an assumption on the data set, for instance normality, is made[11]. As already said, the approximate method is easy to implement and gives an order of magnitude of the results.

The increasing availability of fast computers has made permutation tests more and more accessible and competitive with respect to their approximate or parametric counterparts. In fact, the exhaustive distributions of test statistics can be effectively replaced by combinatorial distributions obtained by the MC method which, more than numeric evaluations, provides statistical estimations without compromising their desirable statistical properties.

(2) *Computational considerations.*

It is worth noticing that the test statistic (see Equation 3.1, p. 36) as well as the compatibility region (see §3.2.5, p. 40) are based on the M-norm, that is, they depend on the inverse of the covariance endomorphism M*cov* of cloud M^I (see §3.2.4, p. 36). From a computational viewpoint, the inversion of the matrix and its use for computing the values of the test statistic can be time-consuming. Computations become much simpler as soon as we choose the orthocalibrated basis of the space (see Chapter 2, §2.3.3, p. 23), then the matrix representation of M*cov* and its inverse is the identity matrix and the squared M-norm of a vector is merely the sum of the squared coordinates (see page 55).

---

[10]Because of increasing computing power, by 2010, p-values based on exact enumeration sometimes exceeded $10,000,000$ samples. A number of $1,000,000$ resamplings is not only recommended but common (Johnston et al., 2007).

[11]When referring to MC methods, some authors use terms such as "approximate permutation test" or even "modified permutation tests"; see, e.g., Dwass (1957).

3.3   One-dimensional Case: Typicality for Mean

The foregoing test for mean point can be applied as is to one-dimensional clouds but it leads to a two-tailed test with a nondirectional conclusion.

In this instance, a directional conclusion can be reached without difficulty. In the case of one-dimensional clouds ($L = 1$), it is usual to work on a numerical variable associated with the reference set I and denoted $x^I = (x^i)_{i \in I}$ together with a numerical variable associated with the group C of n_c individuals and denoted $x^C = (x^c)_{c \in C}$. Thus, by choosing the *Mean* as test statistic, we will be able to conclude by taking into account the sign of the difference between means. The typicality test presented here will amount to comparing the mean of a group to the mean of a reference set (or population).

3.3.1   Sample Set and Sample Space

Recall that the sample set, that is, the set of all n_c-element subsets of I, is indexed by J and that the subset of the n_c individuals of sample j is denoted I_j. Let us denote $x^{I_j} = (x^i)_{i \in I_j}$ the n_c values associated with the subset I_j. The *sample space*, denoted $\mathcal{J}$, is defined as the set of the $\binom{n}{n_c}$ subsets $(x^{I_j})_{j \in J}$.

3.3.2   Test Statistic

The map, denoted M (written in italics), that associates with each sample its mean, defines the *Mean* as a *statistic*.

$$M : \begin{array}{ccc} \mathcal{J} & \to & \mathbb{R} \\ x^{I_j} & \mapsto & \sum_{i \in I_j} x^i / n_c \end{array}$$

The mean of x^{I_j} is denoted m^j and called "mean of sample j".

Distribution of the test statistic M

The distribution of statistic M is the map that assigns to each m^j the proportion of samples whose mean is equal to m^j.

$$\begin{array}{ccc} M(\mathcal{J}) & \to & [0, 1] \\ m^j & \mapsto & p(M = m^j) \end{array}$$

Let $\bar{x} = \sum x^i / n$ denote the mean of x^I and $v = \sum (x^i - \bar{x})^2 / n$ its variance. After the classical theory of sampling in a finite population, one has the following two *properties* (see Cramér, 1946, p. 523):

- the *mean of M* is equal to $\bar{x}$.
- the *variance of M* is equal to $\frac{n - n_c}{n - 1} \times \frac{v}{n_c}$.

Combinatorial p-value

The mean of the group, $\sum x^c/n_c$, is called the *observed mean* and denoted m_{obs}. Then the mean m_{obs} is located in the distribution of M.

To achieve this, we consider the samples whose means are greater than or equal to m_{obs}, that is, which satisfy the property $(M \geq m_{obs})$. The proportion of these samples is denoted $\bar{p} = p(M \geq m_{obs})$ and called *upper typicality level*; it is a directional index of departure of the group of observations from the reference population, with respect to the mean. The *lower typicality level* $\underline{p} = p(M \leq m_{obs})$ is defined accordingly.

The *combinatorial p-value* of the test is defined as the smaller of these two proportions; it is denoted p and called p-value of the test.

$$p = \min\{\bar{p}, \underline{p}\} \quad \text{with} \quad \bar{p} = p(M \geq m_{obs}) \quad \text{and} \quad \underline{p} = p(M \leq m_{obs})$$

The p-value defines the typicality level of the group of observations for the mean with respect to the reference population. The smaller the value of p-value, the lower the typicality.

For any α-level (see p. 31),

- if $p \leq \alpha/2$, the group will be declared *atypical* of the reference population at level $\alpha/2$, on the side of high means if $p = \bar{p}$ or on the side of low means if $p = \underline{p}$;

- if $p > \alpha/2$, the reference population and the group are said to be *compatible at level* α, with respect to the mean.

Remark. When considering the variable $D = (M - \bar{x})/\sqrt{v}$ as a test statistic, we obtain an equivalent test. The statistic D, that is, the variable M scaled with respect to the reference population, is called *Scaled deviation* and is defined by:

$$D: \quad \mathcal{J} \quad \rightarrow \quad \mathbb{R}$$
$$x^{I_j} \quad \mapsto \quad (m^j - \bar{x})/\sqrt{v}$$

$D(x^{I_j})$ is denoted d^j. Letting $d_{obs} = (m_{obs} - \bar{x})/\sqrt{v}$, since $p(M = m^j) = p(D = d^j)$ we obtain $\bar{p} = p(M \geq m_{obs}) = p(D \geq d_{obs})$ and $\bar{p} = p(M \leq m_{obs}) = p(D \leq d_{obs})$, hence statistics M and D lead to the same p-values. Note that the square of D is the particularization of the test statistic D^2 defined above (see Equation 3.1, p. 36) to a one-dimensional cloud (the absolute scaled deviation is the M-distance between points).

3.3.3 Compatibility Interval

As for a multidimensional cloud, given $a \in \mathbb{R}$, let us consider the "shifted cloud" associated with the variable $y^I = (x^i + a)_{i \in I}$. The mean of y^I is $\bar{y} = \bar{x} + a$ and its variance is equal to the variance of the reference population.

The test statistic associated with y^I is the map $M_y : \mathcal{J} \rightarrow \mathbb{R}$ such that

$$M_y : y^{I_j} \mapsto \sum_{i \in I_j} y^i/n_c = m^j + a$$

The upper and lower levels associated with y^I are $\overline{p}_y = p(M_y \geq m_{obs})$ and $\underline{p}_y = p(M_y \leq m_{obs})$, respectively.

Given an α-level, the mean of y^I is said to be *compatible* with m_{obs} at level α if it is not atypical, that is, if $\min\{\overline{p}_y, \underline{p}_y\} > \alpha/2$.

Definition 3.2. *The* $(1 - \alpha)$ *compatibility interval for* m_{obs} *is defined as the set of means* $\overline{y}$ *compatible with* m_{obs} *at level* α.

It verifies the following proposition.

Proposition 3.7. *The* $(1 - \alpha)$ *compatibility interval for* m_{obs} *is*

$$\left[m_{obs} - \overline{m}_\alpha + \overline{x}, \; m_{obs} - \underline{m}_\alpha + \overline{x}\right]$$

with $\overline{m}_\alpha = \max\limits_{j \in J}\{m^j \mid p(M \geq m^j) > \tfrac{\alpha}{2}\}$; $\underline{m}_\alpha = \min\limits_{j \in J}\{m^j \mid p(M \leq m^j) > \tfrac{\alpha}{2}\}$.

Proof. From $M_y(y^{I_j}) = m^j + (\overline{y} - \overline{x})$ we deduce that $M_y(y^{I_j}) \geq m_{obs} \iff m^j \geq m_{obs} - (\overline{y} - \overline{x})$, hence $\overline{p}_y = p(M \geq m_{obs} - (\overline{y} - \overline{x}))$. One has $\overline{p}_y > \alpha/2$ if $m_{obs} - (\overline{y} - \overline{x}) \leq \overline{m}_\alpha$, that is, if $\overline{y} \geq m_{obs} - (\overline{m}_\alpha - \overline{x})$.
Similarly, $\underline{p}_y = p(M \leq m_{obs} - (\overline{y} - \overline{x}))$ hence $\underline{p}_y > \alpha/2$ if $m_{obs} - (\overline{y} - \overline{x}) \geq \underline{m}_\alpha$, that is, if $\overline{y} \leq m_{obs} + \overline{x} - \underline{m}_\alpha$. □

In terms of statistic D, letting $\overline{d}_\alpha = (\overline{m}_\alpha - \overline{x})/\sqrt{v}$ and $\underline{d}_\alpha = (\underline{m}_\alpha - \overline{x})/\sqrt{v}$, the $(1 - \alpha)$ compatibility interval is: $\left[d_{obs} - \overline{d}_\alpha, \; d_{obs} - \underline{d}_\alpha\right]$.

Note that, in general, the distribution of M (or D) is not symmetric, and that $(\overline{x}_\alpha + \underline{x}_\alpha)/2 \neq \overline{x}$; therefore the interval is not centred at the observed value of the test statistic (m_{obs} or d_{obs}).

Target example

To exemplify the one-dimensional case, we take the two clouds of the *Target example* projected onto the horizontal axis. Figure 3.8 shows the two one-dimensional clouds (reference and group clouds) and Table 3.2 gives variables x^I and x^C (coordinates of points of the two clouds).

Figure 3.8
Target example. Target with 10 impacts and target with 4 impacts.

Table 3.2
Target example. Coordinates of points.

x^I	0	3	7	3	6	−4	1	3	5	6

x^C	−1.5	1.5	0.5	−0.5

Descriptive appraisal

One has $\overline{x} = 3$, $v = 10$ and $m_{obs} = 0$. With respect to the reference population, the scaled deviation between the observed mean and the reference mean is $d_{obs} = (0 - 3)/\sqrt{10} = -0.9487$.

Since $|d_{obs}| > 0.4$ (see Chapter 2, p. 22), we conclude that

descriptively the deviation is notable.

Combinatorial typicality test for the mean

Thus, we will now assess the typicality level of the group of observations.

The distribution of the statistic M is shown in Table 3.3 and Figure 3.9.

From Table 3.3, we deduce that $\underline{p} = p(M \leq 0) = 3/210$ and $\overline{p} = p(M \geq 0) = 1$, hence $p = \underline{p} < .025$, hence the conclusion:

> *The observed mean is atypical of the reference mean at level .025 (one-sided), on the side of low values.*

Table 3.3

Taste example. Distribution of the *Mean* expressed in numbers of samples.

m^j	0	0.5	0.75	1	1.25	1.5	1.75	2
	3	4	5	4	10	9	11	10

m^j	2.25	2.5	2.75	3	3.25	3.5	3.75	4
	11	15	12	17	12	14	15	13

m^j	4.25	4.5	4.75	5	5.25	5.5	6
	13	9	9	4	6	3	1

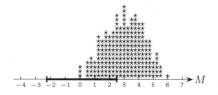

Figure 3.9

Taste example. Stack diagram of the *Mean* distribution and 95% compatibility interval for $m_{obs} = 0$.

Compatibility interval

One has $p(M \geq 5.25) = 10/210 > .025$ and $p(M > 5.25) = 4/210 < .025$, hence $\overline{m}_\alpha = 5.25$. Similarly, one finds $\underline{m}_\alpha = 0.5$. For M, the 95% compatibility interval is equal to $[-2.25, +2.50]$. The reference mean $\overline{x} = 3$ is outside of the 95% compatibility interval ($\overline{x}$ is not compatible with m_{obs} at level .05) and greater than the upper limit of the interval, hence the conclusion:

> m_{obs} *is significantly lower than $\overline{x}$ at level .025 (one-sided).*

For D, the 95% compatibility interval is equal to $[-1.66, -0.16]$.

3.3.4   Approximate Test: Z-test

Consider now the test statistic $Z = \dfrac{M - \overline{x}}{\sqrt{\frac{n - n_c}{n - 1} \times \frac{v}{n_c}}}$, called Z-score. The test statistics *Mean* and *Z-score* lead to the same p-values, that is, they are equivalent statistics. When n_c and $n - n_c$ are both large, M has approximately a normal distribution with mean $\overline{x}$ and variance $\text{Var}\, M = \frac{n - n_c}{n - 1} \times \frac{v}{n_c}$ (see page 45), then Z is approximately a standard normal variable $\mathcal{N}(0, 1)$.

Approximate p-value

Let $z_{obs} = \sqrt{n_c} \times \sqrt{\frac{n - 1}{n - n_c}} \times d_{obs}$ be the observed value of statistic Z.

If $z_{obs} > 0$, the approximate p-value is equal to the proportion of the normal distribution greater than z_{obs} (approximate upper level); if $z_{obs} < 0$, the approximate p-value is equal to the proportion of the normal distribution less than z_{obs} (approximate lower level).

Approximate compatibility interval

Let z_α denote the critical value of the normal distribution at level α (recall that $z_\alpha = 1.96$ for $\alpha = .05$). Letting $h_\alpha = z_\alpha \sqrt{\frac{n-n_c}{n-1}} \sqrt{\frac{v}{n_c}}$ (half-length of the interval), the *approximate* $(1 - \alpha)$ *compatibility interval* is written

$$\left[m_{obs} - h_\alpha, \, m_{obs} + h_\alpha \right]$$

3.4 Combinatorial Typicality Test for Variance

The method of combinatorial typicality test can be used for any test statistic, whether its distribution is known or not. Thus one is always free to choose the test statistic. We could have chosen a statistic based on the geometric distance instead of the Mahalanobis distance but the former does not take into account the shape of the reference cloud. In the one-dimensional case, we could have considered the median as a test statistic even though its combinatorial distribution is unknown. We could also be interested in the dispersions of clouds. We wonder if, in terms of dispersion, the group is atypical of the reference population.

As an example, let us study the typicality in terms of variance. The aim of the test is to compare the variance of the group cloud to that of the reference cloud. More precisely the test answers the question:

"Is the variance of the group cloud higher or lower than the variance of the reference cloud?"

In such a test, we have to take into account the sign of the deviation between variances. The test will be briefly applied to the *Target example*.

3.4.1 Test Statistic

The *variance* of the sample cloud M^{I_j} is equal to $\frac{1}{2} \sum_{i \in I_j} \sum_{i' \in I_j} (M^i M^{i'})^2 / n_c^2$ (see formula p. 12) and denoted v_j.

The map $\quad V : \begin{aligned} \mathcal{J} &\to \mathbb{R}_{\geq 0} \\ M^{I_j} &\mapsto v_j \end{aligned} \quad$ defines the *Variance* as a statistic.

The *distribution of V* is defined by the map

$$\begin{aligned} V(\mathcal{J}) &\to [0,1] \\ v_j &\mapsto p(V = v_j) \end{aligned}$$

Proposition 3.8. *The mean of V is equal to $\frac{n}{n-1} \times \frac{n_c-1}{n_c} \times v$.*

Proof. $v^j = \frac{1}{n_c^2} \sum_{i \in I} \sum_{i' \in I} \delta_i^j \delta_{i'}^j (M^i M^{i'})^2 / 2$, Mean $V = \sum_{j \in J} v_j / \binom{n}{n_c}$. One has $\sum_{j \in J} v_j =$

$\frac{1}{n_c^2} \sum_{i \in I} \sum_{\substack{i' \in I \\ i' \neq i}} (\sum_{j \in J} \delta_i^j \delta_{i'}^j)(M^i M^{i'})^2 / 2 = \frac{1}{n_c^2} \binom{n-2}{n_c-2} \sum_{i \in I} \sum_{i' \in I} (M^i M^{i'})^2 / 2$ (see p.33). One has

$\sum_{i \in I} \sum_{i' \in I} (M^i M^{i'})^2 / 2 = n^2 v$, then Mean $V = \frac{1}{n_c^2} \frac{\binom{n-2}{n_c-2}}{\binom{n}{n_c}} (n^2\, v) = \frac{1}{n_c^2} \frac{n_c(n_c-1)}{n(n-1)} (n^2 v)$. $\square$

Combinatorial p-value

The variance of the group cloud, denoted v_{obs}, is called observed variance and is equal to $\frac{1}{2}\big(\sum_{c \in C} \sum_{c' \in C} (x^c - x^{c'})^2 \big)/n_c^2$. As with the test of a mean, the observed variance is located in the distribution of V.

The typicality upper and lower levels are, respectively, $\overline{p} = p(V \geq v_{obs})$ and $\underline{p} = p(V \leq v_{obs})$. The *combinatorial p-value* is $p = \min\{\overline{p}, \underline{p}\}$.

$$p = \min \big\{ p(V \geq v_{obs}) ;, p(V \leq v_{obs}) \big\}$$

Given an α-level,

- if $p \leq \alpha/2$, the group will be declared *atypical* of the reference population at level $\alpha/2$, on the side of high variances if $p = \overline{p}$ or on the side of low variances if $p = \underline{p}$.

- if $p > \alpha/2$, the reference population and the group are said to be *compatible at level* α, with respect to the variance.

3.4.2   Compatibility Interval

The steps of the characterization of the compatibility interval for variance are the same as for the mean, but the "new" reference cloud will be a cloud obtained by homothety of M^I instead of translation.

Firstly, we consider the "zoomed cloud" P^I defined by $\forall i \in I$, $P^i = G + a\overrightarrow{GM^i}$ $(a > 0)$: the cloud P^I is the image of cloud M^I by the homothety with centre G and scale factor a (see page 240), hence $\forall i, i' \in I$, $P^i P^{i'} = a\, M^i M^{i'}$ and the variance of cloud P^I is equal to $a^2 v$.

Secondly, we associate the statistic V_P with the reference cloud P^I:

$$V_P : P^{I_j} \mapsto \frac{1}{2} \frac{1}{n_c^2} \sum_{i \in I_j} \sum_{i' \in I_j} (P^i P^{i'})^2 = a^2 v_j$$

Thirdly, we determine the *p*-value as the minimum of $\overline{p}_y = p(V_P \geq v_{obs}^2)$ and $\underline{p}_y = p(V_P \leq v_{obs}^2)$.

The variance of the cloud $P^I = (G + a\overrightarrow{GM^i})_{i \in I}$ is said to be *compatible with* v_{obs} at level α if it is not atypical, that is, if $\min\{\overline{p}_y, \underline{p}_y\} > \alpha/2$.

Definition 3.3. *The $(1 - \alpha)$ compatibility interval is the set of the variances of "zoomed clouds" that are compatible with v_{obs} at level α.*

It satisfies the following proposition.

Proposition 3.9. *Letting* $\overline{v}_\alpha = \max_{j \in J}\{v_j \mid p(V \geq v_j) > \alpha/2\}$ *and* $\underline{v}_\alpha = \min_{j \in J}\{v_j \mid p(V \leq v_j) > \alpha/2\}$, *the* $(1 - \alpha)$ *compatibility interval is given by*

$$\left[v_{obs} \times \frac{v}{\overline{v}_\alpha}, \; v_{obs} \times \frac{v}{\underline{v}_\alpha} \right]$$

Proof. From $V_P(P^{I_j}) = a^2 v_j$, we deduce $V_P(P^{I_j}) \geq v_{obs} \iff v_j \geq v_{obs}/a^2$, hence $\overline{p}_y = p(V \geq v_{obs}/a^2)$. If $v_{obs}/a^2 \leq \overline{v}_\alpha$ or $a^2 \geq v_{obs}/\overline{v}_\alpha$, then $\overline{p}_y > \alpha/2$. Similarly, if $a^2 \leq v_{obs}/\underline{v}_\alpha$, then $\underline{p}_y > \alpha/2$. Hence $v_{obs}/\overline{v}_\alpha \leq a^2 \leq v_{obs}/\underline{v}_\alpha$. By replacing a^2 by $\text{Var}\,P^I/v$ we obtain $v_{obs} \times \frac{v}{\overline{v}_\alpha} \leq \text{Var}\,P^I \leq v_{obs} \times \frac{v}{\underline{v}_\alpha}$. $\square$

Target example

We consider here the one-dimensional cloud projected on the horizontal axis. The observed variance $v_{obs} = 1.25$ is much smaller than the reference variance ($v = 10$). Then we conclude that *descriptively, the observed variance is notably lower than the reference one.*

We will now assess a typicality level for the group with respect to *Variance*. The distribution of the statistic is given in Figure 3.10. One has $p = \underline{p} = 3/210 < .025$: *the group of the 4 impacts is atypical at level .025, on the side of small variances.*

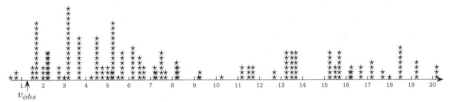

Figure 3.10
Combinatorial distribution of the *Variance* on horizontal axis.

3.5 Combinatorial Inference in GDA

Combinatorial inference is congenial to the descriptive-oriented key idea of geometric data analysis (GDA). Combinatorial typicality tests in GDA are performed on the results of a GDA method (mainly PCA or MCA)[12]. Starting with the principal clouds constructed from a GDA method, we conduct inductive

---

[12]For a presentation of these methods, see Le Roux and Rouanet (2004, Chapters 4 and 5) and Le Roux (2014a, Chapters 6 and 8). Applications to real data sets can respectively be found in Chapters 9 and 11 of these books.

analyses aiming at extending the major descriptive conclusions. In accordance with the descriptive findings, inferences will pertain to clouds projected onto a principal axis, or a principal subspace.

In the sequel, we will focus on the particular case when the reference population is comprised of the individuals that have participated in the determination of principal axes (usually called "active individuals"). The group can be a subset of active individuals characterized by a property—for instance, in the *Central Bankers Study* (Chapter 6, p. 188), the bankers having a specific position-taking—or a group of supplementary individuals—for instance, in the *Parkinson study* (Chapter 6, p. 156), the group of Parkinsonian patients.

In this particular case, we will write the formulas firstly for a principal axis and secondly for a principal subspace. Thus, as we will see below, the numerical formulas written with respect to the principal basis are quite simple.

3.5.1   Principal Reference Cloud

The reference cloud M^I is associated with the n active individuals.

Recall that the direction vectors of principal axes are eigenvectors of the covariance endomorphism Mcov of cloud M^I (see Theorem 2.1, p. 17). The direction vector subspace $\mathcal{L}$ of the affine support $\mathcal{M}$ of the cloud M^I is equipped with the two principal bases: the orthonormal one $(\overrightarrow{\alpha_\ell})_{\ell \in L}$ (with M$cov(\overrightarrow{\alpha_\ell}) = \lambda_\ell \overrightarrow{\alpha_\ell}$) and the "orthocalibrated" one $(\overrightarrow{\beta_\ell})_{\ell \in L}$ with $\overrightarrow{\beta_\ell} = \sqrt{\lambda_\ell}\, \overrightarrow{\alpha_\ell}$,

The *principal coordinate* of point M^i on axis $(G, \overrightarrow{\alpha_\ell})$ (see page 17) is denoted y_ℓ^i, thus $\overrightarrow{GM^i} = \sum_{\ell \in L} y_\ell^i \overrightarrow{\alpha_\ell}$. The mean of the ℓ-th principal variable y_ℓ^I is equal to 0 and the variance is equal to λ_ℓ (eigenvalue).

The coordinates of point M^i over the orthocalibrated basis are equal to $(y_\ell^I/\sqrt{\lambda_\ell})_{\ell \in L}$. The ℓ-th coordinate variable $z_\ell^I = y_\ell^I/\sqrt{\lambda_\ell}$ is standardized (its mean is 0 and its variance is 1).

In GDA, interpretations are based on the *aids to interpretation*, namely the contributions. The *contribution of a point to axis* ℓ is the proportion of the variance of axis ℓ accounted for by the point (Le Roux and Rouanet, 2004, p. 93); the contribution of point M^i to axis ℓ, denoted Ctr_i^ℓ, is equal to $\frac{1}{n}(y_\ell^i)^2/\lambda_\ell = \frac{1}{n}(z_\ell^i)^2$.

We will now present the typicality test for a mean point when the clouds are orthogonally projected onto a principal axis or a principal subspace of the reference cloud.

3.5.2   Principal Axis: Typicality for Mean

Let us consider the orthogonal projections of the reference cloud M^I and of the group cloud M^C onto axis ℓ. The coordinate of the mean point of cloud M^C on axis $(G, \overrightarrow{\alpha_\ell})$ is denoted m_ℓ.

The methodology is the following one.

- Firstly, we *compare descriptively* the observed mean m_ℓ to zero (mean of the reference population) in terms of number of the reference standard deviation ($\sqrt{\lambda_\ell}$), that is, we evaluate the magnitude of the deviation (see §2.3.2, p. 22) by using the *scaled deviation* $d_\ell = (m_\ell - 0)/\sqrt{\lambda_\ell}$.

    - If $|d_\ell| \geq 0.4$, the deviation is *descriptively notable*, then we will assess the typicality level of the group.

    - If $|d_\ell| < 0.4$, the deviation is *descriptively small*, so there is no point in inquiring about its statistical significance.

- Secondly, whenever appropriate, we perform the typicality test for the mean by applying the results of §3.3 (p. 45) and determine the compatibility interval.

Typicality test for the mean

The *sample space*, denoted $\mathcal{J}_\ell$, is the family $(y_\ell^{I_j})_{j \in J}$, with $J = \binom{n}{n_c}$. The mean of sample j is denoted $\overline{y}_\ell^j$ and equal to $\sum\limits_{i \in I_j} y_\ell^i / n_c$.

On sample space $\mathcal{J}_\ell$, we define three equivalent statistics: M_ℓ, D_ℓ, Z_ℓ (see pp. 45, 46 and 48)

$$
\begin{aligned}
M_\ell : \ & y_\ell^{I_j} \ \mapsto \ \overline{y}_\ell^j && \text{(Mean)} \\
D_\ell : \ & y_\ell^{I_j} \ \mapsto \ \overline{y}_\ell^j / \sqrt{\lambda_\ell} = d_\ell^j && \text{(Scaled Deviation)} \\
Z_\ell : \ & y^{I_j} \ \mapsto \ \sqrt{n_c}\sqrt{\frac{n-1}{n-n_c}}\frac{\overline{y}_\ell^j}{\sqrt{\lambda_\ell}} = \sqrt{n_c}\sqrt{\frac{n-1}{n-n_c}}\, d_\ell^j = z_\ell^j && \text{(Z-score)}
\end{aligned}
$$

As noted in §3.3 (pp. 46 and 48), these statistics are equivalent in the sense that they lead to the same combinatorial p-value. Let us consider statistic D_ℓ, the observed value is $d_\ell = m_\ell/\sqrt{\lambda_\ell}$, hence the p-value is

$$
p = \min\left\{ p(D_\ell \geq d_\ell)\,;\, p(D_\ell \leq d_\ell) \right\}
$$

Compatibility interval

If $\overline{d}_\ell^\alpha = \max\limits_{j \in J}\left\{ d_\ell^j | p(D_\ell \geq d_\ell^j) > \alpha/2 \right\}$ and $\underline{d}_\ell^\alpha = \min\limits_{j \in J}\left\{ d_\ell^j | p(D_\ell \leq d_\ell^j) > \alpha/2 \right\}$, the $(1 - \alpha)$ compatibility interval is

$$
\left[d_\ell - \overline{d}_\ell^\alpha \ ;\ d_\ell - \underline{d}_\ell^\alpha \right]
$$

Remarks. (1) The value of statistic $\frac{n_c}{n}D_\ell^2$ for sample j is equal to $\frac{n_c}{n}(d_\ell^j)^2$, which is the contribution Ctr_j^ℓ of the mean point of sample j to axis ℓ. The combinatorial test is therefore linked to the usual aids to interpretation and it can be considered to extend the descriptive concept of contribution of a group of individuals to the variance of an axis.

(2) In the GDA literature, the observed value of Z_ℓ is called *test-value* (see Lebart et al., 2006, p. 291). We will denote it $z_{\ell obs}$, with $z_{\ell obs} = \sqrt{n_c \frac{n-1}{n-n_c}}\, d_\ell$.

(3) In MCA, if the group is a subset of the active individuals, denoted I_c, let us consider the categorical variable with two categories c and c' that is equal to c if $i \in I_c$ and to c' if $i \notin I_c$. Putting this variable as a supplementary variable, one obtains the coordinate of category point c on axis ℓ, which is equal to d_ℓ. Letting $\cos \theta_{c\ell} = \sqrt{\frac{n_c}{n - n_c}} d_\ell$, the observed value of Z_ℓ (i.e., the test-value $z_{\ell obs}$) is equal to $\sqrt{n - 1} \cos \theta_{c\ell}$, whereas $\cos^2 \theta_{c\ell}$ is the quality of representation of the category point c on axis ℓ (see Le Roux, 1998, p. 11–13).

3.5.3 Principal Subspace: Typicality for Mean Point

Let L' denote a subset of L and $\mathcal{M}_{L'} \subseteq \mathcal{M}$ the L'-dimensional principal subspace of cloud M^I passing through the mean point G and generated by the L' principal vectors $(\overrightarrow{\alpha_\ell})_{\ell \in L'}$.

Now, the *reference cloud* is the orthogonal projection of cloud M^I onto $\mathcal{M}_{L'}$; it is denoted $\mathrm{M}^I_{L'}$, with $\overrightarrow{\mathrm{GM}}^i_{L'} = \sum_{\ell \in L'} y^i_\ell \overrightarrow{\alpha_\ell} = \sum_{\ell \in L'} z^i_\ell \overrightarrow{\beta_\ell}$ and its mean point is point G. The *group cloud* is the orthogonal projection of cloud M^C onto $\mathcal{M}_{L'}$; its mean point is the projection of the mean point of M^C onto $\mathcal{M}_{L'}$, it is denoted $\mathrm{C}_{L'}$ with $\overrightarrow{\mathrm{GC}}_{L'} = \sum_{\ell \in L'} m_\ell \overrightarrow{\alpha_\ell} = \sum_{\ell \in L'} d_\ell \overrightarrow{\beta_\ell}$.

First of all, we assess the magnitude of the deviation between points G and $\mathrm{C}_{L'}$ using the M-distance $|\mathrm{GC}_{L'}| = \left(\sum_{\ell \in L'} d_\ell^2 \right)^{1/2}$ (see §2.3.2, p. 22).

- If $|\mathrm{GC}_{L'}| \geq 0.4$, the *deviation is descriptively notable*, then we will assess the typicality level of the group.

- If $|\mathrm{GC}_{L'}| < 0.4$, the *deviation is descriptively small*, so there is no point in inquiring about its statistical significance.

Whenever appropriate, we perform the typicality test by applying the results of §3.2 (p. 32).

Typicality test for Mean point

The *sample space* $\mathcal{J}_{L'}$ is the set of the $\binom{n}{n_c}$ subclouds $(\mathrm{M}^{I_j})_{j \in J}$ projected onto the principal subspace $\mathcal{M}_{L'}$. The mean point of the projection of cloud M^{I_j} is the projection of the mean point H^j onto $\mathcal{M}_{L'}$ that is denoted $\mathrm{H}^j_{L'}$: letting $d^j_\ell = \sum_{i \in I_j} z^i_\ell / n_c$, we have $\overrightarrow{\mathrm{GH}}^j_{L'} = \sum_{\ell \in L'} d^j_\ell \overrightarrow{\beta_\ell}$.

The test statistic, denoted $D^2_{L'}$, is defined by (see Equation 3.1, p. 36):

$$
D^2_{L'} : \quad \mathcal{J}_{L'} \quad \to \quad \mathbb{R}_{\geq 0}
$$
$$
\mathrm{M}^{I_j}_{L'} \quad \mapsto \quad |\mathrm{GH}^j_{L'}|^2 = \sum_{\ell \in L'} (d^j_\ell)^2
$$

The observed value of $D^2_{L'}$ is $|\mathrm{GC}_{L'}|^2 = \sum_{\ell \in L'} d_\ell^2$, and the p-value is $p = p\left(D^2_{L'} \geq \sum_{\ell \in L'} d_\ell^2 \right)$.

Compatibility region

Letting $d_\alpha^2 = \max_{j \in J}\{ \sum_{\ell \in L'} (d_\ell^j)^2 \mid p(D_{L'}^2 \geq \sum_{\ell \in L'} (d_\ell^j)^2) > \alpha\}$, by Proposition 3.6 (p. 42), the $(1-\alpha)$ *compatibility region* for point $\mathrm{C}_{L'}$ is the set of points P of the principal subspace verifying $|\mathrm{PC}_{L'}| \leq d_\alpha$ (with $d_\alpha = \sqrt{d_\alpha^2} \geq 0$).

Geometrically, the $(1-\alpha)$ compatibility region for point $\mathrm{C}_{L'}$ is the set of points of the principal subspace $\mathcal{M}'$ located inside or on the ellipsoid defined by $|\mathrm{PC}_{L'}| = d_\alpha$, that is, the ellipsoid centred at point $\mathrm{C}_{L'}$, whose equation with respect to the orthonormal frame $(\mathrm{G}, (\overrightarrow{\alpha_\ell})_{\ell \in L'})$ is $\sum_{\ell \in L'} (y_\ell - d_\ell)^2 / \lambda_\ell = d_\alpha^2$.

Remarks. (1) The contribution of point H^j to the variance of axis ℓ is equal to $\frac{n_c}{n}(d_\ell^j)^2$, hence $\frac{n_c}{n}(D_{L'}^j)^2 = \sum_{\ell \in L'} \mathrm{Ctr}_j^\ell$. The statistic $D_{L'}^2$ associates with each sample j the sum of its contributions to the L' axes generating the subspace. (2) The test statistic $n_c \frac{n-1}{n-n_c} D^2$ is the multidimensional generalization of the squared Z-score. Its observed value is the sum of the squared test-values of the L' principal axes.

3.6 Computations with R and Coheris SPAD Software

In order to carry out computations, a coordinate system has to be chosen. For the sake of simplicity, the geometric space is equipped with an *orthonormal Cartesian frame*. In other words, the Euclidean distance must be chosen prior to inductive analysis, that is, the first step of the analysis will always be the study of the geometric structure of the reference cloud and the second one will be focused on inferential analysis (see Case Studies in Chapter 6).

An orthonormal basis of the geometric space can be obtained from the initial basis using the Gram–Schmidt orthonormalization procedure (see Theorem 7.2, p. 232), but we advise the use of the *orthonormal principal basis* of the reference cloud obtained from a GDA method (see Chapter 2, p. 17).

In this section, we first present the R script performing typicality test and determining the compatibility region for Mean Point (multidimensional data). Then, we comment on the use of the complete program, with all procedures proposed in this chapter for multidimensional and one-dimensional data, by using an R script interfaced with SPAD, which is a driven-menu software.

3.6.1 R Script

The R script we present computes the p-value of the combinatorial typicality test for the Mean Point (§3.2, p. 32), then the compatibility region (§3.2.5, p. 40). In the following R code, the data management and the display of results are reduced to the minimum. Applied to the *Target example*, it provides the results of pages 38 and 42.

Notations for R scripts. As a rule, the name of a matrix will begin with a capital letter and end with a dot followed by two capital letters designating the two sets indexing the rows and the columns of the matrix (for a column-matrix, the dot will be followed by one capital letter that designates the set indexing the rows). *Examples*: the coordinate matrix of points $(M^i)_{i \in I}$ will be denominated by X.IK and its covariance matrix by Mcov.KK (M to recall cloud M^I and cov to point out "covariance"). The J-column of values of the test statistic D^2 will be denominated by D2.J, etc.

Step 1. The data

The data set consists of the coordinates of the points of the reference cloud and of that of the group cloud over an *orthonormal basis* of the space.

First, we specify the working directory (say "D:/path").

The data file of the reference set is read (the opposite table is that of the *Target example* corresponding to the read.table below) and the cardinality of the reference set (n) as well as the dimensionality of the space (K) are determined. The $n \times K$-matrix of coordinates of the points of the reference cloud is subsequently established (X.IK).

Then, the data file (Target_group.txt) of the group is read; hence the cardinality of the group (n_c) and the $n_c \times K$-matrix of coordinates of the points of the group (X.CK).

I;	X;	Y
$i1$;	0;	-6
$i2$;	3;	-5
$i3$;	7;	-3
$i4$;	3;	-1
$i5$;	6;	0
$i6$;	-4;	1
$i7$;	1;	2
$i8$;	3;	2
$i9$;	5;	5
$i10$;	6;	5

C;	X;	Y
$c1$;	-1.5;	-2.5
$c2$;	1.5;	-0.5
$c3$;	0.5;	1.5
$c4$;	-0.5;	1.5

```
setwd("D:/path")
base <-  read.table(file = "Target_reference.txt", header= TRUE,
                      sep= ";", dec= ".", row.names= 1)
n     <- dim(base)[1]                    # cardinality of reference set
K     <- dim(base)[2]                    # number of variables
X.IK <- as.matrix(base, nrow=n, ncol=K)  # reference coordinates
base <- read.table(file = "Target_group.txt", header= TRUE,
                sep= ";", dec= ".", row.names= 1)
n_c  <- dim(base)[1]                     # cardinality of group
X.CK <- as.matrix(base, nrow=n_c,ncol=K) # group coordinates
if(dim(base)[2] < K)
  stop("the number of columns for group is less than ", K)
```

Step 2. Parameters of the method

The *parameters* of the method are the following ones: notable limit for scaled deviation (see Remark (1) p. 22), a reference α-level, the maximum number of samples to be used (see Remark (1) p. 44) and possibly an integer used to initialize the random number generator[13] for MC method.

---

[13]We may set the random number generator seed by using seed <- 1234 then set.seed(seed): this is done in order to allow researchers to obtain the same result. Most often, the generator seed will not be defined by the user (seed <- NULL), so the results might be different.

```
notable_D  <- 0.4            # notable limit for D
alpha      <- 0.05           # alpha level
max_number <- 100000         # maximum number of samples
seed       <- NULL           # seed for random number generator
```

Step 3. Descriptive analysis

In this step, we compute the scaled deviation (M-distance) between the mean point G of the reference cloud and the mean point C of the group cloud, in order to conclude about the magnitude of the deviation.

We start by proceeding to a *change of basis* for purposes of working in the orthocalibrated principal frame of the affine support $\mathcal{M}$ of the reference cloud (see Remark (2), p. 44). To do this, we compute the covariance matrix (Mcov.KK) of the reference cloud and we proceed to the passage from $\mathcal{V}$ (direction of $\mathcal{U}$ equipped with an orthonormal basis) to $\mathcal{L}$ (direction of $\mathcal{M}$ equipped with the orthocalibrated principal basis; see p. 20). This is done from eigenvectors and eigenvalues of the covariance matrix of cloud M^I. If L is the dimension of the cloud—that is, the number of non-null eigenvalues—the change of basis matrix is a $K \times L$-matrix called BasisChange.KL.

Then, we deduce the matrix of the standardized principal coordinates of points $(\mathrm{M}^i)_{i \in I}$ (Z.IL) (see Chapter 2, Equation 2.6, p. 19) and that of the mean point of the group (ZC.L).

Finally, we compute the squared M-distance between points G and C $(d_{obs}^2 = |\mathrm{GC}|^2$ denoted d2_obs), which is simply the sum of the squared standardized principal coordinates of point C. Then, using the scaled deviation, which is an index of magnitude of the deviation between points G and C (see p. 22), we state the descriptive conclusion.

```
Mcov.KK <- cov.wt(X.IK, method= "ML")$cov    # covariance matrix
eig <- eigen(Mcov.KK, symmetric = TRUE)      # diagonalization
L <- sum(eig$values > 1.5e-8)                # dimensionality of cloud
lambda.L <- eig$values[1:L]                  # non-null eigenvalues
if (L < K) {
   warning("The dimension of the reference cloud is ", L, " < ", K,
           " (number of variables). \n It is advisable to study again",
           " the geometric construction of the reference cloud.")
}
BasisChange.KL <- eig$vectors[ ,1:L] %*% diag(1/sqrt(lambda.L), nrow=L)
Z.IL    <- sweep(X.IK, 2, colMeans(X.IK), "-") %*% BasisChange.KL
ZC.L    <- t(colMeans(X.CK) - colMeans(X.IK)) %*% BasisChange.KL
d2_obs <- sum(ZC.L^2)                # observed value of test statistic
cat(" M-distance D = ", sqrt(d2_obs), "\n",
    " Descriptively, the deviation is ",
    ifelse(d2_obs >= notable_D^2, "notable.",
           "small: there is no point in doing the test."), sep= "")
```

Step 4. The testing procedure

In the case where the scaled deviation is greater than the notable limit, the testing procedure is only performed .

Then, the choice between exhaustive and MC method depends on the number $\binom{n}{n_c}$ of possible samples (see Remark (1) p. 44). If it is less than the maximum number of samples chosen by the user (max_number), then the exhaustive method is used and the $J \times n_C$-matrix identifying the J samples $(I_j)_{j \in J}$ (see §3.2.2, p. 33), called Samples.JC, is constructed. If the MC method is used, samples are determined step by step. The number of samples used in the program is denoted cardJ; it is equal to the minimum of $J = \binom{n}{n_c}$ and of the (maximum) number of samples chosen by the user. Finally we calculate the value of the test statistic for each sample, that is, the column-matrix D2.J, and then we compute the combinatorial p-value (see p. 37).

```
cardJ <- min(choose(n, n_c), max_number)      # number of used samples
D2.J <- matrix(0, nrow= cardJ, ncol= 1)       # test statistic
if (choose(n, n_c) <= max_number) {
  Samples.JC <- t(combn(n, n_c))
  for (j in 1:cardJ)
    D2.J[j]  <- sum(colMeans(as.matrix(Z.IL[Samples.JC[j, ], ],
                                 ncol = L))^2)
} else { set.seed(seed)
  for (j in 1:cardJ)
    D2.J[j]  <- sum(colMeans(as.matrix(Z.IL[sample.int(n, n_c),
                                 ncol = L]))^2)
}
if(d2_obs >= notable_D^2){
  n_sup    <- sum(D2.J >= d2_obs * (1 - 1e-12))
  p_value <- n_sup / cardJ
  cat(" p-value = ", n_sup, "/", cardJ, " = ",
      format(round(p_value, digits= 3), nsmall= 3),"\n", sep= "")
}
```

Step 5. The compatibility region

Then, by applying Proposition 3.6 (p. 42), we determine the scale parameter of the compatibility ellipsoid (d_alpha).

```
rank       <-  cardJ - trunc(cardJ *alpha)
d_alpha    <- sqrt(sort(D2.J)[rank])

cat("\n ", 100*(1 - alpha), "% compatibility region:\n",
    "  principal ellipsoid of the reference cloud ",
    "centred on the group mean point,\n  with scale parameter",
    " d_alpha = ", round(d_alpha, 3), ". \n", sep = "")
```

3.6.2 R Script Interfaced with Coheris SPAD Software

For researchers, it is particularly easy to perform the analyses using the R script that is interfaced with Coheris SPAD software[14]. We will present here the analyses of the *Target example* developed in this chapter. The sequence of analyses is quite simple. It comprises four steps (Figure 3.11).

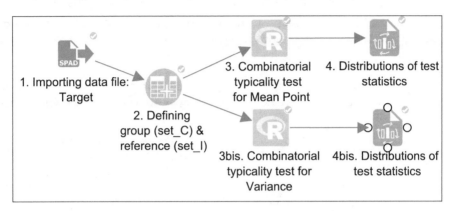

Figure 3.11
Diagram of the Coheris SPAD project.

The *first step* consists in *importing a data file* in a "SPAD data archive" format (many other formats are available). The file is stored inside the project for the sake of the autonomy of the project. Data table is shown in Figure 3.12 (left).

The *second step* consists in adding one or two columns to the data table. The first column (`set_C`) indicates the individuals of the group (designated

Index	ident	X	Y
1	i1	0	-6
2	i2	3	-5
3	i3	7	-3
4	i4	3	-1
5	i5	6	0
6	i6	-4	1
7	i7	1	2
8	i8	3	2
9	i9	5	5
10	i10	6	5
11	c1	-1.5	-2.5
12	c2	1.5	-0.5
13	c3	0.5	1.5
14	c4	-0.5	1.5

Index	ident	X	Y	set_C	set_I
1	i1	0	-6		ref
2	i2	3	-5		ref
3	i3	7	-3		ref
4	i4	3	-1		ref
5	i5	6	0		ref
6	i6	-4	1		ref
7	i7	1	2		ref
8	i8	3	2		ref
9	i9	5	5		ref
10	i10	6	5		ref
11	c1	-1.500	-2.500	grp	
12	c2	1.500	-0.500	grp	
13	c3	0.500	1.500	grp	
14	c4	-0.500	1.500	grp	

Figure 3.12
Data tables after steps 1 and 2.

---

[14]For a general presentation of Coheris SPAD software, see Chapter 1, p. 6.

by a name of category, in our example `grp`) (Figure 3.12, right). The second one—which is necessary only if the reference set corresponds to a subtable—must take the value `ref` for the individuals that belong to the reference set (it is called `set_I` in our example).

Steps 3 and 3bis are the ones of combinatorial typicality methods through the R script for *Mean point* (Step 3) and for *Variance* (Step 3bis).

The *settings* of methods will be defined using the four tabs `Variables`, `Analyses`, `Parameters` and `Outputs` (see Figure 3.13).

- `Variables` (Figure 3.13)

  Firstly, you select the "Combinatorial typicality test" R script (`CIGDA_Comb-v1.R`), which has been previously copied in the folder (defined by the user) of interfaced R script.

  Secondly, you have to select the *numerical variables* on which the analysis is performed (coordinates of points in a geometric space equipped with orthonormal basis) as well as the categorical variables defining the *group* (here `set_I`) and, since here the reference set corresponds to a subtable, the categorical variable defining the *reference set* (here `set_C`).

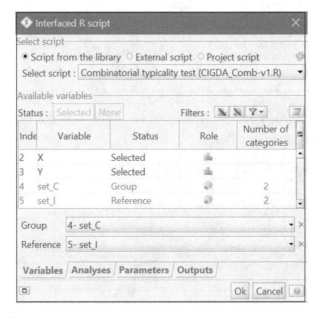

Figure 3.13
The `Variables` dialog box: selection of the R script and of the status of variables for the analyses.

- `Analyses` (Figure 3.14)

  You have to give the name of the category of the group under study (`grp`).

Then, you choose the type of analysis (one-dimensional, multidimensional, or both) and the typicality test (for Mean point or for Variance).

You can perform the test and/or the compatibility region.

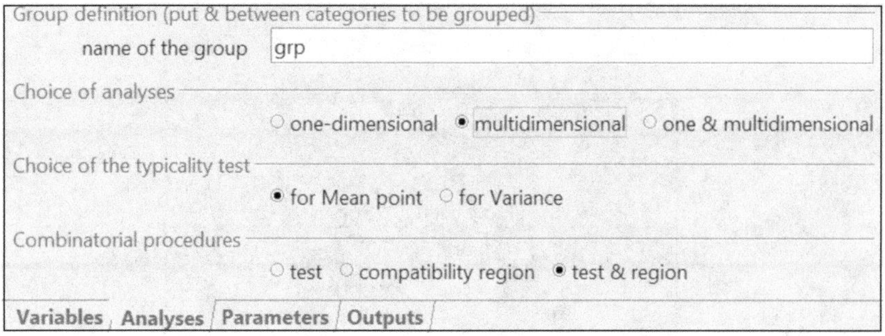

Figure 3.14
The Analyses dialog box.

- Parameters (Figure 3.15)

  All parameters have default values corresponding to the more conventional choices. You can modify these values.

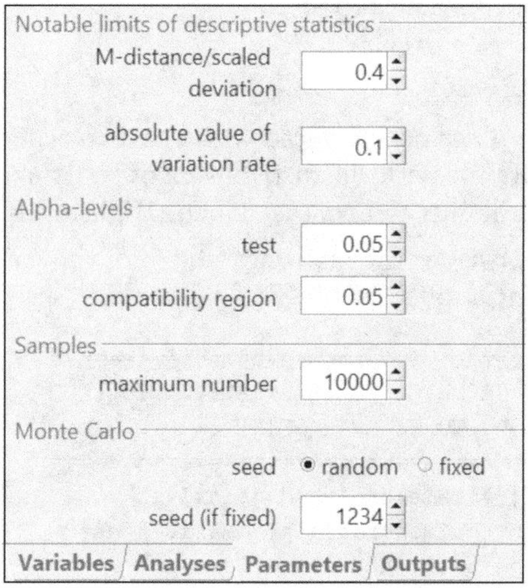

Figure 3.15
The Parameters dialog box.

- Outputs (Figure 3.16)

  The Outputs dialog box permits to choose the number of decimals for

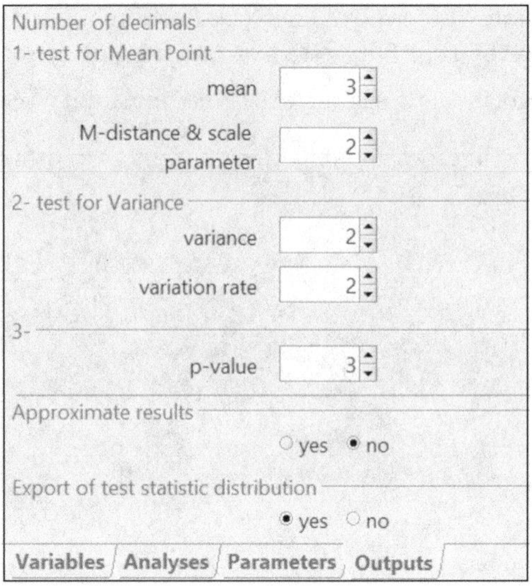

Figure 3.16
The Outputs dialog box.

results, to give or not the results of approximate methods and to export or not the tests statistics to SPAD.

Results of analyses

The results of the R script interfaced with SPAD consist of a text file[15] and, if requested, of a data file with the distribution of test statistics. The results can be accessed from the method context menu (Right-click/Results/Edition).

The program provides the following results of the combinatorial typicality test for *Mean point* applied to the *Target example* (see p. 38).

```
====================================================
COMPARISON OF A GROUP CLOUD TO A REFERENCE CLOUD
             ACCORDING TO MEAN POINT
====================================================
  2 variables (X, Y) are analysed jointly;
  clouds are in a 2-dimensional geometric space.
      size of the reference set:  10
      size of the group (grp): 4

Descriptive analysis
--------------------

  M-distance between mean points: 0.96 >= 0.4
  Descriptively, the deviation is notable.
```

---

[15] In the case of a 2-dimensional space, the compatibility ellipse can be found in an Excel file (Right-click/Results/ Report).

```
Combinatorial inference
-----------------------
   number of possible samples: 210;
   exhaustive method is performed.

~ Combinatorial typicality test
   test statistic: squared M-distance (observed value: 0.93)
   p-value: p = 9/210 = 0.043 <= 0.05
   For mean point, the group is atypical of the reference population,
   at level 0.05.

~ 95% compatibility region: principal ellipse of reference cloud
      centered at the mean point of group cloud and with scale
      parameter 0.96.

Test statistic distribution is exported to SPAD.
```

Distributions of test statistics

The R script provides the values of the test statistic that can be exported to SPAD. From these results, it is possible to construct the distributions using graphical interface of Coheris SPAD (see Step 4 in Figure 3.11).

Concluding Remarks

To answer the question raised at the beginning of the chapter, we studied the deviation between the mean point of the group and that of the reference population. But the typicality test can be defined for any test statistic (we have also presented the variance); thus, researchers have the option of using a wide variety of test statistics.

It is important to note that combinatorial procedures applied in GDA are direct inductive extensions of the descriptive aids to interpretation.

The typicality test is also used in cluster analysis in order to interpret clusters of a partition: for instance comparison of the mean of a variable in a cluster to its mean in the overall set of individuals that have been clustered; this is especially important if the size of the cluster is small. In the case of Euclidean clustering with representation of clusters in the geometric space, we also question the typicality of the subcloud associated with the cluster with respect to the overall cloud.

So, combinatorial typicality tests should be used in GDA much more widely than they are today, and included in software performing GDA methods.

4

Geometric Typicality Test

In this chapter, we present the *geometric typicality test*, in short *geometric test*[1]. This test answers the kind of questions frequently asked by GDA users:

> *"Is the deviation of the mean point of a cloud from a reference point a genuine one, or may it be due to chance?"*

This test could also be termed *typicality test for a mean point of a cloud with respect to a reference point*.

The geometric typicality test will be applied in Chapter 6 to *The Parkinson study* (see p. 156).

4.1 Principle of the Test

To begin with we will present two situations that are exemplary cases for the geometric typicality test, then we will briefly describe the steps of the test.

4.1.1 Paradigmatic Situations

Let us consider the following two situations.

Drug effect

The data set originates from an experiment with a *repeated measures design*. It concerns a group of patients for which a score of performance is registered twice: before (occasion A) and after (occasion B) drug intake.

The question is:

> *Does the treatment have an effect?*

Suppose that the treatment has no effect, then for each patient, the "before" score is just as likely to be larger than the "after" score as it is to be smaller. In other words, a *permutation* within any pair of patient's scores is as likely as the reverse. A way to construct the test is to create all *possible rearrangements* of the data between "before" and "after" scores, keeping the pair of

---

[1]See Le Roux and Rouanet (2004, Chapter 8); Bienaise (2013, Chapter 2).

scores together. Then, a test statistic assessing the magnitude of the difference between scores is computed for each of them. Finally the value of the test
statistic obtained from the observed data is located within the permutation
distribution of the test statistic.

Actually, it is simpler to consider the individual differences of scores, that
is, the *individual effects*, rather than the raw scores themselves. Then, rearranging "before" and "after" scores simply reverses the sign of the differences.

In a *geometric setup*, the two groups of multidimensional observations are
represented by two clouds. Two points are associated with each patient, the
point "before" and the point "after", and the vector going from point "before"
to point "after" defines the individual effect-vector (Figure 4.1-a)[2].

Let I denote a set of patients and $\overrightarrow{e}^i$ the effect-vector of patient $i \in I$. If
the point O represents the null effect, the point $M^i = O + \overrightarrow{e}^i$ is called *effect-
point* of patient i (Figure 4.1-b). The *cloud of effect-points* together with its
mean point G and the reference point O (Figure 4.1-c) will be at the basis of
the geometric typicality test.

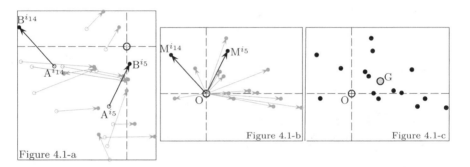

Figure 4.1
The Parkinson study. Construction of the cloud of effect-points. Figure 4.1-a: the two
matched clouds A^I ("before", o) and B^I ("after", •) with effect-vectors; Figure 4.1-b: shifted
effect-vectors and effect-points (Patients $i5$ and $i14$ are in black, the others in grey); Figure
4.1-c: cloud of effect-points.

Geometrically, if the treatment has no effect we may exchange each effect-
point with its symmetrical with respect to point O (point $O + \overrightarrow{e}^i$ is exchanged
with point $O - \overrightarrow{e}^i$).

This principle invites us to construct the *space of possible clouds* as the
set of clouds generated by the symmetry with centre O. Then we choose a
test statistic measuring the "distance" between points. Lastly, we determine
the proportion of the mean points of possible clouds that are more or equally
"distant" from point O than the observed mean point G is and we define it as
the combinatorial *p*-value.

---

[2]The data of Figure 4.1 are those of *The Parkinson study* (see Chapter 6, §6.1).

Target paradigm

Let us consider again the *Target example*. The data set under study is made up of the 10 impact points on a target (see Figure 4.2)[3].

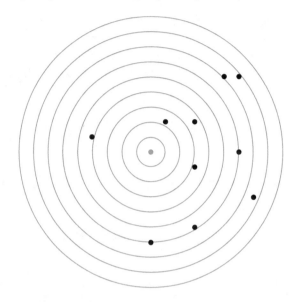

Figure 4.2
Target example. Ten impact points.

The question is the following one:

Is the aiming point the centre of the target?

Regarding the *Target paradigm*, the problem of typicality can be formulated as follows. Suppose that the centre of the target is the "true aiming point", then each impact point can be exchanged with its symmetric with respect to the centre of the target. Doing this for every point generates a *permutation space* of 2^{10} *possible clouds*. Then, considering the mean point of each possible cloud, we obtain a cloud of 2^{10} mean points. To evaluate the "distance" from each mean point to the centre of the target, we choose a *test statistic*. Finally, we look for mean points whose value of the test statistic is more extreme than or as extreme as the observed value (the one for the observed cloud). The proportion of such points defines the combinatorial *p*-value.

For these two examples, starting from the cloud of effect-points or from the cloud of impact points, we will perform the same procedure.

---

[3]Here, we choose a graphic scale which is the double of that of Figure 2.1 (Chapter 2, p. 9), thus almost all graphs of this chapter are at the same scale.

4.1.2   Steps of the Test

Briefly, the steps of the test are the following.

1. *Permutation space*

    Let point P be the reference point (the origin that represents the null effect in Figure 4.1-c, or the centre of the target in Figure 4.2). We change each individual point into its symmetric with respect to point P, one at a time. Starting from a cloud of n points and doing the exchange for every point generates 2^n possible clouds, among which the observed one. The family of possible clouds defines the *permutation space*.

  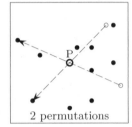

Figure 4.3

Target example. Reference point P = O (centre of the target): Clouds of 10 points with no permuted point (observed cloud) then one and two permuted points (the graphic scale of this figure is the half of the one of Figure 4.2).

2. *Combinatorial p-value*

    A test statistic is chosen in order to measure the "distance" between the mean points of possible clouds and the reference point. Then, the observed value is calculated from the observed data and the *combinatorial p-value* is equal to the proportion of possible clouds whose statistic value is more extreme than or as extreme as the observed one.

3. *Conclusion*

    Taking a reference α-level (see p. 31), we state the conclusion in the following words:

    - If the *p*-value is less than or equal to α, the deviation between the reference point and the observed mean point will be said *statistically significant* (in a combinatorial sense) at level α. Finding a significant deviation is an indication that the deviation between the observed mean point and the reference point is not "due to chance", therefore some explanation is called for.

    - If the *p*-value is greater than α, the deviation between the reference point and the observed mean point will be said *not significant* at level α. We consider that "it may be due to chance", then the two points are said to be *compatible* at level α.

After this short introduction that enlightens the objective of the test, we will outline the general test for Mean point (§4.2) and then develop the particular case of a one-dimensional cloud (§4.3). In Section 4.4, we will briefly present the case of repeated measures design. In Section 4.5, we will evoke other methods that deal with the same problem. Finally, we will give the R script written in connection with the Coheris SPAD software (§4.6).

The *Target paradigm* will be our leading example throughout this chapter.

4.2 Geometric Typicality Test for Mean Point

For this method, the basic data set is made of a *cloud of points* and a *reference point*. In a certain sense, we can say that the test consists in *comparing the mean point of a cloud to a reference point*. Again, the study is conducted in the line of pure geometry and a coordinate system will be chosen only to carry out computations (see §4.6, p. 97).

The set of individuals constitutes a *group of observations*, that is, individuals can be considered as *exchangeable*. It is denoted by I and its cardinality by n. The set I is represented by a cloud of n points in a K-dimensional geometric space $\mathcal{U}$ with direction vector space $\mathcal{V}$ (see §7.4.5, p. 241).

Recall that the mean point of cloud M^I is denoted by G, its affine support by $\mathcal{M}$ and its dimension by L ($L \leq K$) (§2.1, p. 10); recall that the covariance structure of cloud M^I is defined by the endomorphism Mcov (see §2.2.1, p. 15).

$$\mathrm{M}cov : \mathcal{V} \longrightarrow \mathcal{V}$$
$$\vec{u} \longmapsto \sum \frac{1}{n} \langle \overrightarrow{\mathrm{GM}^i} | \vec{u} \rangle \overrightarrow{\mathrm{GM}^i}$$

4.2.1 Permutation Space

Furthermore, let us consider a reference point P. The step 1 of the test, outlined on p. 68, leads us to construct the permutation space made up of 2^n possible clouds of n points. In order to operationalize the construction of the permutation space, we will use the following permutation table.

Permutation table

The set of possible clouds is indexed by the sequence of integers between 1 and 2^n, which is denoted J. In the binary numeral system, the integer $j \in J$ can be represented by the sequence $(\delta_i^j)_{i=1,\dots n}$ of n numbers equal to 0 or 1 such that $j = 1 + \sum_{i=1}^{n} \delta_i^j \times 2^{n-i}$. For $j' = 2^n + 1 - j$, one has $\delta_i^{j'} = 1 - \delta_i^j$ ($i \in I$).

The $J \times I$ permutation table is defined as follows: the entry on row j and column i is $\epsilon_i^j = 1 - 2\delta_i^j$ (see Table 4.1 for $n = 3$). Note that, for $j' = 2^n + 1 - j$, one has $\epsilon_i^{j'} = -\epsilon_i^j$ ($i \in I$).

The family $(\epsilon_i^j)_{i \in I}$ is called *permutation j*.

Table 4.1

Binary numbers for $j = 1$ to 2^3 and permutation table ($n = 3$).

δ_i^j	$i=1$	$i=2$	$i=3$		ϵ_i^j	$i=1$	$i=2$	$i=3$
$j = 1$	0	0	0			+1	+1	+1
$j = 2$	0	0	1			+1	+1	−1
$j = 3$	0	1	0			+1	−1	+1
$j = 4$	0	1	1			+1	−1	−1
$j = 5$	1	0	0			−1	+1	+1
$j = 6$	1	0	1			−1	+1	−1
$j = 7$	1	1	0			−1	−1	+1
$j = 8$	1	1	1			−1	−1	−1

Permutation space

Let P^{ij} be the point defined by $\overrightarrow{\mathrm{PP}^{ij}} = \epsilon_i^j \overrightarrow{\mathrm{PM}^i}$. Thus, if $\epsilon_i^j = +1$ (or $\delta_i^j = 0$), point P^{ij} is M^i and is called a "non-permuted point"; if $\epsilon_i^j = -1$ (or $\delta_i^j = 1$), point P^{ij} is symmetrical of M^i about the point P and is called a "permuted point" (see Figure 4.3, p. 68).

A cloud of n points $(\mathrm{P}^{ij})_{i \in I}$ is associated with each permutation $j \in J$; it is denoted P^{Ij}. In particular, if $j = 1$ one has $\forall i \in I$, $\epsilon_i^j = +1$: no point is permuted and if $j = 2^n$ one has $\forall i \in I$, $\epsilon_i^j = -1$: all points are permuted. Given $j \in J$ and $j' = 2^n + 1 - j$, one has $\forall i \in I, \epsilon_i^{j'} = -\epsilon_i^j$: clouds P^{Ij} and $\mathrm{P}^{Ij'}$ are symmetrical about the point P.

The family of the 2^n clouds $(\mathrm{P}^{Ij})_{j \in J}$ defines the *permutation P-space*, which is denoted $\mathcal{J}_{\mathrm{P}}$.

In what follows, we will study several permutation spaces depending on the choice of point P. One of them is very particular: it is constructed by taking the mean point G of the cloud M^I as a reference point. The others are defined by taking any point $\mathrm{P} \neq \mathrm{G} \in \mathcal{M}$ as a reference point.

4.2.2   Permutation Clouds

We will begin with studying in detail the G-permutation space that plays a key role in the construction of the test and of the compatibility region. Indeed, properties of any P-permutation space can be deduced from those of the G-permutation space.

Permutation G-cloud

Taking G as a reference point, the point G^{ij} is such that $\overrightarrow{\mathrm{GG}^{ij}} = \epsilon_i^j \overrightarrow{\mathrm{GM}^i}$. Given $j \in J$, the cloud of the n points $(\mathrm{G}^{ij})_{i \in I}$ is denoted G^{Ij} and the family of the 2^n clouds $(\mathrm{G}^{Ij})_{j \in J}$ defines the *permutation G-space*, denoted $\mathcal{J}_{\mathrm{G}}$.

The mean point of cloud G^{Ij} is $G^j = \sum_{i \in I} G^{ij}/n$; it is such that

$$\overrightarrow{GG^j} = \sum_{i \in I} \epsilon_i^j \overrightarrow{GM^i}/n \qquad (4.1)$$

The family of the 2^n mean points $(G^j)_{j \in J}$ is called *permutation G-cloud* and denoted G^J.

Given $j \in J$, if $j' = 2^n + 1 - j$, then $\epsilon_i^{j'} = -\epsilon_i^j$ and $\overrightarrow{GG^{j'}} = \sum_{i \in I} \epsilon_i^{j'} \overrightarrow{GM^i}/n =$

$-\sum_{i \in I} \epsilon_i^j \overrightarrow{GM^i}/n = -\overrightarrow{GG^j}$: points G^j and $G^{j'}$ are symmetrical about the point G: the permutation G-cloud is *symmetrical about the point* G (see Figure 4.4, for the *Target example*).

Figure 4.4
Target example. Permutation G-cloud with the mean point G (the grey levels of points depend on the overlapping).

Remark. The G-cloud of the *Target example* has horizontal and vertical patterns of points. This is because some points of cloud M^I have the same abscissas (three points), or ordinates (two pairs of points): two permutations that only differ for two of these points produce the same abscissa or ordinate, etc. But, in general, for a cloud M^I with no group of points that are lined up, the permutation G-cloud does not present such patterns.

• *Partition of the permutation G-cloud*

Given $j \in J$, the subset of I that consists of non-permuted individuals is denoted I_j^+:

$$I_j^+ = \{i \in I \text{ such that } \epsilon_i^j = +1\}$$

Its cardinality, denoted n_j^+, is an integer within the range 0 to n. Given an index set C with $n + 1$ elements, the set of integers from 0 to n will be denoted $(n_c)_{c \in C}$ $(0 \leq n_c \leq n)$. The subset of elements of J whose number of non-permuted individuals is equal to n_c is denoted J_c $(J_c = \{j \in J \mid n_j^+ = n_c\})$; its cardinality is equal to $\binom{n}{n_c}$. The family of the $n + 1$ subsets $(J_c)_{c \in C}$ constitutes a partition of J.

For instance, for $n = 3$ (see Table 4.1, p. 70), the partition of J is made of 4 parts, namely, $J_0 = \{j8\}$, $J_1 = \{j4, j6, j7\}$, $J_2 = \{j2, j3, j5\}$ and $J_3 = \{j1\}$.

A partition of the permutation G-cloud into $n + 1$ subclouds $(G^{J_c})_{c \in C}$ is associated with the partition $(J_c)_{c \in C}$ of J. For $n_c = 0$ (no point is permuted) and $n_c = n$ (all points are permuted), subclouds are reduced to point G. In the other cases, the subcloud G^{J_c} consists in $\binom{n}{n_c}$ points $(G^j)_{j \in J_c}$ (Figure 4.5).

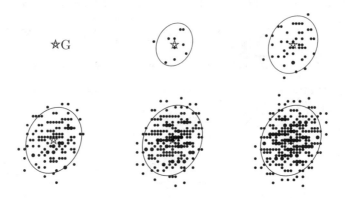

Figure 4.5

Target example. For $n_c = 0, 1, 2, 3, 4, 5$, subclouds G^{J_c} (point sizes are proportional to the superimposition) with their concentration ellipses and the mean point G (star). For $n_c = 9, 8, 7, 6$ subclouds are symmetrical about point G.

The subcloud G^{J_c} is linked to the *sample cloud* of size n_c defined in Chapter 3 (see §3.2.3, p. 33). In the present chapter, the sample cloud will be denoted by H^{J_c} (instead of H^J) in order to mark that its cardinality is equal to n_c.

The following proposition establishes the relation between the subcloud G^{J_c} and the sample cloud H^{J_c}.

Proposition 4.1 (Homothety property). *The subcloud G^{J_c} of G^J is the image of the sample cloud H^{J_c} by the homothety of centre G and scale factor $2n_c/n$.*

$$\forall j \in J_c, \ \overrightarrow{GG^j} = \tfrac{2n_c}{n} \overrightarrow{GH^j}$$

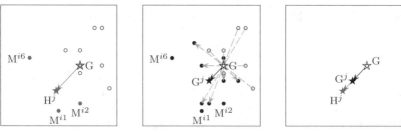

Figure 4.6

Target example. For $I_j^+ = \{i1, i2, i6\}$ ($n_c = 3$): on the left, subcloud $M^{I_j^+}$ (black points) with its mean point H^j; in the middle, cloud G^{I_j} (black points) with its mean point G^j ($\bigstar$), then the homothety property (the graphic scale is half of that of Figure 4.2).

Proof. $\forall j \in J_c$, G is the barycentre of point $H^j = \frac{1}{n_c} \sum\limits_{j \in I_j^+} M^i$ (weighted by n_c) and

point $H'^j = \frac{1}{n-n_c} \sum\limits_{j \notin I_j^+} M^i$ (weighted by $n - n_c$) with $n_c \overrightarrow{GH^j} + (n - n_c)\overrightarrow{GH'^j} = \overrightarrow{0}$

(barycentric property, p. 11). Yet $\overrightarrow{GG^j} = \sum\limits_{i \in I} \epsilon_i^j \overrightarrow{GM^i}/n = \sum\limits_{i \in I_j^+} \overrightarrow{GM^i}/n - \sum\limits_{i \notin I_j^+} \overrightarrow{GM^i}/n$

$= \frac{n_c}{n}\overrightarrow{GH^j} - \frac{n-n_c}{n}\overrightarrow{GH'^j} = \left(\frac{n_c}{n} + \frac{n-n_c}{n} \times \frac{n_c}{n-n_c} \right)\overrightarrow{GH^j}$ (see Figure 4.6). $\square$

From the homothety property and Proposition 3.4 (p. 35) we deduce the following corollary.

Corollary 4.1. *$\forall c \in C$ and $\forall j \in J_c$ the M-distance between points G^j and G is less than or equal to 1:* $|GG^j| \leq 2\sqrt{\frac{n_c(n-n_c)}{n^2}} \leq 1.$

Geometrically, points of the permutation G-cloud are all located inside (or on) the indicator ellipsoid of cloud M^I.

Lemma 4.1. *The mean point of the subcloud G^{J_c} is point G.*

Proof. By Proposition 3.1 (p. 34), the mean point of the sample subcloud H^{J_c} is the point G. Subcloud G^{J_c} is the image of H^{J_c} by the homothety of centre G, hence the property. $\square$

The following proposition results from the previous lemma as well as the symmetry property of the permutation G-cloud.

Proposition 4.2. *The mean point of the permutation G-cloud is point G.*

- *Covariance structure of the permutation G-cloud*

The covariance endomorphism of cloud G^J (see Definition 2.4, p. 15) is denoted Gcov and is such that:

$$\forall \overrightarrow{u} \in V, \text{ Gcov} : \overrightarrow{u} \mapsto \sum \frac{1}{2^n} \langle \overrightarrow{GG^j} | \overrightarrow{u} \rangle \overrightarrow{GG^j}$$

We will now establish the relation between the covariance endomorphisms of clouds G^J and M^I. For that purpose, we begin by studying the covariance endomorphism of the subcloud G^{J_c} and we deduce that of cloud G^J.

Lemma 4.2. *The covariance endomorphism of the subcloud G^{J_c}, denoted $Gcov_{J(c)}$, is equal to $\frac{1}{n-1} \frac{4n_c(n-n_c)}{n^2}$ Mcov.*

$$Gcov_{J(c)} = \frac{1}{n-1} \frac{4n_c(n-n_c)}{n^2} \text{ Mcov}$$

Proof. This lemma is an immediate consequence of the homothety property (p. 72) and of Proposition 3.2 (p. 34). $\square$

Theorem 4.1. *The covariance endomorphism of the permutation G-cloud, denoted Gcov, is equal to Mcov/n.*

$$Gcov = \frac{1}{n}\text{Mcov}$$

Proof. From the partition $(G^{J_c})_{c \in C}$ of cloud G^J, we deduce the between-C and the within-C clouds (see §2.4, p. 25). By Theorem 2.3 (p. 28), the covariance endomorphism of cloud G^J is the sum of that of the between-C cloud and of that of the within-C cloud. The mean points of subclouds $(G_{J_c})_{c \in C}$ are all equal to G (see Proposition 4.1, p. 73), hence the between-C cloud is a one-point cloud (point G), and its covariance endomorphism is null. After Equation 2.12 (p. 27), the within-C covariance endomorphism is equal to $\frac{1}{2^n} \sum_{c \in C} \binom{n}{n_c} \mathrm{Gcov}_{J(c)}$. From

Lemma 4.2, we deduce $\mathrm{G}_{J(C)} = \frac{1}{2^n} \sum_{n_c=0}^{n} \binom{n}{n_c} \frac{4}{n-1} \frac{n_c(n-n_c)}{n^2} \mathrm{Mcov}$, hence $\mathrm{G}_{J(C)} =$

$\frac{1}{n} \frac{4}{2^n} \left[\sum_{n_c=1}^{n-1} \frac{n_c}{n} \frac{n-n_c}{n-1} \binom{n}{n_c} \right] \mathrm{Mcov} = \frac{1}{n} \frac{1}{2^{n-2}} \left[\sum_{n_c=1}^{n-1} \binom{n-2}{n_c-1} \right] \mathrm{Mcov} = \frac{1}{n} \mathrm{Mcov}.$ $\square$

For $n_c = 0$ and $n_c = n$, the covariance endomorphisms of subclouds G^{J_c} are null since subclouds are one-point clouds. For $0 < n_c < n$, the covariance endomorphisms of subclouds $(G^{J_c})_{c \in C}$ are proportional to that of cloud M^I, the proportionality coefficient being equal to $\frac{4}{n-1} \frac{n_c(n-n_c)}{n^2}$. Therefore, the principal ellipsoids of the $C + 1$ subclouds $(G^{J_c})_{c \in C}$ are homothetic to that of cloud M^I and the homothety scale factor is an increasing function of n_c for $n_c \leq n/2$ and a decreasing function for $n/2 \leq n_c \leq n$ (see Figure 4.5, p. 72).

Permutation P-cloud

Recall (see p. 70) that $\forall j \in J, \forall i \in I, \overrightarrow{PP}^{ij} = \epsilon_i^j \overrightarrow{PM}^i$ and that the family of the 2^n clouds $(P^{Ij})_{j \in J}$ defines the permutation P-space $\mathcal{J}_P$. The 2^n mean points $(P^j)_{j \in J}$ of these clouds define the *permutation* P-*cloud*, denoted P^J, with

$$\forall j \in J, \ P^j = \sum_{i \in I} P^{ij}/n \quad \text{or} \quad \overrightarrow{PP}^j = \sum_{i \in I} \epsilon_i^j \overrightarrow{PM}^i/n$$

For $j \in J$ and $j' = 2^n + 1 - j$, we have $\epsilon_i^{j'} = -\epsilon_i^j$, hence $\overrightarrow{PP}^{j'} = \sum_{i \in I} \epsilon_i^{j'} \overrightarrow{PM}^i/n$

$= - \sum_{i \in I} \epsilon_i^j \overrightarrow{PM}^i/n = -\overrightarrow{PP}^j$. Points P^j and $P^{j'}$ are symmetrical about the point P. Consequently, the permutation P-cloud is *symmetrical about the point* P and its mean point is point P.

The properties of the permutation P-cloud can be deduced from those of the permutation G-cloud as a consequence of the following proposition, which exhibits the relation between the two clouds.

Proposition 4.3. *Letting* $\bar{\epsilon}^j = \sum_{i \in I} \epsilon_i^j/n$, *the permutation* P-*cloud satisfies the*

following property:

$$\forall j \in J, \ \overrightarrow{PP}^j = \overrightarrow{GG}^j + \bar{\epsilon}^j \overrightarrow{PG}$$

Proof. Given any $j \in J$, one has $\overrightarrow{PP}^j = \frac{1}{n} \sum_{i \in I} \epsilon_i^j \overrightarrow{PM}^i = \frac{1}{n} \sum_{i \in I} \epsilon_i^j (\overrightarrow{PG} + \overrightarrow{GM}^i) =$

$\left(\frac{1}{n} \sum_{i \in I} \epsilon_i^j \right) \overrightarrow{PG} + \frac{1}{n} \sum_{i \in I} \epsilon_i^j \overrightarrow{GM}^i = \bar{\epsilon}^j \overrightarrow{PG} + \overrightarrow{GG}^j.$ $\square$

• *Partition of the permutation* P-*cloud*

A partition of the permutation P-cloud into $n+1$ subclouds $(P^{J_c})_{c \in C}$ is associated with the partition $(J_c)_{c \in C}$ of J defined on page 71. As for the permutation G-cloud, we first establish the properties of subclouds $(P^{J_c})_{c \in C}$, then we study those of cloud P^J.

Proposition 4.4. *The subcloud* P^{J_c} *is the image of subcloud* G^{J_c} *by the translation of vector* $2\frac{n-n_c}{n} \overrightarrow{GP}$:

$$P^j = G^j + 2\frac{n-n_c}{n} \overrightarrow{GP}$$

Proof. $\forall j \in J_c, \overline{\epsilon}^j = \frac{n_c - (n-n_c)}{n} = 2\frac{n_c}{n} - 1$. From Chasles's relation, $\overrightarrow{PP^j} = \overrightarrow{PG} + \overrightarrow{GP^j}$ and from Proposition 4.3, $\overrightarrow{PP^j} = \overrightarrow{GG^j} + \overline{\epsilon}^j \overrightarrow{PG}$, hence $P^j = G^j + 2\frac{n-n_c}{n} \overrightarrow{GP}$: □

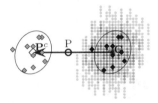

Figure 4.7
Target example. Permutation G-cloud and, for $n_c = 1$, subclouds G^{J_c} (black diamonds) and P^{J_c} (grey diamonds) with their mean points and their concentration ellipses.

Using the relation between the two clouds, we deduce that:

1. the mean point P^c of the subcloud P^{J_c} is the image of point G by translation of vector $2\frac{n-n_c}{n} \overrightarrow{GP}$, that is, $P^c = G + 2\frac{n-n_c}{n} \overrightarrow{GP}$;

2. the $n+1$ points $(P^c)_{c \in C}$ are on the line passing through points G and P, consequently, the cloud $(P^c)_{c \in C}$ is one-dimensional;

3. for all $j \in J_c$, $\overrightarrow{P^c P^j} = \overrightarrow{GG^j}$.

• *Covariance structure of the permutation* P-*cloud*

For the purpose of determining the covariance endomorphism of the permutation P-cloud, we will use the between-within decomposition of the permutation P-cloud. We will show that the covariance endomorphism of P^J is proportional to the generalized covariance endomorphism of cloud M^I with respect to point P (see Definition 2.5, p. 16).

Lemma 4.3. *The covariance endomorphism of the between-C cloud, denoted* $Pcov_C$, *is such that:*

$$\forall \overrightarrow{u} \in \mathcal{V}, Pcov_C(\overrightarrow{u}) = \frac{1}{n} \langle \overrightarrow{PG} | \overrightarrow{u} \rangle \overrightarrow{PG}$$

Proof. We have $\text{Pcov}_C(\vec{u}) = \frac{1}{2^n} \sum\limits_{c \in C} \binom{n}{n_c} \langle \overrightarrow{\text{PP}^c} | \vec{u} \rangle \overrightarrow{\text{PP}^c}$ (see Equation 2.10, p. 26).

Yet $\overrightarrow{\text{PP}^c} = \overrightarrow{\text{PG}} + \overrightarrow{\text{GP}^c}$ (Chasles's relation), hence $\overrightarrow{\text{PP}^c} = \overrightarrow{\text{PG}} + 2\frac{n-n_c}{n}\overrightarrow{\text{GP}} = (1 - 2\frac{n_c}{n})\overrightarrow{\text{GP}}$ and $\text{Pcov}_C(\vec{u}) = \sum\limits_{c \in C} \binom{n}{n_c}(1 - 2\frac{n_c}{n})^2 \langle \overrightarrow{\text{PG}} | \vec{u} \rangle \overrightarrow{\text{PG}}/2^n$.

$$\sum_{c \in C} \binom{n}{n_c}(1 - \frac{2n_c}{n})^2 = \sum_{c \in C} \binom{n}{n_c}\left(1 - 4\frac{n_c(n-n_c)}{n^2}\right) = \sum_{k=0}^{n} \binom{n}{k}\left(1 - 4\frac{k(n-k)}{n^2}\right) = 2^n -$$
$$\frac{n}{n^2} \sum_{k=1}^{n-1} \binom{n}{k}k(n-k) = 2^n - 4\frac{n-1}{n} \sum_{k=1}^{n-1} \binom{n-2}{k-1} = 2^n - 4\frac{n-1}{n}2^{n-2} = \frac{2^n}{n}/n. \quad \square$$

The following lemma is a direct consequence of the property of translation.

Lemma 4.4. *The covariance endomorphism of the subcloud* $\text{P}^{J_c} \subset \text{P}^J$, *denoted* $\text{Pcov}_{J(c)}$, *is equal to that of the subcloud* $\text{G}^{J_c} \subset \text{G}^J$.

$$\forall c \in C, \ \text{Pcov}_{J(c)} = \text{Gcov}_{J(c)}$$

Figure 4.8
Target example. On the left, the P-cloud with the mean points and the concentration ellipses of the 11 subclouds $(\text{P}^{J_c})_{c \in C}$; on the right the P-cloud with its concentration ellipse (the grey levels of points depend on the overlapping).

Proposition 4.5. *The covariance endomorphism of the permutation P-cloud, denoted* Pcov, *is proportional to the generalized covariance endomorphism of cloud* M^I *with respect to point P:*

$$\text{Pcov} = \frac{1}{n}\text{Mcov}\text{P}$$

Proof. The between-within decomposition of the covariance endomorphism of cloud P^J writes (Theorem 2.3, p. 28): $\text{Pcov} = \text{Pcov}_C + \text{Pcov}_{J(C)}$.
We have: $\text{Pcov}_{J(C)} = \sum\limits_{c \in C} \binom{n}{n_c}\text{Pcov}_{J(c)}/2^n = \sum\limits_{c \in C} \binom{n}{n_c}\text{Gcov}_{J(c)}/2^n = \text{Gcov} = \frac{1}{n}\text{Mcov}$,
$\forall \vec{u} \in \mathcal{V}, \text{Pcov}_C(\vec{u}) = \frac{1}{n}\langle \overrightarrow{\text{PG}} | \vec{u} \rangle \overrightarrow{\text{PG}}$, hence $\text{Pcov}(\vec{u}) = \frac{1}{n}(\text{Mcov}(\vec{u}) + \langle \overrightarrow{\text{PG}} | \vec{u} \rangle \overrightarrow{\text{PG}})$.
Using Proposition 2.5 (p. 16), we deduce the property. $\quad \square$

Corollary 4.2. *The M-covariance endomorphism of the permutation P-cloud, denoted* M.Pcov, *is such that*

$$\forall \vec{u} \in \mathcal{L}, \ \text{M.Pcov}(\vec{u}) = \frac{1}{n}\left(\vec{u} + [\overrightarrow{\text{PG}} | \vec{u}]\overrightarrow{\text{PG}}\right)$$

Proof. If $\mathcal{L}$ is equipped with the M-scalar product, applying Proposition 4.5, we obtain $\mathrm{M.P}cov = \frac{1}{n}\mathrm{M.M}cov^\mathrm{P}$. Then, $\forall \vec{u} \in \mathcal{L} : \mathrm{M.P}cov(\vec{u}) = \frac{1}{n}\left(\mathrm{M.M}cov(\vec{u}) + [\overrightarrow{\mathrm{PG}}|\vec{u}]\overrightarrow{\mathrm{PG}}\right)$. By Proposition 2.13, (p. 21), $\mathrm{M.M}cov = id_L$, hence the property. $\square$

4.2.3   Test Statistic

From now on, we will consider the restrictions of covariance endomorphisms to the direction vector subspace $\mathcal{L} \subseteq \mathcal{V}$ of the affine support $\mathcal{M}$ of cloud M^I.

In order to make the test procedure operational, we have to define a test statistic. We choose a test statistic which is a magnitude index of the deviation between points taking into account the shape of the permutation P-cloud.

The mean point of the permutation P-cloud is point P and its covariance endomorphism Pcov is proportional to $\mathrm{M}cov^\mathrm{P}$ (Proposition 4.5, p. 76), that is why we choose a test statistic that depends on the *squared generalized Mahalanobis norm* with respect to point P (see Definition 2.7, p. 21).

More precisely, the test statistic, denoted D_P^2, is such that:

$$D_\mathrm{P}^2 : \mathcal{J}_\mathrm{P} \to \mathbb{R}_{\geq 0}$$
$$\mathrm{P}^{Ij} \mapsto |\mathrm{PP}^j|_\mathrm{P}^2 \qquad (4.2)$$

with $|\mathrm{PP}^j|_\mathrm{P}^2 = \langle \mathrm{M}cov^{\mathrm{P}-1}(\overrightarrow{\mathrm{PP}^j})|\overrightarrow{\mathrm{PP}^j}\rangle$, that we simply denote $d_\mathrm{P}^2[j]$.

By Theorem 2.2 (p. 21), we express the values of the test statistic (squared generalized Mahalanobis norm $|\cdot|_\mathrm{P})^2$ as a function of the squared Mahalanobis norm $(|\cdot|^2)$.

$$d_\mathrm{P}^2[j] = |\mathrm{PP}^j|_\mathrm{P}^2 = |\mathrm{PP}^j|^2 - \frac{[\overrightarrow{\mathrm{PG}}|\overrightarrow{\mathrm{PP}^j}]^2}{1+|\mathrm{PG}|^2} \qquad (4.3)$$

The observed value of the statistic is equal to $|\mathrm{PG}|_\mathrm{P}^2$ and denoted $d_{\mathrm{P}obs}^2$. Using the reciprocity formula (p. 21), one has:

$$d_{\mathrm{P}obs}^2 = \frac{|\mathrm{PG}|^2}{1+|\mathrm{PG}|^2} \qquad (4.4)$$

Permutation distribution of the test statistic

Given $j \in J$, let $p(D_\mathrm{P}^2 = d_\mathrm{P}^2[j])$ denote the proportion of $j \in J$ that satisfies the property $(D_\mathrm{P}^2 = d_\mathrm{P}^2[j])$. The *distribution* of statistic D_P^2 is defined by the map from $D_\mathrm{P}^2(\mathcal{J}_\mathrm{P})$ to $[0,1]$, which associates with each value $d_\mathrm{P}^2[j]$ the proportion $p(D_\mathrm{P}^2 = d_\mathrm{P}^2[j])$.

$$D^2(\mathcal{J}_\mathrm{P}) \to [0,1]$$
$$d_\mathrm{P}^2[j] \mapsto p(D_\mathrm{P}^2 = d_\mathrm{P}^2[j])$$

This distribution is discrete and will most often be represented by a histogram (for the *Target example*, see Figure 4.10, p. 80).

Combinatorial p-value

We want to know how extreme the observed value $d^2_{P_{obs}}$ is over the permutation distribution of D^2_P. We look for points of the permutation cloud P^J for which the value of the statistic is greater than or equal to the observed one. This proportion defines the *combinatorial p-value* and is denoted p.

$$p = p(D^2_P \geq d^2_{P_{obs}})$$

Geometric interpretation of the p-value

Let us consider the family of the principal ellipsoids of the permutation P-cloud. The κ-ellipsoid with centre P (mean point of cloud P^J) going trough point G is such that $\kappa = \sqrt{n}\,|PG|_P$. The p-value can be interpreted as the proportion of the points of the permutation P-cloud located on or outside this κ-ellipsoid (see Figure 4.9 for the *Target example*).

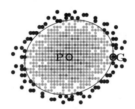

Figure 4.9
Target example. Permutation P-cloud (2^{10} points) with the principal ellipse of the cloud passing through point G (the reference point P is the centre of the target); 86 points of the permutation cloud (black points) are on or outside the ellipse.

Remarks on the choice of test statistic.

In order to assess the importance of the deviation from the points of the permutation cloud to the reference point, we have chosen the generalized Mahalanobis norm. This statistic is a Euclidean norm that leads to interesting properties about the characterization of the compatibility region (see Definition 4.1, p. 80). It is a *descriptive statistic* which does not depend on the length unit and which is linked to the covariance structure of the cloud.

We could have chosen the squared norm associated with Pcov as a test statistic. This statistic is *permutationally equivalent*[4] to D^2_P (since it is equal to nD^2_P) but it is not a descriptive statistic.

Proposition 4.6. $\forall P \in \mathcal{M}$, *the mean of the statistic D^2_P is equal to L/n.*

$$\forall P \in \mathcal{M},\ \text{Mean}\, D^2_P = L/n$$

---

[4] For a discussion of the concept of permutationally equivalent statistics, see, e.g., Pesarin (2001, p. 43).

Proof. Mean $D_P^2 = \sum |PP^j|_P^2/2^n = \frac{1}{2^n} \sum \langle Mcov^{P-1}(\overrightarrow{PP^j}) | \overrightarrow{PP^j} \rangle$; $Mcov^P = nPcov$, hence Mean $D_P^2 = \frac{1}{n} \sum \frac{1}{2^n} \langle Pcov^{-1}(\overrightarrow{PP^j}) | \overrightarrow{PP^j} \rangle$. Mean D_P^2 is equal to $\frac{1}{n}$ times the variance of cloud P^j calculated by using the Mahalanobis distance associated with $Pcov$, which is equal to L (Corollary 2.3, p. 21), hence the property. $\square$

Target example

Let $\overrightarrow{\epsilon_1}$ and $\overrightarrow{\epsilon_2}$ be the unit vectors of the horizontal and vertical axes of the plane. With respect to the orthonormal frame $(O, \overrightarrow{\epsilon_1}, \overrightarrow{\epsilon_2})$, the coordinates of G are $(3,0)$, and the matrix associated with $Mcov$ is the covariance matrix $\mathbf{V} = \begin{bmatrix} 10 & 2 \\ 2 & 13 \end{bmatrix}$ (see p. 14). The components of $\overrightarrow{GP}$ are $(-3,0)$.

Descriptive appraisal

The squared M-distance between points G and P is

$$|GP|^2 = \frac{1}{126} \times \begin{bmatrix} -3 & 0 \end{bmatrix} \begin{bmatrix} 13 & -2 \\ -2 & 10 \end{bmatrix} \begin{bmatrix} -3 \\ 0 \end{bmatrix} = \frac{117}{126} = 0.9286.$$

The M-distance between two points is an indicator of the magnitude of the deviation between points (see Chapter 2, p. 22). Since $|GP| = 0.96 \geq 0.40$, we conclude:

Descriptively, the deviation between P and G is of notable magnitude.

Geometric typicality test

This finding triggers the following natural query from the researcher (informally stated):

"Is the observed deviation a genuine one, or might it be due to chance?"

In order to answer this question, we will compare the mean point of the 10 impact-points to the centre of the target taken as a reference point.

The statistic defined by Equation 4.2 (p. 77) depends on endomorphism $Mcov^O$ (the reference point is O), which is equal to $\mathbf{V}_O = \begin{bmatrix} 19 & 2 \\ 2 & 13 \end{bmatrix}$ (see p. 19). Hence, $\mathbf{V}_O^{-1} = \frac{1}{243} \begin{bmatrix} 13 & -2 \\ -2 & 19 \end{bmatrix}$ and the observed value of the test statistic D_O^2 is

$$d_{Oobs}^2 = \frac{1}{243} \times \begin{bmatrix} 3 & 0 \end{bmatrix} \begin{bmatrix} 13 & -2 \\ -2 & 19 \end{bmatrix} \begin{bmatrix} 3 \\ 0 \end{bmatrix} = \frac{117}{243} = 0.48.$$

The distribution of the test statistic D_O^2 is depicted in Figure 4.10 (p. 80).

The combinatorial *p*-value is equal to $86/1024 = 0.084$; it is greater than the conventional level $\alpha = .05$. Hence, the deviation from point G to point O is not significant at level .05: points G and O are said to be compatible at level .05. Then, we assume that the location of point G "may be due to chance".

To answer the question raised on page 67, we will say that:

Data are compatible with the hypothesis that the aiming point is the centre of the target area.

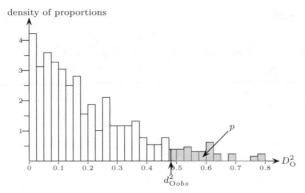

Figure 4.10
Target example. Reference point: O. Distribution of test statistic D_O^2 with (interpolated) p-value (shaded area) $p = 0.084$ ($d_{O obs}^2 = 0.48$).

4.2.4 Compatibility Region

Now we will consider as a reference point not just point P but every point of the affine support of the cloud. Recall (see p. 68) that, given an α-level, a point is said to be compatible with point G if the associated p-value is greater than α. Hence the following definition of the $(1 - \alpha)$ *compatibility region*[5].

Definition 4.1. *The $(1 - \alpha)$ compatibility region for point G is defined as the set of points* $P \in \mathcal{M}$ *compatible with point G at level* α.

$$\{P \in \mathcal{M} \text{ such that } p(D_P^2 \geq |PG|_P^2) > \alpha\}$$

Remark. The p-value depends on points to which the deviation from the reference point is non-null. This means that removing all individuals that are located at point P (null deviation to P) from the analysis provides the same result. For instance, for the "drug effect" situation, all individuals for whom the treatment has no effect can be discarded from the analysis.

On the contrary, it is worth observing that null deviations from point P play their part in the determination of the compatibility region as well as all other deviations. This remark pleads for the construction and the interpretation of the compatibility region.

The compatibility region has no analytical form: it depends on the cloud, more precisely, on its shape and the distribution of its points in the space. To determine it, we proceed in two steps. We begin by characterizing the compatibility interval on a line, then we show that, in the space, the compatibility region can be adjusted by a principal ellipsoid of the cloud.

---

[5] In combinatorial inference, we say "compatibility region" (or "compatibility interval") rather than "confidence region", which is the term used in traditional inference (see note p. 40). Other authors say "permutation confidence interval" (Pesarin, 2001; Good, 2011).

Compatibility interval along a line

Given a line $\mathcal{D} \subset \mathcal{M}$ passing through point G, and a permutation $j \in J$, we first determine the set of points $P \in \mathcal{D}$ such that $|PP^j|_P^2 \geq |PG|_P^2$. Then, we establish the $(1 - \alpha)$ compatibility interval along line $\mathcal{D}$.

Lemma 4.5. *Given a line $\mathcal{D}$ in $\mathcal{M}$ passing through point G, there exists, for each permutation $j \in J$, a closed line segment of $\mathcal{D}$ including point G that is bounded by points D_1^j and D_2^j such that:*

$$P \in [D_1^j, D_2^j] \Leftrightarrow |PP^j|_P^2 \geq |PG|_P^2$$

Proof. Let $\overrightarrow{\beta}$ be a M-unit vector and $\mathcal{D}$ the line passing through point G and with direction vector $\overrightarrow{\beta}$. The M-orthogonal decomposition of $\overrightarrow{GG^j}$ onto $\mathcal{D}$ and $\mathcal{D}^\perp$ writes $\overrightarrow{GG^j} = u^j \overrightarrow{\beta} + v^j \overrightarrow{\beta'}$ (with $|\overrightarrow{\beta}| = |\overrightarrow{\beta'}| = 1$, $[\overrightarrow{\beta} | \overrightarrow{\beta'}] = 0$). From Proposition 4.3 (p. 74), we have $\overrightarrow{PP^j} = \overrightarrow{GG^j} + \bar{\epsilon}^j \overrightarrow{PG}$. Then, letting $\overrightarrow{GP} = x \overrightarrow{\beta}$, $\overrightarrow{PP^j} = (u^j - \bar{\epsilon}^j x) \overrightarrow{\beta} + v^j \overrightarrow{\beta'}$.

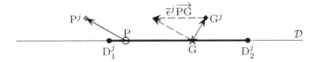

Figure 4.11
Plane generated by points G, P and G^j ($\bar{\epsilon}^j = -0.8$).

In the sequel of the proof we will omit the superscript j for u^j, v^j and $\bar{\epsilon}^j$. $|GP|^2 = x^2$; $|PP^j|^2 = (u - \bar{\epsilon}x)^2 + v^2$; $[\overrightarrow{PP^j} | \overrightarrow{PG}] = -(u - \bar{\epsilon}x)x$. Using formulae 4.3 and 4.4 (p. 77), we deduce $|PP^j|_P^2 = (u - \bar{\epsilon}x)^2 + v^2 - \frac{(u-\bar{\epsilon}x)^2 x^2}{1+x^2}$ and $|PG|_P^2 = \frac{x^2}{1+x^2}$. Therefore $|PP^j|_P^2 = \frac{(u-\bar{\epsilon}x)^2}{1+x^2} + v^2$. Letting $|GG^j|^2 = d^2 = u^2 + v^2$, we obtain

$$|PP^j|_P^2 - |PG|_P^2 = \frac{(u - \bar{\epsilon}x)^2}{1 + x^2} + v^2 - \frac{x^2}{1 + x^2} = \frac{-(1 - v^2 - \bar{\epsilon}^2)x^2 - 2\bar{\epsilon}ux + d^2}{1 + x^2}$$

We look for points P verifying $|PP^j|_P^2 - |PG|_P^2 \geq 0$, or, equivalently, for x verifying

$$(v^2 + \bar{\epsilon}^2 - 1)x^2 - 2\bar{\epsilon}ux + d^2 \geq 0$$

By Corollary 4.1 (p. 73) we have $d^2 \leq 4\frac{n_j(n-n_j)}{n^2}$. From $\bar{\epsilon} = 2\frac{n_j}{n} - 1$, we deduce $4\frac{n_j(n-n_j)}{n^2} = 1 - \bar{\epsilon}^2$, hence $v^2 \leq d^2 \leq 1 - \bar{\epsilon}^2$ and $v^2 + \bar{\epsilon}^2 - 1 \leq 0$. Therefore, if $v^2 \neq 1 - \bar{\epsilon}^2$, the coefficient of x^2 is strictly negative and the constant term (d^2) is positive; hence the two roots of the quadratic polynomial are

$$x_1 = -\frac{\bar{\epsilon}u + \sqrt{\bar{\epsilon}^2 u^2 + d^2(1 - v^2 - \bar{\epsilon}^2)}}{1 - v^2 - \bar{\epsilon}^2} \quad \text{and} \quad x_2 = -\frac{\bar{\epsilon}u - \sqrt{\bar{\epsilon}^2 u^2 + d^2(1 - v^2 - \bar{\epsilon}^2)}}{1 - v^2 - \bar{\epsilon}^2} .$$

with $x_1 < 0$ and $x_2 > 0$. The quadratic polynomial is positive for all values of x between x_1 and x_2, therefore $|PP^j|_P^2 - |PG|_P^2$ is positive for all points P of the line segment delimited by the two points $D_1^j = G + x_1 \overrightarrow{\beta}$ and $D_2^j = G + x_2 \overrightarrow{\beta}$.

Particular case. If $v^2 = 1 - \bar{\epsilon}^2$, from properties $d^2 \leq 1 - \bar{\epsilon}^2$ and $v^2 \leq d^2$, we deduce $v^2 = d^2 = 1 - \bar{\epsilon}^2$, hence $u = 0$ (the projection of point G^j onto line $\mathcal{D}$ is point G). In this case, $|PP^j|_P^2 - |PG|_P^2 = \frac{d^2}{1+x^2}$ is positive or null for all x: the property is verified for any point $P \in \mathcal{D}$, hence $x_1^j = -\infty$ and $x_2^j = +\infty$. $\square$

Theorem 4.2. *Let $\mathcal{D}$ be the line with* M-*unit direction vector $\overrightarrow{\beta}$ passing through point* G. *Given an α-level, let us denote $\widetilde{x}_1$ (resp. $\widetilde{x}_2$) the smaller value $(x_1^j)_{j \in J}$ (resp. the greater value $(x_2^j)_{j \in J}$) such that the proportion of $j \in J$ verifying $x_1^j \leq \widetilde{x}_1$ (resp. $x_2^j \geq \widetilde{x}_2$) is greater than α. The two points $D_1 = G + \widetilde{x}_1 \overrightarrow{\beta}$ and $D_2 = G + \widetilde{x}_2 \overrightarrow{\beta}$ determine the compatibility interval for point* G *at level α along line $\mathcal{D}$.*

Proof. Given $x > 0$ and $P = G + x \overrightarrow{\beta}$, the proportion of permutations j such that $|PP^j|_P^2 - |PG|_P^2 \geq 0$ is equal to the proportion of points $D_2^j = G + x_2 \overrightarrow{\beta}$ such that $x_2^j \leq x$. If $x \leq \widetilde{x}_2$, this proportion is greater than α, hence point P is compatible with point G along the line $\mathcal{D}$. By the same reasoning, every point $P = G + x \overrightarrow{\beta}$ with $0 > x \geq \widetilde{x}_1$ is compatible with point G.   □

Adjusted compatibility region

This theorem allows us to produce an exact and algebraically defined interval of compatibility for each one-dimensional subspace of the affine support of the cloud. In the L-dimensional space, the compatibility region has no analytical form. However the compatibility region can be adjusted by a principal ellipsoid of the cloud.

Given a point P and a permutation $j \in J$, note that, by using relations 4.3 and 4.4, the property $|PP^j|_P^2 \geq |PG|_P^2$ writes $|PP^j|^2 - \frac{[\overrightarrow{PG}|\overrightarrow{PP^j}]^2}{1+|PG|^2} \geq \frac{|PG|^2}{1+|PG|^2}$, thus, it depends on the M-scalar product. Therefore, the compatibility region, as defined on page 80, will be studied in the Mahalanobis space, that is, in the subspace $\mathcal{M}$ of $\mathcal{U}$ associated with the direction vector subspace $\mathcal{L} \in \mathcal{V}$ equipped with the M-scalar product, which is denoted $[\cdot|\cdot]$.

In the Mahalanobis space, the covariance endomorphism of the cloud M^I is Id_L (identity of $\mathcal{L}$, see Proposition 2.13, p. 21), that of the permutation G-cloud is Id_L/n (Theorem 4.1, p. 73) and that of the permutation cloud P^J is M.Pcov with $\forall \overrightarrow{u} \in \mathcal{L}$, M.Pcov$(\overrightarrow{u}) = \frac{1}{n}\left(\overrightarrow{u} + [\overrightarrow{PG}|\overrightarrow{u}]\overrightarrow{PG}\right)$ (Corollary 4.2, p. 76). By Theorem 7.4 (p. 234), we deduce the following property.

Proposition 4.7. *The vector $\overrightarrow{PG}$ is eigenvector of the* M-*covariance endomorphism of the permutation* P-*cloud associated with eigenvalue $(1+|PG|^2)/n$ and the subspace of $\mathcal{L}$ which is* M-*orthogonal to $\overrightarrow{PG}$ is eigenspace associated with eigenvalue $1/n$ with multiplicity $L - 1$.*

Consider two points P and Q that belong to the κ-ellipsoid of cloud M^I, then $|PG| = |QG| = \kappa$. We deduce from Proposition 4.7 that the M-covariance endomorphisms of the permutation P-cloud and of the permutation Q-cloud have the same eigenvalues, namely $(1 + \kappa^2)/n$ and $1/n$ (with multiplicity $L - 1$); the eigenvectors are, respectively, $\overrightarrow{GP}$ and $\overrightarrow{GQ}$ and the $(L - 1)$-dimensional eigenspaces are, respectively, orthogonal to $\overrightarrow{GP}$ and to $\overrightarrow{GQ}$. Thus, the concentration ellipsoids of the two permutation clouds are identical up to a plane rotation (see graph on the left in Figure 4.12, p. 83). On the one hand, their projections on the plane spanned by points P, Q and G can be deduced from each other thanks to a rotation with centre G and with angle defined by vectors $\overrightarrow{GP}$ and $\overrightarrow{GQ}$ (graphs on the right in Figure 4.12); on the other hand, their projections on the subspace of $\mathcal{M}$ orthogonal to this plane are identical.

By Proposition 2.4 (p. 21), we deduce that $|PG|^2_P = |QG|^2_Q = \frac{\kappa^2}{1+\kappa^2}$; therefore the two proportions $p\big(D^2_P \geq \frac{\kappa^2}{1+\kappa^2}\big)$ and $p\big(D^2_Q \geq \frac{\kappa^2}{1+\kappa^2}\big)$ are nearly equal ("nearly equal" rather than "equal" due to "worries of discreteness"). As a consequence, the compatibility region is approximately an ellipsoid of cloud M^I. In any case (for values of n not too small) we will adjust the compatibility region by a principal ellipsoid of cloud M^I (see Figure 4.13, p. 84).

From a technical standpoint, the *adjusted compatibility region* will be defined from the κ values obtained from the L principal lines of cloud M^I to which the κ values of lines randomly selected in the L-dimensional space may be added. Note that, in order to determine the $(1 - \alpha)$ compatibility region, the size n of the cloud must verify $\alpha \times 2^{n-1} \geq 1$ (for $\alpha = 0.05$, $n \geq 6$).

Target example

In the space equipped with the Mahalanobis distance, Figure 4.12 shows:

(1) on the left, the G-permutation cloud with the "concentration circle" with centre G and radius $2\sqrt{1/10}$; points P and Q with $|GP| = |GQ| = 1$ and the concentration ellipses of the permutation clouds associated with P and Q; lengths of semi-axes are equal to $2\sqrt{2/10}$ and $2\sqrt{1/10}$;

(2) in the middle, the permutation P-cloud with its principal ellipse passing through point G;

(3) on the right, the permutation Q-cloud with its principal ellipse passing through point G.

The proportions of points that are not inside the ellipses (*p*-values) are, respectively, equal to $68/1024$ and $70/1024$ (slightly different due to "worries of discreteness").

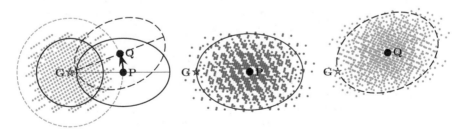

Figure 4.12
Target example. Mahalanobis space. (1) On the left, points P and Q with $|GP| = |GQ| = 1$, circle with centre G and radius 1 (grey dashed line), concentration ellipses of permutation clouds G^J (circle in solid line), P^J (ellipse in solid line) and Q^J (ellipse in dashed line); (2) in the middle, P-cloud P^J with its principal ellipse passing through point G; (3) on the right, Q-cloud Q^J with its principal ellipse passing through point G.

In the geometric space, we present the 95% compatibility region determined by using 180 lines (Figure 4.13, on the left) and its adjustment by a κ-ellipse whose κ is the mean of 360 values associated with the 180 lines (Figure 4.13, on the right). The mean is equal to 1.08, the minimum is equal to 1.014 and the maximum is equal to 1.162.

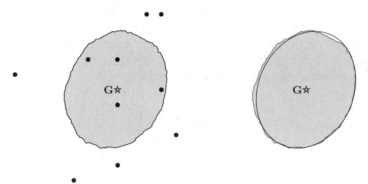

Figure 4.13

Target example. On the left, cloud M^I and the 95% compatibility region; on the right, the 95% compatibility region adjusted by the κ-ellipse of cloud M^I with $\kappa = 1.08$.

Monte Carlo method. The remarks about exhaustive vs. Monte Carlo developed on page 44 are in order here. So we will only comment on the methods developed in this chapter.

The computations based on the enumeration of all possible permutations[6] can be easily performed up to $n = 20$. When the cardinality of the permutation space is large, the Monte Carlo method is used to approximate the permutation distribution, calculate the p-value and determine the compatibility region by randomly drawing permutations among the 2^n ones. A useful strategy is to start with 10,000 random permutations and continue with larger numbers only if p is small enough and makes it worthwhile to pursue the analysis, e.g., $p < 0.10$.

We should emphasize that simulations are carried out by means of *resampling from the permutation space* and that the procedure is substantially different from the well-known *bootstrap techniques* (see §4.5, p. 92).

Computational considerations (see also p. 97). It is worth noticing that the test statistic as well as the compatibility region are based on the M-norm, that is, they depend on the inverse of the covariance endomorphism M*cov* of cloud M^I. As for the combinatorial typicality test (see p. 44) computations become much simpler as soon as we choose the orthocalibrated principal basis of the space (see Chapter 2, §2.3.3, p. 23); then the matrix representation of M*cov* and its inverse is the identity matrix and the squared M-norm of a vector is merely the sum of the squared coordinates, which substantially reduces computing time.

---

[6]In fact, due to the symmetry of the permutation distribution, the computations are made by using 2^{n-1} permutations (see §4.6, p. 97).

4.2.5 Approximate Test and Compatibility Region

If n is large enough, the permutation P-cloud can be fitted by an L-dimensional normal distribution (L is the dimension of the affine support of the cloud), and the distribution of the test statistic nD_P^2 can be approximated by a chi-squared distribution with L degrees of freedom (denoted χ_L^2).

The observed value of the statistic D_P^2 is $d_{P_{obs}}^2 = |PG|_P^2 = \frac{|PG|^2}{1+|PG|^2}$ (see Formula 4.4, p. 77), hence the combinatorial *p-value* can be approximated by

$$\widetilde{p} = p(\chi_L^2 \geq n \times \tfrac{|PG|^2}{1+|PG|^2})$$

The M-covariance endomorphisms of permutation clouds associated with two points on the same ellipsoid of cloud M^I can be deduced from each other by a rotation, so, for moderate n, we can fit the permutation clouds by similar normal distribution. Given an α-level and denoting $\chi_L^2[\alpha]$ the critical value of the chi-squared distribution with L degrees of freedom, *the approximate compatibility region* is the set of points P verifying $n\,d_{P_{obs}}^2 < \chi_L^2[\alpha]$, or equivalently $|PG| < \kappa$ with (owing to Formula 4.4) $\kappa^2 = \frac{\chi_L^2[\alpha]/n}{1-\chi_L^2[\alpha]/n}$.

The $(1-\alpha)$ *approximate compatibility region* is delimited by the κ-ellipsoid of the cloud.

Target example

The reference point is the centre of the target (origin point O). The distribution of the statistic nD_O^2 is graphically depicted in Figure 4.14 with its fitting by the χ^2 distribution with 2 d.f.

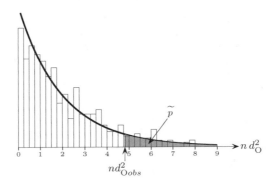

Figure 4.14
Target example. Permutation distribution of nD_O^2 and approximation χ^2 (with 2 d.f.).

One has $n\,d_{O_{obs}}^2 = n|OG|_O^2 = 10 \times 0.6939^2 = 4.815$, hence the approximate *p*-value $\widetilde{p} = p(\chi_2^2 \geq nd_{O_{obs}}^2) = p(\chi_2^2 \geq 4.815) = .090$ (to be compared with the exact *p*-value .084).

The approximate test leads to the same conclusion as the exact test.

The scale parameter of the approximate 95% compatibility ellipse is equal to $(\frac{5.9915/10}{1-5.9915/10})^{1/2} = 1.223$, a value larger than the adjusted value obtained previously (1.084).

Figure 4.15
Target example. Exact(in grey) and approximate (in black) 95% compatibility region.

4.3   One-dimensional Case: Typicality for Mean

In the case of a one-dimensional cloud, the basic data are a set I of n individuals constituting a group of observations together with a numerical variable $x^I = (x^i)_{i \in I}$. We will denote $\bar{x}$ the mean of the variable and v its variance.

Our aim is to study the typicality of the mean $\bar{x}$ with respect to a reference value u. The foregoing test can be applied as is to a one-dimensional cloud, though we wish a conclusion taking into account the sign of the deviation from the reference value to the mean of the group of observations. So, we choose a test statistic based upon mean rather than distance (D or equivalently D^2 used previously).

4.3.1   Permutation Space

As in the general case, if u is the reference value, then each value x^i could be exchanged with its symmetric with respect to u, that is, with $2u - x^i$. If we perform the exchange in all, some, or none of the n individuals, we get a set of 2^n variables $(u^{Ij})_{j \in J}$. Using the permutation table (see §4.2.1, p. 69), the value u^{ij} writes $u^{ij} = u + \epsilon_i^j(x^i - u)$.

The family of the 2^n variables $(u^{Ij})_{j \in J}$ is called *permutation space* and denoted $\mathcal{J}_u$.

4.3.2 Test Statistic

The test statistic, denoted E_u, is such that:

$$E_u : \mathcal{J}_u \to \mathbb{R}$$
$$u^{Ij} \mapsto \tfrac{1}{n} \sum_{i \in I} u^{ij} - u$$

The value $E_u(u^{Ij})$ is denoted e_u^j. Letting $\bar{\epsilon}^j = \sum_{i \in I} \epsilon_i^j / n$ and $\bar{x}^j = \sum_{i \in I} \epsilon_i^j x^i / n$, one has: $e_u^j = \tfrac{1}{n} \sum_{i \in I} u^{ij} - u = \tfrac{1}{n} \sum_{i \in I} \epsilon_i^j (x^i - u) = \bar{x}^j - \bar{\epsilon}^j u$. The *observed value* of the statistic, denoted e_{obs}, is such that $e_{obs} = \bar{x} - u$ (deviation from the reference value to the observed mean).

Distribution of the test statistic

The distribution of statistic E_u is the map $E_u(\mathcal{J}_u) \to [0,1]$ that assigns with each value e_u^j the proportion $p(E_u = e_u^j)$.

$$E_u(\mathcal{J}_u) \to [0,1]$$
$$e_u^j \mapsto p(E_u = e_u^j)$$

- *The distribution of E_u is symmetrical with respect to 0*

  Proof. With each $j \in J$ is associated $j' = 2^n + 1 - j$ such that $(\epsilon_i^{j'} = -\epsilon_i^j)_{i \in I}$ (see p. 69); therefore $e_u^{j'} = \tfrac{1}{n} \sum_{i \in I} \epsilon_i^{j'} (x^i - u) = -\tfrac{1}{n} \sum_{i \in I} \epsilon_i^j (x^i - u) = -e_u^j$, hence the symmetry of the distribution. □

- *The mean of the statistic is null.*

  This property is a consequence of the symmetry one.

- *The variance of E_u is equal to $\tfrac{1}{n}\left(v + (\bar{x} - u)^2\right)$.*

  Proof. $\sum_{j \in J} (e_u^j)^2 = \tfrac{1}{n^2} \sum_{j \in J} \left(\sum_{i \in I} \epsilon_i^j (x^i - u) \right)^2 = \tfrac{1}{n^2} \sum_{i \in I} \sum_{i' \in I} \left(\sum_{j \in J} \epsilon_i^j \epsilon_{i'}^j \right)(x^i - u)(x^{i'} - u)$.
  $\sum_{j \in J} (\epsilon_i^j)^2 = 2^n$; for $i \neq i'$, $\sum_{j \in J} \epsilon_i^j \epsilon_{i'}^j = 0$ (since $\epsilon_i^j \epsilon_{i'}^j$ is 2^{n-1} times equal to 1 and 2^{n-1} times to -1, thus the sum is null). $\operatorname{Var} E_u = \tfrac{1}{2^n} \sum_{j \in J} (e_u^j)^2 = \tfrac{1}{n^2} \sum_{i \in I} (x^i - u)^2$, hence, by applying the Huygens's formula (p. 12), the property. □

Combinatorial p-value

In order to determine the p-value, the observed mean is located in the distribution of E_u. Since the distribution is symmetrical with respect to 0, the p-value will be the proportion of values $(\bar{x}^j)_{j \in J}$ that are greater than or equal to the absolute value of e_{obs}, that is:

$$p = p(E_u \geq |\bar{x} - u|)$$

Given an α-level (see note p. 31),

- if $p \leq \alpha/2$, the group of observations will be declared *atypical* of the reference value at level $\alpha/2$, on the side of high means if $\overline{x} > u$ or on the side of low means if $\overline{x} < u$;

- if $p > \alpha/2$, the reference value and the mean of the group of observations are said to be *compatible at level* α.

4.3.3 Compatibility Interval

As for a multidimensional cloud, we will now consider as a reference value not just u but every value $u \in \mathbb{R}$. Given an α-level, the set of values u which are compatible with $\overline{x}$ at level α define the $(1 - \alpha)$ compatibility interval.

$$u \in \mathbb{R} \text{ is compatible with } \overline{x} \text{ at level } \alpha \iff p(E_u \geq |\overline{x} - u|) > \alpha/2$$

Theorem 4.3. *Consider the family* $(u^j = (\overline{x} - \overline{x}^j)/(1 - \overline{e}^j))_{j \in J}$. *Let us denote* u_1 *(resp.* u_2*) the smaller (resp. the greater) value* $(u^j)_{j \in J}$ *such that the proportion of* $j \in J$ *verifying* $u^j \leq u_1$ *(resp.* $u^j \geq u_2$*) is superior to* $\alpha/2$, *then the* $(1 - \alpha)$ *compatibility interval is the closed interval* $[u_1, u_2]$.

Proof. Given $j \in J$, if $\overline{x} > u$, the property $(e_u^j \geq |e_{obs}|)$ is equivalent to $\overline{x}^j - \overline{e}^j u \geq \overline{x} - u$ or $(\overline{x} - \overline{x}^j)/(1 - \overline{e}^j) \leq u$, that is, $u^j \leq u$. Since the proportion of $j \in J$ verifying $u^j \leq u_2$ is superior to $\alpha/2$, one has $p(E_u \geq e_{obs})$ for all values $u \leq u_2$. Similarly, if $\overline{x} > u$, $p(E_u \leq e_{obs}) > \alpha/2$ for all values $u \leq u_1$, hence the interval. □

In general, this interval is not centred at the observed mean $\overline{x}$.

Remarks. Instead of E_u, we could have taken the *Mean* $(E_u + u)$, the *Scaled deviation* $(D_u = E_u/\sqrt{v})$, or the *Z-score* $(Z_u = (E_u - \text{Mean } E_u)/\sqrt{\text{Var } E_u})$ as test statistic. These statistics are equivalent since the properties $(E_u \geq e_{obs})$, $(D_u \geq d_{obs})$ and $(Z_u \geq z_{obs})$ are equivalent; therefore they lead to the same proportions, hence to the same p-values.

As a result of the symmetry of the distribution, we can take the absolute value of E_u as a statistic. Then, the p-value is $\frac{1}{2}p(|E_u| \geq |\overline{x} - u|)$. This remark points to only enumerate half of possible permutations (see §4.6).

Target example

To exemplify the one-dimensional case, we take the cloud of the *Target example* projected onto the horizontal axis.

Figure 4.16 shows the one-dimensional cloud and Table 4.2 gives variables x^I (coordinate of points of the cloud).

Figure 4.16
Target example. Target with 10 impacts.

Table 4.2
Target example. Coordinates of points.

| x^I | 0 | 3 | 7 | 3 | 6 | −4 | 1 | 3 | 5 | 6 |

Descriptive appraisal

One has $\bar{x} = 3$ and $u = 0$, hence $e_{obs} = \bar{x}_u = 3$. Since $v = 10$, the scaled deviation between the observed mean and the reference value is $d_{obs} = \frac{3-0}{\sqrt{10}} = 0.95$. Since $d_{obs} > 0.4$ (see Chapter 2, p. 22), we conclude that

descriptively the deviation is notable.

Thus, we will now assess the typicality level of the group of observations.

Combinatorial typicality test for the Mean

The distribution of the test statistic is shown in Table 4.3 and Figure 4.17.

Table 4.3
Distribution of E_u for $E_u \geq 0$.

0	0.2	0.4	0.6	0.8
60	54	54	56	46

1.0	1.2	1.4	1.6	1.8
44	46	34	28	32

2.0	2.2	2.4	2.6	2.8
22	14	18	12	4

3.0	3.2	3.6	3.8
8	6	2	2

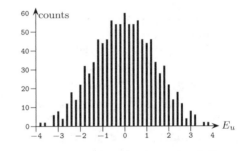

Figure 4.17
Bar diagram of the distribution of E_u.

From Table 4.3, we deduce that $p = p(E_u \geq 3) = 18/1024 = 0.018 < .025$, and the 95% compatibility interval is $[0.6, 5.333]$, hence the conclusion:

The mean of the group is atypical of the reference value at level .025 on the side of high values.

4.3.4 Approximate Test and Compatibility Interval

If n is large enough, the distribution of E_u can be approximated by the normal distribution with mean 0 and variance $\text{Var}\, E_u = (v + (\bar{x} - u)^2)/n$. Hence the approximate test is based on the test statistic $Z_u = E_u/\sqrt{\text{Var}\, E_u}$ whose distribution can be approximated by a standard normal distribution.

Let $z_{obs} = \sqrt{n}\, e_{obs}/\sqrt{v + e_{obs}^2}$ be the observed value of statistic Z_u. The *approximate p-value* is equal to the proportion of the normal distribution greater than $|z_{obs}|$.

Let z_α be the critical value of the normal distribution at level α (two-sided)[7]. Letting $h_\alpha = \frac{z_\alpha \sqrt{v/n}}{\sqrt{1 - z_\alpha^2/n}}$, the approximate compatibility interval at level α is $\left[\bar{x} - h_\alpha, \bar{x} + h_\alpha\right]$.

---

[7]Recall that $z_\alpha = 1.96$ for $\alpha = 0.05$

4.4 The Case of a Design with Two Repeated Measures

A design with two repeated measures is a privileged case in which the geometric typicality test applies. As an example, there is the paradigmatic situation of drug effect mentioned at the beginning of this chapter (see p. 65).

Recall that, in this case, the data consist in a set I of n subjects who are observed twice: before and after treatment. If the observations are multidimensional (several variables are observed), two points M^{it_1} and M^{it_2} can be associated with subject i in a geometric space. Then the basic data set consists of n pairs of points $(M^{it_1}, M^{it_2})_{i \in I}$. Our aim is to compare the mean point of the cloud $(M^{it_1})_{i \in I}$ to that of cloud $(M^{it_2})_{i \in I}$. In other words, we wonder if the deviation between the two points is significantly different from the null deviation or not.

As previously said (page 66), it is simpler to consider the individual differences, that is, for each subject i, the "effect of treatment" represented by the effect-vector $\overrightarrow{M^{it_1} M^{it_2}}$. Given a point O representing the null effect, the point $D^i = O + \overrightarrow{M^{it_1} M^{it_2}}$ is associated with subject i; it is called effect-point. Then rearranging "before" and "after" simply reverses the sign of the individual effects or amounts to exchanging the effect-point D^i with its symmetrical with respect to the reference point O.

Now the problem of the null effect consists in comparing the mean point of the relevant cloud D^I to the reference point by performing a geometric test.

> **Remarks.** Obviously, observations within each subject (statistical individual) are dependent because they are measured on the same unit on different occasions, whereas the n subjects are assumed to be exchangeable.
>
> The one-sample matched-pairs problem, where independent units are paired according to some known covariates, are formally equivalent to that of paired observations.
>
> This test can also be seen as a homogeneity test of two groups in the particular case of a design with two repeated measures, which will be further developed in Chapter 5 (§5.5, p. 133).

We will extensively present two examples. The first one is one-dimensional and concerns the famous Student's example (Student, 1908, p. 20); it will be studied in the next section. The second one deals with multidimensional data; it concerns the *Parkinson Study* and will be studied in Chapter 6 (§6.1, p. 156).

4.4.1 Student's Example

The sleep of 10 patients was measured without hypnotic and after treatment with two drugs: drug A and drug B. Patients took drug A in a first period and drug B in a second period. In each period, the average number of hours of sleep gained by the use of the drug is registered. Thus, the data consist in two scores for each patient, hence a *repeated measures design*.

The *question* is:

Does drug B produce a larger effect than drug A?

The scores are tabulated in Table 4.4 together with the means and the standard deviations (SD) and graphically represented in Figure 4.18. The one-dimensional cloud is depicted in Figure 4.19.

Table 4.4
Student's example. Number of hours of sleep gained and individual effects.

Patient	$s1$	$s2$	$s3$	$s4$	$s5$	$s6$	$s7$	$s8$	$s9$	$s10$	means	SDs
Drug A	+0.7	−1.6	−0.2	−1.2	−0.1	+3.4	+3.7	+0.8	0	+2.0	+0.75	1.70
Drug B	+1.9	+0.8	+1.1	+0.1	−0.1	+4.4	+5.5	+1.6	+4.6	+3.4	+2.33	1.90
Difference	+1.2	+2.4	+1.3	+1.3	0	+1.0	+1.8	+0.8	+4.6	+1.4	+1.58	1.17

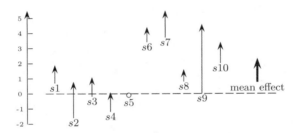

Figure 4.18
Student's example. Individual effects and mean effect.

Figure 4.19
Student's example. Cloud of individuals effect-points and mean effect-point.

Descriptive analysis

All individual effects are positive (null for subject $s5$); the mean effect is 1.58 hour. The variance is $v = 1.3616$ (SD = 1.167); hence the *scaled mean effect* is equal to $1.58/1.167 = 1.35$.

Everyone will agree that the mean effect, which is more than one standard deviation, is large. The descriptive conclusion—pertaining to the patients who are examined—is that drug B is (far) superior to drug A.

Descriptively, the mean effect is positive and large in magnitude.

Inductive analysis

At this point, we have established that drug B is descriptively superior to drug A. We now try to extend the descriptive conclusion by using the geometric typicality test in order to answer the natural query:

> *Is the observed effect of drug a genuine one or might it be due to chance?*

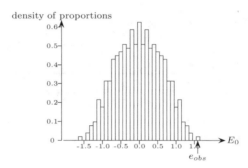

Figure 4.20
Student's example. Histogram of the distribution of the test statistic E_O.

In this example, the data are extremal since nine differences are strictly positive and one is null, hence $p = 1/2^9 = .002 < 0.025$. In conclusion, the group of observations is atypical of the reference value at level $.025$ on the side of the high values.

> *The data are in favour of the superiority of drug B.*

The 95% compatibility interval for the mean is equal to $[0.833, 2.467]$.

4.5   Other Methods

Now, we will provide a brief overview of some other methods that also answer the question of comparison of the mean point of a group to a reference point.

For instance, the bootstrap and the Pesarin's combination of univariate tests are two methods which are not intrinsically geometric but stay in the same spirit as the typicality test since they are "statistics without probability". We will also present the traditional Hotelling's test and its statistical framework.

Recall that in this book, we confine ourselves to the permutation approach in a geometric scope, so we do not present these procedures in detail.

4.5.1 Bootstrap Method

The bootstrap was introduced by Efron, in 1979, as a technique for assessing the statistical accuracy of estimates, especially in nonparametric settings. It may also be used to test statistical hypotheses (see Efron and Tibshirani, 1993). Specifically, at least in their simpler form, bootstrap inferences are based on a distribution of the chosen statistic calculated from a large number of n-samples which are randomly drawn *with replacement* from the observed data. In the case of the multivariate one-sample issue, the bootstrap hypothesis test is based on the Hotelling's T^2 statistic (Hotelling, 1951).

The method[8] applied to a cloud of points can be described as follows.

1. Draw B *resamples* of size n *with replacement* from the set I.

   Given a resample $b \in B$, the sequence of the n elements extracted from set I is denoted by I^{*b} and an element of I^{*b} by i^*. The cloud associated with I^{*b}, denoted $C^{I^{*b}}$, is a subcloud of M^I with repeated points according to the sequence[9] I^{*b}. The mean point of cloud $C^{I^{*b}}$ is denoted C^b and its covariance endomorphism restricted to its support is denoted $C^b cov$.

2. Compute the squared $C^b cov$-Mahalanobis distance between the mean point of each resample cloud[10] and the reference point P, which is denoted $d^{*2}[b]$, hence the statistic denoted D^{*2}:

$$D^{*2} \ : \ C^{I^{*b}} \mapsto d^{*2}[b] = \langle (C^b cov)^{-1} (\overrightarrow{PC^b}) | \overrightarrow{PC^b} \rangle$$

3. Calculate the observed value of the statistic that is equal to the squared Mahalanobis distance attached with the cloud M^I between the observed mean point G and the reference point P, that is, $|PG|^2$.

4. Determine the bootstrap p-value, which is the proportion of resamples verifying $d^{*2}[b] \geq |PG|^2$.

$$p_{boot} = p(D^{*2} \geq |PG|^2)$$

From the bootstrap distribution of D^{*2} a *bootstrap confidence region* can be determined by using the method of "simple percentile" (see Efron and Tibshirani, 1993, Chapter 13). If κ_α^* is the $100 \times (1-\alpha)$-th percentile of the bootstrap distribution, the bootstrap confidence region for point G can be defined as the set of points P verifying $|PG| \leq \kappa_\alpha^*$. Thus the region is defined by the κ_α^*-ellipsoid of cloud M^I.

---

[8] We follow the approach of Efron and Tibshirani (1993, p. 226) for the univariate one-sample problem using the Student's t statistic.

[9] With each resample b corresponds a system of weights $(\varpi_i)_{i \in I}$ with $0 \leq \varpi_i \leq n$ and $\sum \varpi_i = n$, therefore the cloud $C^{I^{*b}}$ can be seen as the weighted cloud $(M^i, \varpi_i)_{i \in I}$.

[10] A resample cloud with only one point will be excluded from the distribution.

Remarks

(1) The permutation test exploits the special symmetry that exists under the "null hypothesis" to create a permutation distribution of the test statistic. In contrast, the bootstrap explicitly estimates the distribution under the null hypothesis in order to estimate the p-value. The two tests are intrinsically different.

(2) The major disadvantage of the bootstrap test is that the test statistic changes for each bootstrap resample, hence B matrix inversions must be done, and the computing time can become long. Futhermore, the distances are not measured with the same scale: *it is not a geometric method.*

The bootstrap technique has been used to provide statements regarding the accuracy of relative locations of points in space, see, e.g., Lebart et al. (2006); Lebart (2007) for PCA and MCA and Weinberg et al. (1984) for INDSCAL.

4.5.2   Combining Univariate Tests

The method of nonparametric combination (NPC) of dependent permutation tests is due to Pesarin (2001).

In the NPC methodology, each variable is tested one by one and the obtained p-values are combined using a test statistic called *combining function*. For instance, if for variable k the combinatorial p-value is denoted p_k, the Fisher's omnibus combining function writes $-2 \sum \log(p_k)$.

In the sequel, we compare the mean $\overline{x}_k$ of each variable x_k^I to a reference value u_k. The method is the one described in §4.3 but the test statistic is $D_k = |E_{u_k}|$ (for the definition of E_{u_k}, see §4.3, p. 86). The observed value of the statistic D_k, denoted d_k, is equal to $|\overline{x}_k - u_k|$ (absolute value of the difference between $\overline{x}_k$ and u_k).

Recall that the set of the 2^n permutations is denoted J.

1. Compute, for each variable k, the combinatorial p-value $p_k = p(D_k \geq d_k)$ and using the combination function, calculate $Lp = -2 \sum \log(p_k)$.

2. Compute, for each permutation $j \in J$ and each variable k, the value of the test statistic D_k, that is, $d_k[j] = |\overline{x}_k^j - \overline{\epsilon}^j u_k|$; then calculate the pseudo p-values $p_k^j = p(D_k \geq d_k[j])$ and, using the combination function, calculate $Lp^j = -2 \sum_{k \in K} \log(p_k^j)$.

3. Estimate the combined p-value of the global test as the proportion of values Lp^j that are greater than or equal to Lp.

Remark. In the geometric typicality test, one single overall test statistic is available (based on Mahalanobis distance related to cloud M^I); then, in terms of computational complexity, permutation solutions become equivalent to simple univariate procedures.

4.5.3 Hotelling's Test

Referring the cloud to an *orthonormal Cartesian frame*, we denote $\mathbf{V}$ the covariance matrix (dividing the sums of squares and products by n) which is associated with the covariance endomorphism M*cov*, and $\mathbf{S} = \frac{n}{n-1}\mathbf{V}$ the covariance matrix corrected with $n-1$ d.f. (dividing the sums of squares and products by $n-1$).

Let us take the origin point O of the Cartesian frame as a *reference point*. We call *observed effect* the vector that represents the deviation from the reference point O to the mean point G. This vector is denoted $\overrightarrow{d}_{obs}$ with $\overrightarrow{d}_{obs} = \overrightarrow{OG}$. The coordinates of point G (or of vector $\overrightarrow{d}_{obs}$) is represented by a column-matrix denoted $\mathbf{d}_{obs}$ or simply $\mathbf{d}$. The squared M-distance between points O and G (see Chapter 2, p. 20) is denoted $|OG|^2$, with

$$|OG|^2 = \mathbf{d}^\top \mathbf{V}^{-1} \mathbf{d} = \frac{n-1}{n}\mathbf{d}^\top \mathbf{S}^{-1} \mathbf{d}$$

Statistical modelling. The intuitive rationale for inference procedures developed in multivariate statistics is that an observed point is just one outcome among a set of possible outcomes, e.g., the mean-point G is supposed to be the realization of a *Mean-point* variable G (italics) varying in some *sample space* specified by the statistical modelling. We adopt the convenient "trueness metaphor", that is, the *Mean-point* variable G will be thought to vary around an idealized point, called a "true mean point" and denoted Γ, to which the inference pertains: $G = \Gamma + \overrightarrow{e}$, $\overrightarrow{e}$ being an *error-vector* term.

Accordingly, in terms of vector-deviations from point O, the observed effect is supposed to be the realization of a vector-variable $\overrightarrow{d} = \overrightarrow{\delta} + \overrightarrow{e}$, where $\overrightarrow{\delta}$ denotes the *true vector-effect*. In the inference procedure, the true mean-point Γ (parameter) is estimated by the observed mean-point G, and the true effect $\overrightarrow{\delta}$ by the observed effect $\overrightarrow{d}_{obs}$ (see opposite figure).

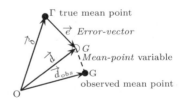

In this section, the dimension of the cloud[11] will be denoted r.

In the *usual parametric model*, the cloud of points is an i.i.d. sample from an r-dimensional normal distribution centred at the true mean point (main parameter) and with covariance matrix Σ (secondary parameter) that is estimated by $\mathbf{V}$ (maximum-likelihood estimation) or by $\mathbf{S}$ (unbiased estimation). The sampling properties involve the *Wishart* ($\mathcal{W}$), chi-square (χ^2) and *Snedecor* (F) distributions (see, e.g., Anderson (2003, Chapter 5), Rao (1973, Chapter 8)).

Let us denote $\boldsymbol{d}$ and $\boldsymbol{S}$ (or $\boldsymbol{V}$) the statistics whose observed values are $\mathbf{d}$

---

[11] In the notation of preceding sections, r is equal to K (dimension of the Euclidean space), or L (dimension of the support of the cloud), or L' (projected cloud).

and S (or V). Then , for fixed δ and Σ regarded as fixed parameters (see, e.g., Rouanet and Lecoutre, 1983):

1. d and S are independent;

2. d is distributed as $\mathcal{N}(\delta, \frac{1}{n}\Sigma)$ (r-dimensional normal distribution with mean δ and covariance matrix $\frac{1}{n}\Sigma$);

3. $(n-1)\,S$ is distributed as $\mathcal{W}_{r,n-1}(\Sigma)$ (Σ-scaled r-dimensional Wishart with r and $n-1$ d.f.).

Under this sampling model, the null hypothesis $\mathcal{H}_0$ that the true mean-point is O, or, equivalently, that the true effect is null, can be tested by the classical Hotelling's test (see, e.g., Anderson, 2003, pp. 171–178) based on the following property:

Under $\mathcal{H}_0$, the statistic $T^2 = (n-1)|\overrightarrow{d}|^2$ is distributed as $\frac{n-1}{n-r}\,r\,F_{r,n-r}$ where $F_{r,n-r}$ is the central F distribution with r and $n-r$ d.f.

The *p-value* of Hotelling's test, that is, the sampling probability that, under $\mathcal{H}_0$, the M-norm of the vector-effect exceeds its observed value, is

$$\widetilde{p} = p(F_{r,n-r} \geq \tfrac{n-r}{r}\,|\mathrm{OG}|^2)$$

The *confidence region* for the *Mean* at level α is defined by:

$$|\mathrm{OG}|^2 < \tfrac{r}{n-r}\,F_{r,n-r}[\alpha] \tag{4.5}$$

The confidence region is *geometrically interpreted* as the interior of a κ-ellipsoid of the cloud M^I, with $\kappa^2 = \tfrac{r}{n-r}\,F_{r,n-r}[\alpha]$ (if $n-r$ is large, one has $\kappa^2 \leq \chi_r^2[\alpha]/(n-1)$).

Saporta and Hatabian (1986) proposed these formulas to draw, in principal planes, confidence ellipses around points representing categories of categorical variables for PCA and MCA.

Particular case: one-dimensional cloud

The group $(x^i)_{i \in I}$ is supposed to be an i.i.d. sample from a normal distribution with mean μ and variance σ^2. The parameters of the model are estimated from the corresponding descriptive statistics, that is, $\overline{x}$ for μ and v for σ^2. Then the statistic M (*Mean*) is normally distributed with mean δ and variance σ^2/n, and for any σ^2, the variable $t = \sqrt{n-1}(M-\mu)/\sqrt{v}$ (Student's ratio) is distributed as Student's t-distribution with $(n-1)$ d.f., hence the p-value is equal to

$$p\left(t_{n-1} \geq \sqrt{n-1}\,\tfrac{\overline{x}-u}{\sqrt{v}}\right)$$

The confidence interval at level α is, letting $h_\alpha = t_{n-1}[\alpha]\sqrt{\tfrac{v}{n-1}}$, such that:

$$[\overline{x} - h_\alpha \,;\, \overline{x} + h_\alpha]$$

4.6 Computations with R and Coheris SPAD Software

In this section, we will first describe the R script performing geometric typicality test and determining the compatibility region for the mean point of a cloud. Then, we will comment on the use of the full program including all the permutation procedures presented in this chapter, by using an R script interfaced with SPAD, which is a driven-menu software.

As already noted in Chapter 3 (p. 55), we carry out computations using coordinates of points over any *orthonormal basis* of the space.

4.6.1 R Script

The following R script computes the *p*-value and the compatibility region for the mean point of a multi- or one-dimensional cloud (see §4.2, p. 69).

Step 1. The data

First of all, we specify the working directory (say "D:/path"). Then, we read the data file of the coordinates of the n points of cloud M^I with respect to an orthonormal frame with origin point O (the data file of the *Target example* corresponding to the read.table below is named Target.txt; see p. 56).

The number of points (n) as well as the dimensionality of the space (K) are determined from the data base. The $I \times K$-matrix of coordinates of the n points is subsequently established (X.IK). The coordinates of the reference point P are in X_P (by default, it is the origin point of the frame).

```
setwd("D:/path")
base <-  read.table(file = "Target.txt", header= TRUE,
                   sep= ";", dec= ".", row.names= 1)
n      <- dim(base)[1]                        # number of points
K      <- dim(base)[2]                        # number of variables
X.IK   <- as.matrix(base, nrow= n, ncol= K)   # matrix of coordinates
X_P    <- rep(0,K)                   # coordinates of reference point
```

Step 2. The parameters

The *parameters* are the limit for notable scaled deviation (see p. 22), a reference α-level, the maximum number of permutations to be used for computations, the number of lines used for determining the adjusted compatibility ellipsoid (see p. 82), and possibly an integer (seed) used to initialize the random number generator for MC method (see footnote 13, p. 56).

```
notable_D  <- 0.4         # notable limit for D
alpha      <- 0.05        # alpha level
max_number <- 1e+06       # maximum number of samples
n_dir      <- 200         # number of lines for compatibility region
seed       <- NULL        # seed for random number generator
```

Step 3. Descriptive analysis

For a descriptive appraisal of magnitude of the deviation between the mean point G of the cloud and the reference point P, we compute the M-distance between points G and P (see p. 22).

> This index, as well as the test statistic and the compatibility region, depends on the inverse of the covariance endomorphism of the cloud M^I. Therefore, in order to optimize the computational time (see Remark (2), p. 44), we proceed to a change of bases from the initial orthonormal basis of $\mathcal{V}$ (direction of $\mathcal{U}$) to the orthocalibrated principal basis of $\mathcal{L}$ (direction of the support of cloud M^I) whose dimensionality is L (see p. 20), that is, from eigenvectors and eigenvalues of the covariance matrix (since the basis of $\mathcal{V}$ is orthonormal).

The covariance matrix of cloud M^I is the $K \times K$-matrix denoted `Mcov.KK`. If L is the number of non-null eigenvalues (dimensionality of $\mathcal{L}$), the change of basis matrix is the $K \times L$ matrix denoted `BasisChange.KL`. Then, the $I \times L$-matrix of the standardized principal coordinates of vectors $(\overrightarrow{GM^i})_{i \in I}$ is denoted `Z_GM.IL`, and the L-column of those of vector $\overrightarrow{PG}$ is denoted `Z_PG.L` (see Formula 2.6, p. 19). The squared M-norm of $\overrightarrow{PG}$ (`norm2_PG`) is simply the sum of squares of the standardized principal coordinates of $\overrightarrow{PG}$. Then we state a descriptive conclusion.

```
Mcov.KK <- cov.wt(X.IK, method = "ML")$cov
eig <- eigen(Mcov.KK, symmetric = TRUE)
L <- sum(eig$values > 1.5e-8) ; lambda.L <- eig$values[1:L]
BasisChange.KL <- eig$vectors[ ,1:L] %*% diag(1/sqrt(lambda.L), nrow=L)
Z_GM.IL   <- sweep(X.IK, 2, colMeans(X.IK), "-") %*% BasisChange.KL
Z_GP.L    <- t(BasisChange.KL) %*% t(t(colMeans(X.IK) - X_P))
norm2_PG <- sum(Z_GP.L^2)                      # squared M-distance PG
cat(" M-distance D = ", round(sqrt(norm2_PG),3), "\n",
    " Descriptively, the deviation is ",
    ifelse(norm2_PG >= notable_D^2, "notable.",
           "small: there is no point in doing the test."), sep= "")
```

Step 4. The testing procedure

Firstly, we have to choose between exhaustive and Monte Carlo methods for enumeration of permutations (see p. 84), and then determine the associated set of points of the permutation G-cloud (see §4.2.2, p. 70). If the maximum number of permutations chosen by the user (`max_number`) is greater than 2^n, the exhaustive method is performed, if not we proceed to MC method. Algorithms are slightly different according to the method. In both cases, because of the symmetry property, we only enumerate half of the permutations (`cardJ`). For the exhaustive method, we construct half of the permutation table (see §4.2.1, p. 69) that is called `Epsilon.JI`. Then using Equation 4.1 (p. 71), we establish the corresponding matrix (`Z_GGj.JL`) of coordinates of vectors $\overrightarrow{GG^j}$

with respect to the orthocalibrated basis. For the MC method, random permutations are defined one by one (since the permutation table can be very large), and the coordinates of the corresponding vector $\overrightarrow{GG}^j$ are computed; in this case the matrix Z_GGj.JL is filled inside the loop on j.

```
if (2^(n-1) <= max_number){
  cardJ <- 2^(n-1)
  Epsilon.JI <- matrix(1L, nrow= cardJ, ncol= n)
  for(i in 1:(n-1))
    Epsilon.JI[ , i+1] <- rep(c(1L,-1L), each=2^(n-i-1), times=2^(i-1))
    Epsilon.J <- t(t(rowSums(Epsilon.JI)))/n
    Z_GGj.JL  <- Epsilon.JI %*% Z_GM.IL/n
    rm(Epsilon.JI)
} else {
  cardJ     <-  max_number; set.seed(seed)
  Epsilon.J <- matrix(0L, nrow= cardJ, ncol= 1)
  Z_GGj.JL  <- matrix(0,  nrow= cardJ, ncol= L)
  for (j in 1:cardJ) {
    epsilon_j.I   <- t(sample(c(-1L, 1L), size= n, replace= TRUE))
    Epsilon.J[j]  <- sum(epsilon_j.I)/n
    Z_GGj.JL[j, ] <- epsilon_j.I %*% Z_GM.IL/n
  }
}
```

Remark. We recommend—therefore we perform—the exhaustive permutation method whenever the number of resampled values requested by the user is greater than the number of permutations (oversampling). The limitation of the exact method is due to the size of the Epsilon.JI matrix that contains $n \times 2^{n-1}$ elements. For n=25, this number is equal to 419,430,400 (the size of Epsilon.JI is around 1.7 gigabytes); the number of permutations is 2^{24}, that is, around 17 millions, hence a long computing time! When the number of permutations is about one million ($n \leq 21$), the exhaustive method can be performed (see Remark (1), p. 44).

Secondly, by applying Proposition 4.3 (p. 74) we determine the matrix of the coordinates of the cardJ vectors $\overrightarrow{PP}^j$ (Z_PPj.JL) and we compute the squared M-norm of these vectors (norm2_PPj.J). Then, applying Formula 4.3 (p. 77), we deduce the values of the test statistic (d2P.J). By Formula 4.4 (p. 77), we compute the observed value of the test statistic (d2P_obs) and finally the p-value.

```
if (norm2_PG >= notable_D^2) {
  Z_PPj.JL <- Z_GGj.JL + Epsilon.J %*% t(Z_GP.L)
  norm2_PPj.J <- rowSums(Z_PPj.JL^2)
  d2P.J <- norm2_PPj.J - (Z_PPj.JL %*% Z_GP.L)^2 / (1 + norm2_PG)
  rm(Z_PPj.JL)
  d2P_obs    <- norm2_PG/(1 + norm2_PG)
  n_sup      <- sum(d2P.J >= d2P_obs * (1 - 1e-12))
  p_value    <- n_sup/cardJ
```

```
  if(K>1){
    cat(" p-value = ", 2*n_sup, "/", 2*cardJ, " = ",
        format(round(p_value, 3), nsmall= 3),"\n", sep= "")
    } else {
    cat(" p-value = ", n_sup, "/", 2*cardJ, " = ",
        format(round(p_value, 3), nsmall= 3),"\n", sep= "")
  }
}
```

Step 5. The compatibility region

The scale parameter κ of the adjusted compatibility ellipsoid is computed from n_dir random lines with M-unit direction vectors $\vec{\beta}$ (beta.L), using Theorem 4.2 (p. 82).

1. The squared M-norms of vectors $\overrightarrow{GG}^j$ are computed and put in the column-matrix C.J. The set of indexes for which the coefficient $|GG^j|^2 + (\bar{\epsilon}^j)^2 - 1$ is null is denoted Anul (see p. 81).

2. For each line $\mathcal{D}$ defined by the random vector beta.L (loop on d), the coordinates of $(\overrightarrow{GG}^j)_{j \in J}$ along $\mathcal{D}$ are computed, hence column-matrix U.J.

3. By using Lemma 4.5 (p. 81), for each $j \in J$, the interval $[x_1^j, x_2^j]$ is determined, hence the column-matrices X1.J and X2.J.

4. By Theorem 4.2 (p. 82), we obtain two values of κ for each axis hence the $2 * $ n_dir values of κ (kappa.D) whose mean is the adjusted κ value.

```
set.seed(seed); rank_inf <- trunc(alpha * cardJ) + 1
C.J <- t(t(rowSums(Z_GGj.JL^2)))
Anul <- which(abs(C.J + Epsilon.J^2 - 1)  < 1e-12)
if ( K== 1) n_dir <- 1
kappa.D <- rep(0, 2 * n_dir)
for (d in 1: n_dir){
  xy.L <- runif(L, -1, 1) ; beta.L <- t(t(xy.L / sqrt(sum(xy.L^2))))
  U.J <- Z_GGj.JL %*% beta.L
  A.J <- C.J - U.J^2 + Epsilon.J^2- 1
  B.J <- -Epsilon.J * U.J
  DeltaP.J <- abs(B.J^2 - A.J*C.J)
  X1.J <- (-B.J + sqrt(DeltaP.J))/A.J
  X2.J <- (-B.J - sqrt(DeltaP.J))/A.J
  X1.J[Anul] <- -Inf ; X2.J[Anul] <- Inf
  X1_sorted  <- sort(X1.J); X2_sorted  <- sort(X2.J)
  kappa.D[2*d - 1] <- abs(X1_sorted[rank_inf])
  kappa.D[2*d] <- X2_sorted[cardJ + 1 - rank_inf]
}
if (K > 1){
  cat("\n Adjusted ", 100*(1 - alpha), "% compatibility region:\n",
      " principal ellipsoid of the cloud with scale parameter",
      " kappa = ", round(mean(kappa.D), 3), " \n",
      " (mean of ", 2*max(L,n_dir)," kappa values in the range ",
```

```
        floor(min(kappa.D)*1000)/1000, " to ",
        ceiling(max(kappa.D)*1000)/1000, "). \n", sep= "")
} else {
    x1 <- colMeans(X.IK) -  beta.L * kappa.D[1] * sqrt(Mcov.KK[1,1])
    x2 <- colMeans(X.IK) +  beta.L * kappa.D[2] * sqrt(Mcov.KK[1,1])
    cat("\n ", 100*(1 - alpha), "% compatibility interval [",
        round( min(x1,x2),3)," ; ", round( max(x1,x2), 3),"]", sep= "")
}
```

4.6.2   R Script Interfaced with Coheris SPAD Software

For users, it is easier to perform the analyses using the R script interfaced with Coheris SPAD software[12], which is a menu-driven solution.

We will present here the analyses of the *Target example* that are developed in this chapter. The sequence of analyses is quite simple. It comprises two steps (see diagram in Figure 4.21).

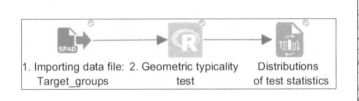

Index	Id	X	Y	C
1	i1	0	-6	c1
2	i2	3	-5	c1
3	i3	7	-3	c3
4	i4	3	-1	c3
5	i5	6	0	c3
6	i6	-4	1	c1
7	i7	1	2	c3
8	i8	3	2	c3
9	i9	5	5	c2
10	i10	6	5	c2

1. Importing data file: Target_groups 2. Geometric typicality test   Distributions of test statistics

Figure 4.21
Diagram of the Coheris SPAD project and data table.

The *step 1* is the *import of the data set*. Here the file is a "SPAD data archive" file (but many different formats of import are possible) which is stored in the project for the sake of the autonomy of the project. Data table is shown in Figure 4.21. The *step 2* is the one of the geometric typicality test for *Mean point* through the R script.

The settings are defined using four tabs (see Figure 4.22): Variables, Analyses, Parameters and Outputs.

- Variables (Figure 4.22)

  Firstly, you have to select the "Geometric typicality test" R script that was copied in the folder (defined by the user) of the interfaced R scripts (see Figure 4.22). Then, you have to select the *numerical variables* on which the analysis is performed (coordinates of points with respect to an orthonormal basis). If you want to perform the geometric method on a

---

[12]For a general presentation of Coheris SPAD software, see Chapter 1, p. 6.

subset of data, you have to name the factor (categorical variable) that permits to select the subset.

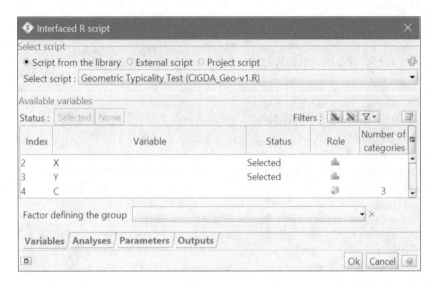

Figure 4.22
`Variables` dialog box: selection of the R script and of the status of variables.

- `Analyses` (Figure 4.23)

  You must choose the data to be analysed, the reference point and the type of analysis (one-dimensional, multidimensional analysis or both). You can

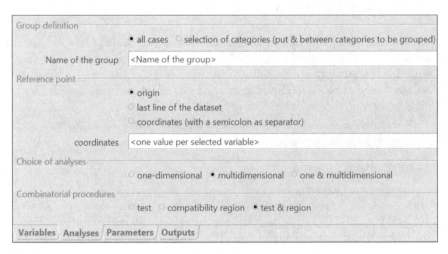

Figure 4.23
`Analyses` dialog box.

perform the test and/or determine the (adjusted) compatibility region (in the analysis of a subset, the category of the factor must be specified).

- Parameters (Figure 4.24)

  All parameters have default settings corresponding to the more conventional choices. These values can be modified.

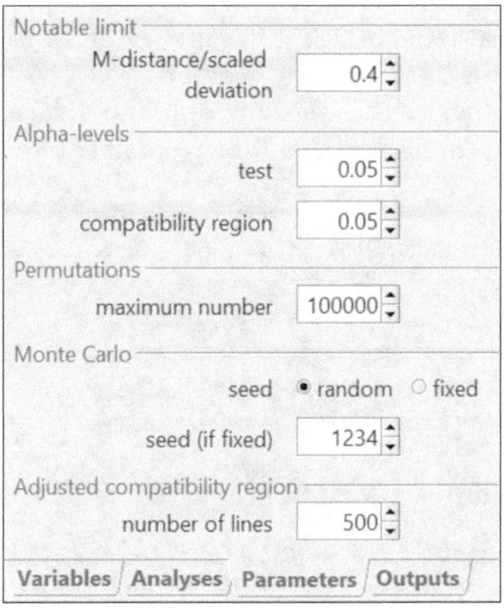

Figure 4.24
Parameters dialog box.

- Outputs (Figure 4.25)

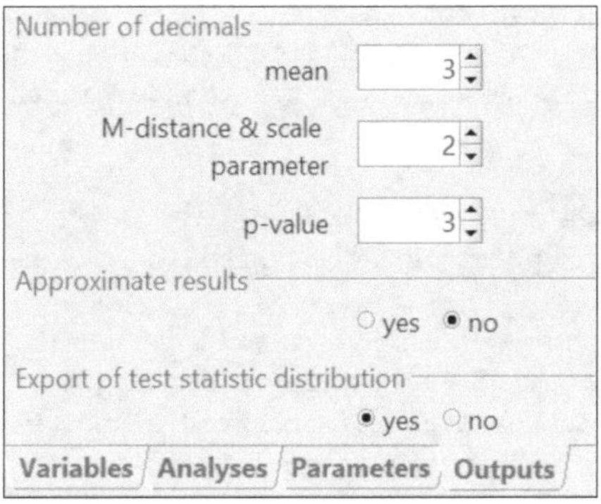

Figure 4.25
Outputs dialog box.

The outputs dialog box permits to choose the format of the results (number of decimals), to give or not the results of approximate methods, and to export or not distributions of test statistics to SPAD.

Results of analyses

The results of the R script consist of a text file[13] and, if requested, of the values of the test statistic distributions. The text file can be accessed from the context menu of the method (Right-click – Results – Edition). The distributions can be exported by using "data export" methods.

The program provides the following results of the combinatorial typicality test for *Mean point* applied to the *Target example* (see `Results/Edition`).

```
================================================================
COMPARISON OF THE MEAN MOINT OF A CLOUD TO A REFERENCE POINT
================================================================
2   variables ( X, Y ) are analysed jointly;
cloud is in a   2-dimensional geometric space.
   size of the group of observations = 10
   reference point:      (0.000; 0.000)

Descriptive analysis
--------------------
   observed mean point: (3.000; 0.000)
   M-distance between observed mean point and reference point: 0.964>0.4
   Descriptively, the deviation is notable.

Combinatorial inference
-----------------------
   Number of possible clouds: 1024;
   exhaustive method is performed.

~ Geometric typicality test
   test statistic: squared generalized Mahalanobis distance with respect
   to reference point between mean point and reference point (observed
   value: 0.48)

   p-value = 86/1024 = 0.084 > 0.05
   The observed mean point is not atypical of the reference point,
   at level 0.05.

~ adjusted 95% compatibility region is defined by
   the principal kappa-ellipse of observed cloud, with kappa = 1.08
   (mean of 1000 values in the range 1.01 to 1.16)

Test statistic distributions are exported to SPAD.
```

---

[13]In the case of a 2-dimensional space, the compatibility ellipse can be found in an Excel file (Right-click/Results/ Report).

Distributions of test statistics

The R script computes the values of test statistics and, if requested, exports them to SPAD. From these results, it is possible to construct the distributions using graphical interface of SPAD software.

Concluding Remarks

As can be seen, our target paradigm extends the classical permutation test to several dimensions, in the Fisher–Pitman tradition (see, e.g., Cox and Hinkley, 1974). Yet, the applicability of such tests is often thought to be confined to the situations involving the so-called "physical act" of randomization (see, e.g., Edgington, 2007). Such a conception severely restricts their range of application. In the combinatorial framework, no randomization is assumed, insofar as the basic concept is a *proportion* (not a probability) of possible data sets (Rouanet et al., 1986).

The geometric typicality test will be used in the case studies that will be presented in Chapter 6.

5

Homogeneity Permutation Tests

In this chapter, we introduce homogeneity permutation tests along an approach similar to that used in the two preceding chapters.

The homogeneity situations can be described as follows: considering several groups of univariate or multivariate observations, and some statistic of interest, the *homogeneity problem* is raised and intuitively expressed thus:

Can the groups be merged? Or, are they heterogeneous?

or, more specifically,

How can a homogeneity level be assessed, according to some statistic of interest?

The chapter is organized as follows. First of all, we present examples of the homogeneity problem (§5.1), and then the principles of combinatorial homogeneity tests (§5.2). Then, we deal with the homogeneity of independent groups (§5.3). After that, we develop the particular case of two independent groups, which is quite frequently encountered in real-life case studies, by focusing not only on the test procedure but also on the definition of a compatibility region (§5.4). Then, we briefly study the homogeneity for a repeated measures design (§5.5). Then, we evoke other methods that deal with the comparison of means of several groups (§5.6). Finally, we give the R script for the homogeneity of two groups and the guide for performing analyses by using the R script interfaced with Coheris SPAD software (§5.7).

5.1 The Homogeneity Problem

Let us consider, as motivating examples, the following three situations.

Pedagogy

In order to compare two kinds of pedagogy of mathematics teaching, two groups of pupils were constituted. Pupils in one of the two groups learn mathematics with a modern pedagogy and the others with a traditional pedagogy. At the end of the course, an exam on combinatorics is given to the pupils.

Is there an effect of pedagogy on the results in combinatorics examination?

Visual acuity

In a comparison of visual acuity of deaf and hearing children, eye movement rates were taken on deaf and hearing children. A clinical psychologist believes that deaf children have greater visual acuity than hearing children. The larger a child's eye movement rate is, the more visual acuity the child possesses.

> *Is there a difference between deaf and hearing children with respect to visual acuity?*

Globalisation and the members of the French Parliament

A questionnaire on globalisation was submitted to the members of the French Parliament (MPs). The Multiple Correspondence Analysis of the questionnaire provides a bidimensional cloud of the respondents (see Chapter 6, §6.2). The MPs are divided into parliamentary groups, mainly Right, Centre and Left. The questions, which will be studied in Chapter 6.2, are:

> *Do the groups differ regarding globalisation? Do the right-wing group and the left-wing group have different representations of globalisation?*

The preceding situations exemplify the homogeneity problem. As in the case of the typicality problem, it is tempting to do some conventional significance tests. Yet again, no randomness is assumed in the data generating process and no unverifiable assumptions are made.

To get combinatorial homogeneity tests, we will take the classical *permutation tests*, or *Fisher–Pitman tests*, initiated by Fisher (1925, 1935) and Pitman (1937); for a brief historical account, see Edgington (2007, pp. 13–17).

5.2   Principle of Combinatorial Homogeneity Tests

To say that several groups of observations are homogeneous amounts to considering that the subdivision into groups may be ignored, that is, any observation belonging to a group might as well have belonged to any one of the groups. This exchangeability principle invites us to consider the *baseline data set* obtained by disregarding the subdivision into groups, and then to construct all possible data sets obtained by reallocating the observations of this baseline data set to the groups in all possible ways. Technically, this leads to applying a permutation group to the data set by taking into account the data structure, against which the observed data set is situated.

The permutation group used to generate the set of possible data sets depends on the *design structure*. Hereafter, we study the case of *independent groups* design for several groups and then especially for two groups. Then, we briefly describe the case of *repeated measures design*.

Experimental designs

In an *independent groups design* (or between-subjects design), each subject is assigned to only one condition of the independent variable. There may be several groups of subjects, but each group only takes part in one condition of the independent variable and does not repeat anything (see, for instance, the *Pedagogy* situation described above).

In a *repeated measures design* (or within-subjects design), each subject is assigned to each condition. The experimental group consists of exactly the same subjects repeating the same task but under different conditions (see, for instance, the *Parkinson study*, Chapter 6, p. 156).

The *steps of homogeneity tests* are the same as those of typicality tests.

1. *Permutation set*

   The permutation group is applied to the baseline data set in order to generate all *possible data sets* that have the same *design structure* as the observed one.

2. *Combinatorial p-value*

   A statistic is chosen and then is calculated for each possible data set as well as for the observed one. The proportion of possible data sets for which the value of the statistic is more extreme than, or as extreme as, the observed one defines the combinatorial p-value.

   This proportion is an evaluation of the level of homogeneity of groups: The smaller the p-value, the lower the homogeneity.

3. *Conclusion*

   Given a reference level α (called α-level), we state the conclusion as:

   - if the p-value is less than or equal to α, the groups are said to be *heterogeneous* at level α, for the property of interest;

   - if the p-value is greater than α, the groups cannot be declared heterogeneous at level α.

5.3 Homogeneity of Independent Groups: General Case

In this section, we study the general case of C' independent groups of interest among C ($2 \leq C' \leq C$).

The question is: *Are the C' groups heterogeneous?*

There are three different cases.

- If $C' = C$, the C' groups determine a partition of the observations, then we say that we carry out a *global comparison*.

- If $C' < C$, groups $c \notin C'$ are brought together in one group denoted c_r. The set of groups is $C' \cup \{c_r\}$, we say that we carry out a *partial comparison*.

- If the data set is restricted to the C' groups under comparison, we say that we perform a *specific comparison*.

Remarks. In most texts dealing with homogeneity of independent groups, only two groups are considered (the so-called *two-sample problem*, see, e.g., Efron and Tibshirani, 1993, p. 202) and most often involve only one response variable. Our purpose is more general.

On the one hand, we consider more than two groups and, in the case of the comparison of two groups among more than two ($C > 2$), we not only study the *specific comparison* by restricting the data set to the two groups but also the *partial comparison* by taking account of all groups.

On the other hand, the *observations are points* of a geometric cloud, that is, they come from multivariate data.

5.3.1   Nested Design

Let us denote I the set indexing the observations and n its cardinality; the set I is called *set of individuals*[1]. We consider the structure in which the set of individuals is partitioned into groups indexed by C ($C \geq 2$)[2]. The cardinality of group $c \in C$ is denoted n_c with $\sum n_c = n$ and called *size of group c*.

In *planning experiment*, "factor"[3] I is said to be nested in "factor" C (conditions), if each element of I occurs in conjunction with only one category of C, that is, there exists a surjection $f : I \to C$. Given c, the subset $f^{-1}(c) \subset I$, called group c, is denoted $I<c>$ and its cardinality is n_c. The surjection induces a partition of I into C groups denoted $I<C>$ (read "I nested in C"), which is named "nesting $I<C>$". A nesting $I<C>$, whose group sizes are equal to $(n_c)_{c \in C}$, is called *nesting $I<C>$ of type n_C*.

A *cloud of points* in a geometric space $\mathcal{U}$ is associated with the set I; it is denoted M^I. Denoting $\mathcal{V}$ the Euclidean vector space associated with the geometric space $\mathcal{U}$, recall (see Chapter 2, Definition 2.4, p. 15) that the covariance endomorphism of cloud M^I, denoted Mcov, is defined by

$$\mathrm{Mcov} : \mathcal{V} \to \mathcal{V}$$
$$\vec{u} \mapsto \sum_{i \in I} \tfrac{1}{n} \langle \overrightarrow{\mathrm{GM}^i} | \vec{u} \rangle \overrightarrow{\mathrm{GM}^i}$$

The partition of I into C groups induces the *partition of cloud* M^I into C subclouds $(\mathrm{M}^{I<c>})_{c \in C}$. Recall that the cloud of the C mean points

---

[1]See note 3, p. 31.

[2]Recall that, in general, we denote the cardinality of a finite set as the set itself.

[3]An *experimental factor* has nothing to do with *principal variables*, often called "factors". This is why we never use the term "factors" or "factorial axes" to name the principal variables or the principal axes of a cloud (see Chapter 2, p. 17).

$(\sum_{i\in I<c>} M^i/n_c)_{c\in C}$ weighted by $(n_c)_{c\in C}$ is the between-C cloud that we denote $G^C = (G^c)_{c\in C}$; its mean point is the mean point of cloud M^I, which is denoted G (see Chapter 2, §2.4, p. 26).

Target example

To exemplify the combinatorial homogeneity test in a geometric setup, we use the *Target example* and study the partition of I into three groups c_1, c_2, c_3, with $I<c_1>= \{i_1, i_2, i_6\}$, $I<c_2>= \{i_9, i_{10}\}$, $I<c_3> = \{i_3, i_4, i_5, i_7, i_8\}$, as shown in the following figure[4].

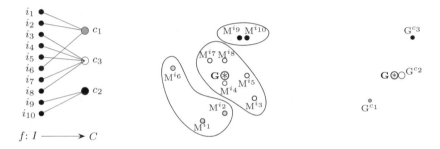

Figure 5.1
Target example. Surjection $f : I \to C$. Initial cloud M^I with the partition into three subclouds and the mean point G (⊛). Between-C cloud with the mean point G (⊛).

5.3.2 Permutation Nesting Set

The *permutation nesting set* is the set of all *possible nestings* of type n_C, that is, of all different allocations of the n individuals to C groups whose sizes are $(n_c)_{c\in C}$. Its cardinality is equal to the multinomial coefficient:

$$\frac{(\sum n_c)!}{\prod n_c!}$$

The permutation nesting set is indexed by the set J. For each $j \in J$, there is a surjection $f_j : I \to C$ that defines the partition of I into C groups such that, for $c \in C$, the cardinality of $f_j^{-1}(c)$ is n_c. Given j, the subset $f_j^{-1}(c)$ of I is denoted by $I<c>_j$ and the partition $f_j^{-1}(C) = (f_j^{-1}(c))_{c\in C}$ of I into C groups by $I<C>_j$. The element $I<c>_j$ of the nesting set will often be called "nesting j".

Given $c \in C$, we define the following equivalence relation on J:

$$j \sim j' \iff f_j^{-1}(c) = f_{j'}^{-1}(c)$$

---

[4]Obviously, this example was chosen for teaching purposes. In fact, it makes it possible to show in detail the steps of the method and to perform exact calculations. Applications to real data sets can be found in Chapter 6.

The quotient set of J by the equivalence relation is denoted J/c. The cardinality of J/c is equal to $\binom{n}{n_c}$ and all equivalence classes have the same cardinality, namely $(n - n_c)!/\prod_{c' \neq c} n_{c'}! = n_c!(n - n_c)!/\prod n_c!$.

5.3.3 Permutation Nesting Space

The C subclouds $(M^{I<c>_j})_{c \in C}$ of cloud M^I associated with nesting $I<C>_j$ determine a partition of cloud M^I of type n_C, which is denoted $M^{I<C>_j}$. The family of the $J = (\sum n_c)!/(\prod n_c!)$ partitioned clouds defines the *permutation nesting space*:

$$\mathcal{J} = (M^{I<C>_j})_{j \in J}$$

Given $j \in J$ and $c \in C$, the mean point of the subcloud $M^{I<c>_j}$ of cloud M^I is denoted G^{cj} with $G^{cj} = \sum_{i \in k>_j} M^i/n_c$ and its weight is n_c.

Target example

A possible nesting $I<C>_j$, with $I<c_1>_j = \{i_1, i_2, i_3\}$, $I<c_2>_j = \{i_9, i_{10}\}$ and $I<c_3>_j = \{i_4, i_5, i_6, i_7, i_8\}$ is shown in Figure 5.2 as well as the three mean points $G^{c_1 j}$, $G^{c_2 j}$ and $G^{c_3 j}$.

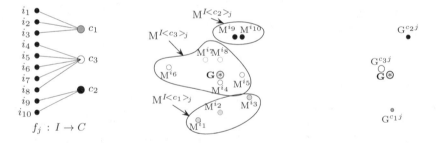

Figure 5.2
Target example. A possible nesting $I<C>_j$. Initial cloud M^I with the partition associated with nesting j and the mean point G (✱). Between-Cj cloud with the mean point G (✱).

Proposition 5.1. *For any $c \in C$, the mean point of cloud G^{cJ} is the mean point G of cloud M^I and its covariance endomorphism is proportional to Mcov.*

$$\forall c \in C, \text{ Mean } G^{cJ} = G; \quad \frac{1}{J}\sum_{j \in J} \langle \overrightarrow{GG}^{cj} | \overrightarrow{u} \rangle \overrightarrow{GG}^{cj} = \frac{1}{n-1} \times \frac{n - n_c}{n_c} \times \text{Mcov}(\overrightarrow{u})$$

Proof. The cloud $G^{cJ} = (G^{cj})_{j \in J}$, all the weights of which are equal to n_c, can be replaced by the equally-weighted cloud $G^{J/c}$, the common value of weights being $\left((n - n_c)!/\prod_{c' \neq c} n_{c'}!\right) \times n_c$. This cloud is exactly the n_c-sample cloud studied in Chapter 3 (p. 33), hence we deduce its mean and its covariance endomorphism. □

The cloud $G^{C'J} = (G^{cj})_{c \in C', j \in J}$ is the union of the C' clouds $(G^{cJ})_{c \in C'}$. From the above proposition, the mean point of each cloud G^{cJ} is point G, hence the between-C' cloud is a one-point cloud, namely point G. Consequently, using Theorem 2.3 (p. 28) the covariance endomorphism of cloud $\mathrm{G}^{C'J}$ is equal to that of the within-C' cloud, which is the weighted mean of those of subclouds $(\mathrm{G}^{cJ})_{c \in C'}$ (see Chapter 2, Equation 2.12, p. 27). Hence, from Proposition 5.1, we deduce that the covariance endomorphism of cloud $\mathrm{G}^{C'J}$ is proportional to Mcov: this calls us to choose a test statistic based upon the M-distance between mean points.

> *From now on, all covariance endomorphisms are restricted to the direction vector subspace $\mathcal{L} \subseteq \mathcal{V}$ of the affine support of cloud M^I.*

In what follows, we detail the study of the *partial comparison* of C' groups ($C' \leq C$); the properties regarding a global comparison can readily be obtained from those regarding a partial comparison.

5.3.4 Test Statistic

In order to make the test procedure operational, we have to choose a test statistic. As previously mentioned, the covariance endomorphism of cloud $\mathrm{G}^{C'J}$ is proportional to that of cloud M^I, therefore, it seems better to choose a test statistic based upon the M-distance than upon the geometric distance between points.

Recall that the M-distance (Mahalanobis distance) between points A and B is denoted $|AB|$ with $|AB|^2 = \langle \mathrm{M}cov^{-1}(\overrightarrow{AB}) | \overrightarrow{AB} \rangle$ (see Chapter 2, p. 20).

Definition of the test statistic

The variance of the C' mean points of subclouds $(\mathrm{M}^{I<c>})_{c \in C'}$ of cloud M^I is called *between-C' variance of cloud M^I.*

The map that, with each element of the nesting space, associates its between-C' M-variance (see Chapter 2, Formula 2.9, p. 26), is denoted V_M and defines V_M as a statistic.

$$V_\mathrm{M} : \mathcal{J} \quad \to \mathbb{R}_{\geq 0}$$
$$\mathrm{M}^{I<C>_j} \mapsto \frac{1}{2} \sum_{c \in C'} \sum_{c' \in C'} \frac{n_c n_{c'}}{(n')^2} |\mathrm{G}^{cj}\mathrm{G}^{c'j}|^2$$

with $n' = \sum\limits_{c \in C'} n_c$. The value of the statistic for nesting j is denoted $v_\mathrm{M}[j]$.

The between-C' M-variance of cloud $\mathrm{M}^{I<C>}$ is called *observed value of* V_M and denoted $v_{\mathrm{M}obs}$.

$$v_{\mathrm{M}obs} = \frac{1}{2} \sum_{c \in C'} \sum_{c' \in C'} \frac{n_c n_{c'}}{(n')^2} |\mathrm{G}^{c}\mathrm{G}^{c'}|^2$$

Permutation distribution of the test statistic

Given $j \in J$, let $p(V_\mathrm{M} = v_\mathrm{M}[j])$ denote the proportion of possible nestings for which the value of the test statistic is equal to $v_\mathrm{M}[j]$. The *frequency distribution* of statistic V_M is defined by the map from $V_M(\mathcal{J})$ to $[0, 1]$ that associates the proportion $p(V_M = v_M[j])$ with each value $v_M[j]$.

$$V_M(\mathcal{J}) \to [0, 1]$$
$$v_M[j] \mapsto p(V_M = v_M[j])$$

The permutation distribution of V_M is discrete and is most often represented by a histogram (see, for example, Figure 5.3, p. 115).

Proposition 5.2. *The mean of statistic V_M is equal to $L \times \frac{C'-1}{n-1} \times \frac{n}{n'}$.*

Proof. If G^j denotes the mean point of the C' weighted points $(G^{jc}, n_c)_{c \in C'}$, the M-variance of the cloud $(G^{jc})_{c \in C'}$ is equal to $\sum\limits_{c \in C'} \frac{n_c}{n'} |G^j G^{jc}|^2$, that is,

to $\sum\limits_{c \in C'} \frac{n_c}{n'} |GG^{jc}|^2 - |GG^j|^2$ (Huyghens's formula, p. 12). Hence, Mean $V_M =$

$\frac{1}{J} \sum\limits_{j \in J} \left(\sum\limits_{c \in C'} \frac{n_c}{n'} |G^j G^{jc}|^2 \right) = \sum\limits_{c \in C'} \frac{n_c}{n'} \times \frac{1}{J} \sum\limits_{j \in J} |GG^{jc}|^2 - \frac{1}{J} \sum\limits_{j \in J} |GG^j|^2$. By Proposition

3.5 (p. 38), one has $\frac{1}{J} \sum\limits_{j \in J} |GG^{jc}|^2 = \frac{L}{n_c} \times \frac{n-n_c}{n-1}$ $(c \in C)$ and, since cloud G^J corre-

sponds to samples of size n', we have $\frac{1}{J} \sum\limits_{j \in J} |GG^j|^2 = \frac{L}{n'} \times \frac{n-n'}{n-1}$. Then, we deduce

Mean $V_M = \sum\limits_{c \in C'} \left(\frac{n_c}{n'} \frac{L}{n_c} \times \frac{n-n_c}{n-1} \right) - \frac{L}{n'} \times \frac{n-n'}{n-1} = \frac{L}{n'(n-1)} \left(nC' - n' - (n - n') \right).$ □

For a global comparison, $C' = C$ and $n' = n$, hence Mean $V_M = L \times \frac{C-1}{n-1}$.

Combinatorial p-value

We want to know how extreme the observed value v_{Mobs} is over the distribution of V_M. To that end, we look for possible nestings that satisfy the property[5] $(V_M \geq v_{Mobs})$. The proportion of those nestings is the *combinatorial p-value* that defines the level of homogeneity of the groups.

$$p = p(V_M \geq v_{Mobs})$$

Target example

Recall (see p. 14) that, if the plane of the target is referred to two rectangular axes (horizontal and vertical) passing through point O (centre of the target), the coordinates of the mean point G are $(3, 0)$, the matrix associated with the covariance endomorphism Mcov is $\mathbf{V} = \begin{bmatrix} 10 & 2 \\ 2 & 13 \end{bmatrix}$ (covariance matrix) and its inverse is $\mathbf{V}^{-1} = \frac{1}{126} \begin{bmatrix} 13 & -2 \\ -2 & 10 \end{bmatrix}$. The coordinates of mean point G^{c_1} are $(-1/3, -10/3)$, those of G^{c_2} are $(5.5, 5)$, those of G^{c_3} are $(4, 0)$, and those of the mean point G are $(3, 0)$. The squared *geometric distances* from these mean points to point G are $(GG^{c_1})^2 = (-1/3 - 3)^2 + (-10/3 - 0)^2 = 200/9$, $(GG^{c_2})^2 = 31.25$, and $(GG^{c_3})^2 = 1$. The squared M-*distances* are: $|GG^{c_1}|^2 =$

$\frac{1}{126} \times \begin{bmatrix} -\frac{10}{3} & \frac{10}{3} \end{bmatrix} \begin{bmatrix} 13 & -2 \\ -2 & 10 \end{bmatrix} \begin{bmatrix} -\frac{10}{3} \\ \frac{10}{3} \end{bmatrix} = 1.6755$, $|GG^{c_2}|^2 = 2.2321$ and $|GG^{c_3}|^2 = 0.1032$.

---

[5] A nesting $j \in J$ is said to satisfy the property $(V_M \geq v_{Mobs})$ if $v_M[j] \geq v_{Mobs}$.

Global comparison

The between-C variance is $\text{Var}_C = \frac{3}{10} \times 200/9 + \frac{2}{10} \times 31.25 + \frac{5}{10} \times 1 = 13.42$, the total variance is $V_{\text{cloud}} = 23$, hence $\eta^2 = 13.42/23 = 0.58$ (see Chapter 2, p. 28). By any standard, such a value can be deemed to be large. Thus, the conclusion:

Descriptively, the global difference between groups is large.

Now, we perform the *homogeneity test* in order to extend the descriptive conclusion. The observed value of the test statistic is $v_{\text{Mobs}} = \frac{3}{10} \times 1.6755 + \frac{2}{10} \times 2.2321) + \frac{5}{10} \times 0.1032 = 1.00$. The number of possible nestings is equal to 2,520 and those for which the value of the test statistic is greater than or equal to the observed value (v_{Mobs}) is equal to 37, hence the combinatorial p-value is $p = 37/2520 = .015 < .05$ (see Figure 5.3).

The data are in favour of the heterogeneity of the three groups, at level .05.

Partial comparison

We now study the partial comparison between the two groups c_1 and c_2. The variance of the cloud of two points $(\text{G}^{c_1}, \text{G}^{c_2})$ is $\frac{n_{c_1} n_{c_2}}{(n_{c_1}+n_{c_2})^2}(\text{G}^{c_1}\text{G}^{c_2})^2$ (Formula 2.2, p. 12). The coordinates of $\overrightarrow{\text{G}^{c_1}\text{G}^{c_2}}$ are $(5.5 + \frac{1}{3}, 5 + \frac{10}{3}) = (17.5/3, 25/3)$, hence $\text{Var}_{C'} = \frac{3 \times 2}{5^2}((17.5/3)^2 + (25/3)^2) = 74.5/3$. The proportion of the overall variance taken account by the partial comparison is equal to $\frac{3+2}{10} \times (74.5/3)/23 = 0.540$ (see Formula 2.14, p. 28).

Descriptively, the difference between groups c_1 and c_2 is large.

The squared M-distance between mean points G^{c_1} and G^{c_2} is $|\text{G}^{c_1}\text{G}^{c_2}|^2 =$

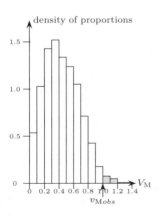 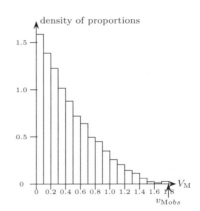

Figure 5.3
Target example. Histograms of the distribution of test statistic V_{M} for the global comparison (left) and for the partial comparison of groups c_1 and c_2 (right). The surface area under the curve past v_{Mobs} (shaded grey area) is equal to the (interpolated) p-value.

$\frac{1}{126} \times \begin{bmatrix} \frac{17.5}{3} & \frac{25}{3} \end{bmatrix} \begin{bmatrix} 13 & -2 \\ -2 & 10 \end{bmatrix} \begin{bmatrix} \frac{17.5}{3} \\ \frac{25}{3} \end{bmatrix} = 7.479$, and the observed value of the test statistic is $v_{\mathrm{Mobs}} = \frac{3 \times 2}{5^2} \times 7.479 = 1.795$.

The distribution of the test statistic is shown in Figure 5.3 (p. 115). The p-value is $p = 1/2520$.

5.3.5   Approximate Test

Let S_{M}^2 denote the statistic defined by $S_{\mathrm{M}}^2 = \frac{n'}{n}(n-1)V_{\mathrm{M}}$. By adjusting the distribution of S_{M}^2 by a χ^2 distribution with $L(C'-1)$ degrees of freedom, we obtain an approximate p-value.

$$\tilde{p} = p\left(\chi^2_{L(C'-1)} \geq \frac{n'}{n}(n-1)\,v_{\mathrm{Mobs}}\right)$$

5.4   Homogeneity of Two Independent Groups

Now, we shall proceed to study in greater depth the partial comparison of two groups of interest that are denoted c_1 and c_2, thus $C' = \{c_1, c_2\}$. The individuals that do not belong to any of the two groups of interest are gathered in one group denoted c_r (possibly empty). The cardinalities of groups (group sizes) are respectively denoted n_{c_1}, n_{c_2} and n_{c_r}. We let $n = n_{c_1} + n_{c_2} + n_{c_r}$, $n' = n_{c_1} + n_{c_2}$ ($n' \leq n$), hence $n_{c_r} = n - n'$.

Letting $C = C' \bigcup \{c_r\}$, we will study the nesting $I{<}C{>}$ of type $n_C = (n_{c_1}, n_{c_2}, n - n')$, which induces the partition of cloud M^I into three subclouds: $\mathrm{M}^{I{<}c_1{>}}$, $\mathrm{M}^{I{<}c_2{>}}$ and $\mathrm{M}^{I{<}c_r{>}}$.

5.4.1   Test Procedure

The *principle of the test* is exactly that described in §5.2 (p. 108).

If we proceed to a *partial comparison* of the two groups, the *permutation nesting set* is the set of all possible nestings of type $(n_{c_1}, n_{c_2}, n - n')$.

If we proceed to a *specific comparison* of the two groups, the *permutation nesting set*—whose construction is extremely simple since it consists in exchanging the individuals of the two groups in all possible ways—is the set of all possible nestings of type (n_{c_1}, n_{c_2}).

The remainder of the test procedure is always the same. We choose a *test statistic*, which is calculated for each possible nesting in order to obtain the combinatorial distribution of the test statistic. Then, the observed value of the test statistic is located in the distribution, hence the *combinatorial p-value*.

In what follows, we deal with the general case of the partial comparison of two groups.

Permutation nesting set and permutation nesting space

The permutation nesting set is the set of possible nestings of type $n_C = (n_{c_1}, n_{c_2}, n - n')$; it is indexed by the set J and its cardinality is equal to $\frac{n!}{n_{c_1}! n_{c_2}! (n-n')!}$. The nesting space $\mathcal{J}$ is constructed as in §5.3.3 (p. 112)

$$\mathcal{J} = (\mathrm{M}^{I<C>_j})_{j \in J}$$

The *effect of interest* depends only on the two groups c_1 and c_2.

Test statistic

The test statistic defined in §5.3.4 (p. 113), that is, the between-C' M-variance, is replaced by an equivalent but simpler test statistic, namely the squared M-distance between the two mean points associated with groups c_1 and c_2. Furthermore, the latter allows a *geometric interpretation* of the p-value and the construction of a *compatibility region*.

For nesting j, the deviation $\mathrm{G}^{c_2 j} - \mathrm{G}^{c_1 j}$ between the mean points of the two subclouds $\mathrm{M}^{I<c_1>_j}$ and $\mathrm{M}^{I<c_2>_j}$ is denoted by $\overrightarrow{d^j}$, and its M-norm by $d[j]$. The test statistic, denoted D_{M}^2, is defined by

$$
\begin{aligned}
D_{\mathrm{M}}^2 : \mathcal{J} &\to \mathbb{R}_{\geq 0} \\
\mathrm{M}^{I<C>_j} &\mapsto d^2[j] = |\mathrm{G}^{c_1 j} \mathrm{G}^{c_2 j}|^2
\end{aligned}
$$

The observed deviation $\mathrm{G}^{c_2} - \mathrm{G}^{c_1}$ is denoted by $\overrightarrow{d_{obs}}$, and its M-norm by d_{obs}; the observed value of the test statistic D_{M}^2 is d_{obs}^2.

> **Remarks.** Statistics V_M and D_{M}^2 are proportional ($V_M = \frac{n_{c_1} n_{c_2}}{n'^2} D_{\mathrm{M}}^2$), thus they are permutationally equivalent, that is, they provide the same p-value.
>
> We could have chosen the M-distance between mean points, instead of its square. However, even if the M-distance and its square define statistics that are permutationally equivalent, on the one hand, the relations between the diverse Mahalanobis distances, which intervene thereafter, may be written simply in terms of squares, and on the other, the distribution of a squared Mahalanobis distance can be approximated by a chi-squared distribution.

Permutation distribution of test statistic D_{M}^2

Given $j \in J$, let $p(D_{\mathrm{M}}^2 = d^2[j])$ be the proportion of possible nestings that satisfy the property $(D_{\mathrm{M}}^2 = d^2[j])$. The *distribution* of D_{M}^2 is defined by the map that associates the proportion $p(D_{\mathrm{M}}^2 = d^2[j])$ with each value $d^2[j]$.

$$
\begin{aligned}
D_{\mathrm{M}}^2(\mathcal{J}) &\to [0,1] \\
d^2[j] &\mapsto p(D_{\mathrm{M}}^2 = d^2[j])
\end{aligned}
$$

This distribution is discrete and often represented by a histogram.

Proposition 5.3. *The mean of statistic D_{M}^2 is equal to $L \times \frac{1}{n-1} \times \frac{nn'}{n_{c_1} n_{c_2}}$.*

Proof. This property results from the relation $V_M = \frac{n_{c_1} n_{c_2}}{n'^2} D_M^2$ and from Proposition 5.2 (p. 114). $\square$

Combinatorial p-value

We want to know how extreme the observed value of the test statistic is over the permutation distribution. We look for possible nestings whose value of the test statistic is greater than or equal to the observed value. This proportion defines the *combinatorial p-value*, denoted p.

$$p = p\big(D_M^2 \geq d_{obs}^2\big)$$

Theorem 5.1 (Equivalence property). *In the case of the specific comparison of two groups, the homogeneity test is equivalent to the combinatorial typicality test for which the reference population is the union of the two groups and the group to be tested is one of the two groups (Chapter 3, §3.2, p. 32).*

Proof. In this case, one has $\overrightarrow{G^{c_1}G^{c_2}} = -\frac{n}{n_{c_2}}\overrightarrow{GG}^{c_1}$ and $\overrightarrow{G^{c_1j}G^{c_2j}} = -\frac{n}{n_{c_2}}\overrightarrow{GG}^{c_1j}$, hence the two properties $(|G^{c_1j}G^{c_2j}|^2 \geq |G^{c_1}G^{c_2}|^2)$ and $(|GG^{c_1j}|^2 \geq |GG^{c_1}|^2)$ are equivalent. After the equivalence relation on J defined on page 112, for all $j \in J/c_1$, the points G^{c_1j} and the point H^j of the n_{c_1}-sample cloud M^{I_j} are the same, hence the theorem. $\square$

5.4.2   Cloud of Deviation Points

For each nesting $j \in J$, the test statistic depends on the deviation $\overrightarrow{d^j}$ from the mean point G^{c_1j} to the mean point G^{c_2j}. Thus, following the same approach as that presented in Chapter 4 (p. 66), we choose a point O that represents the null deviation, and we consider the cloud of the $J = \frac{n!}{n_{c_1}! n_{c_2}! (n-n')!}$ *deviation-points* $(D^j)_{j \in J}$ defined by

$$\forall j \in J, \, D^j = O + \overrightarrow{d^j}$$

The forthcoming properties can be defined and interpreted from the cloud D^J to which we add the observed deviation-point $D_{obs} = O + \overrightarrow{d_{obs}}$.

Proposition 5.4. *The mean point of cloud D^J is point O.*

Proof. $\sum \overrightarrow{OD^j}/J = \frac{1}{J}\sum(\overrightarrow{OG^{c_2j}} - \overrightarrow{OG^{c_1j}})$. By Proposition 5.1 (p. 112), one has Mean $G^{c_1J} = G = $ Mean G^{c_2J}, therefore $\sum \overrightarrow{OG^{c_1j}}/J = \sum \overrightarrow{OG^{c_2j}}/J = \overrightarrow{0}$. $\square$

Target example

In Figure 5.4 (p. 119), we consider two groups[6] $I{<}c_1{>}= \{i_1, i_2, i_6\}$ (grey points) and $I{<}c_2{>}= \{i_9, i_{10}\}$ (black points) whose sizes are 3 and 2. The cloud of the $10!/(3!2!5!) = 2520$ deviation-points is shown in Figure 5.5.

---

[6]See Note 4, p. 111.

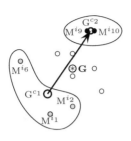

Figure 5.4
Target example. Cloud M^I and the two sub-clouds with the deviation-vector $\overrightarrow{G^{c_1}G^{c_2}}$.

Figure 5.5
Cloud of deviation points with points O and D_{obs} (black points).

Lemma 5.1. *Let* G^j *denote the mean point of the two weighted points* $(G^{c_1 j}, n_{c_1})$ *and* $(G^{c_2 j}, n_{c_2})$, *one has the relation:*

$$\forall \overrightarrow{u} \in \mathcal{L}, \sum_{c \in C'} \frac{n_c}{n'} \langle \overrightarrow{GG}^{cj} | \overrightarrow{u} \rangle \overrightarrow{GG}^{cj} = \frac{n_{c_1} n_{c_2}}{(n')^2} \langle \overrightarrow{d}^j | \overrightarrow{u} \rangle \overrightarrow{d}^j + \langle \overrightarrow{GG}^j | \overrightarrow{u} \rangle \overrightarrow{GG}^j$$

Proof. The covariance endomorphism of the cloud of two points $(G^{c_1 j}, n_{c_1})$ and $(G^{c_2 j}, n_{c_2})$ is such that $\overrightarrow{u} \mapsto \frac{n_{c_1} n_{c_2}}{(n')^2} \langle \overrightarrow{d}^j | \overrightarrow{u} \rangle \overrightarrow{d}^j$ ($\overrightarrow{u} \in \mathcal{L}$) (see Formula 2.3, p. 16). From Proposition 2.5 (p. 16), taking point G as a reference point, we obtain $\sum_{c \in C'} \frac{n_c}{(n')^2} \langle \overrightarrow{GG}^{cj} | \overrightarrow{u} \rangle \overrightarrow{GG}^{cj} = \frac{n_{c_1} n_{c_2}}{(n')^2} \langle \overrightarrow{d}^j | \overrightarrow{u} \rangle \overrightarrow{d}^j + \langle \overrightarrow{GG}^j | \overrightarrow{u} \rangle \overrightarrow{GG}^j$. □

Proposition 5.5. *The covariance endomorphism of cloud of deviation points* D^J, *denoted* Dcov, *is proportional to that of cloud* M^I. *Letting* $\gamma = \frac{n_{c_1} n_{c_2}}{nn'}$:

$$\text{Dcov} = \frac{1}{n-1} \times \frac{1}{\gamma} \text{Mcov}$$

Proof. By definition, $\forall \overrightarrow{u} \in \mathcal{L}$, $\text{Dcov}(\overrightarrow{u}) = \frac{1}{J} \sum_{j \in J} \langle \overrightarrow{OD}^j | \overrightarrow{u} \rangle \overrightarrow{OD}^j = \frac{1}{J} \sum_{j \in J} \langle \overrightarrow{d}^j | \overrightarrow{u} \rangle \overrightarrow{d}^j$. From the preceding lemma, we deduce:
$\langle \overrightarrow{d}^j | \overrightarrow{u} \rangle \overrightarrow{d}^j = \frac{n'}{n_{c_2}} \langle \overrightarrow{GG}^{c_1 j} | \overrightarrow{u} \rangle \overrightarrow{GG}^{c_1 j} + \frac{n'}{n_{c_1}} \langle \overrightarrow{GG}^{c_2 j} | \overrightarrow{u} \rangle \overrightarrow{GG}^{c_2 j} - \frac{(n')^2}{n_{c_1} n_{c_2}} \langle \overrightarrow{GG}^j | \overrightarrow{u} \rangle \overrightarrow{GG}^j$.
Hence $\text{Dcov}(\overrightarrow{u}) = \frac{n'}{n_{c_2}} \left(\frac{1}{J} \sum_{j \in J} \langle \overrightarrow{GG}^{c_1 j} | \overrightarrow{u} \rangle \overrightarrow{GG}^{c_1 j} \right) + \frac{n'}{n_{c_1}} \left(\frac{1}{J} \sum_{j \in J} \langle \overrightarrow{GG}^{c_2 j} | \overrightarrow{u} \rangle \overrightarrow{GG}^{c_2 j} \right) - \frac{(n')^2}{n_{c_1} n_{c_2}} \left(\frac{1}{J} \sum_{j \in J} \langle \overrightarrow{GG}^j | \overrightarrow{u} \rangle \overrightarrow{GG}^j \right)$. By Proposition 5.1 (p. 112), the covariance endo-morphism of cloud $G^{c_1 J}$ is equal to $\frac{1}{n-1} \frac{n - n_{c_1}}{n_{c_1}}$ Mcov, that of cloud $G^{c_2 J}$ to $\frac{1}{n-1} \frac{n - n_{c_2}}{n_{c_2}}$ Mcov and that of cloud G^J (corresponding to the union of groups c_1 and c_2) to $\frac{1}{n-1} \frac{n - n'}{n'}$ Mcov. So, we deduce $\text{Dcov} = \left(\frac{n'}{n_{c_2}} \frac{1}{n-1} \frac{n - n_{c_1}}{n_{c_1}} + \frac{n'}{n_{c_1}} \frac{1}{n-1} \frac{n - n_{c_2}}{n_{c_2}} - \frac{(n')^2}{n_{c_1} n_{c_2}} \frac{1}{n-1} \frac{n - n'}{n'} \right)$ Mcov $= \frac{1}{n-1} \frac{n'}{n_{c_1} n_{c_2}} n$ Mcov. □

Remark. The former proposition brings us to choose a test statistic based upon the covariance structure of the cloud of deviation points in order to take its shape into account. The test statistic D_{M}^2 satisfies this requirement.

Geometric interpretation of the combinatorial p-value

The combinatorial p-value is equal to the proportion of nestings $j \in J$ that verify $d^2[j] \geq d^2_{obs}$. In geometric terms, the combinatorial p-value is the proportion of points D^j such that $|\overrightarrow{OD^j}|^2 \geq |\overrightarrow{OD_{obs}}|^2$. The points D^j that verify the property are located on or outside the principal ellipsoid of cloud M^I centred at point O passing through point D_{obs} (see Figure 5.7).

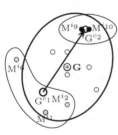

Figure 5.6
Target example. Cloud M^I with its concentration ellipse and the two subclouds and the deviation-vector $\overrightarrow{G^{c_1}G^{c_2}}$.

Figure 5.7
Cloud of deviation points with its concentration ellipse (solid line). One point is on, no point is outside the principal ellipse passing through point D_{obs} (dashed line): $p = 1/2450$.

5.4.3 Compatibility Region

Recall that the two groups of interest are c_1 and c_2 whose cardinalities are n_{c_1} and n_{c_2}. Then, we consider the nesting $I{<}C{>}$ of type n_C with $n_C = (n_{c_1}, n_{c_2}, n - n')$. The homogeneity test amounts to comparing the observed deviation $\overrightarrow{d_{obs}}$ between the two groups c_1 and c_2 to the null deviation $\overrightarrow{0}$. The purpose of the compatibility region is to define a set of deviations that can be said to be "compatible" with the observed deviation $\overrightarrow{d_{obs}}$.

To make the construction of the compatibility region operational, we proceed in three steps. Let us first of all briefly present the three steps before going into detail about them.

Step A. *Reference cloud defined on the nesting $I{<}C{>}$ of type n_C.*

This step consists in constructing a *reference cloud* defined on $I{<}C{>}$, whose within-C covariance endomorphism is equal to that of cloud $M^{I{<}C{>}}$ and whose deviation of interest is null.

Step B. *Cloud defined on the nesting $I{<}C{>}$ associated with a vector $\overrightarrow{e}$.*

Let $\overrightarrow{e}$ be a vector in $\mathcal{L}$. A *cloud* on nesting $I{<}C{>}$ is deduced from the reference cloud by means of translations of the two subclouds of interest such that the deviation of interest is equal to $\overrightarrow{d_{obs}} - \overrightarrow{e}$.

Step C. *Compatibility region.*

We apply the test of homogeneity for the partial comparison of groups c_1 and c_2 to the cloud defined in the previous step. Thus, the deviation $\overrightarrow{d}_{obs} - \overrightarrow{e}$ is compared to the null deviation (see §5.4.1, p. 116), hence the following definition of the compatibility region.

Definition 5.1. *Given an α-level, a deviation $\overrightarrow{e}$ is compatible with $\overrightarrow{d}_{obs}$ at level α if, for cloud $\mathrm{E}^{I<C>}$, the p-value of the test of homogeneity of the groups c1 and c2 is greater than α.*

Definition 5.2. *The $(1 - \alpha)$ compatibility region is the set of deviations $\overrightarrow{e} \in \mathcal{L}$ compatible with $\overrightarrow{d}_{obs}$ at level α.*

We will now detail the steps of the construction of the compatibility region.

Step A. Reference Cloud

As mentioned above, the reference cloud serves as a basis for constructing the compatibility region, so we will study it in detail.

It is denoted $\mathrm{R}^{I<C>}$ and must meet the following properties:
1. the mean point of the reference cloud is point G;
2. the deviation between the mean points of the two subclouds $\mathrm{R}^{I<c_1>}$ and $\mathrm{R}^{I<c_2>}$ is null;
3. the covariance endomorphisms of subclouds $(\mathrm{R}^{I<c>})_{c \in C}$ are equal to those of subclouds $(\mathrm{M}^{I<c>})_{c \in C}$.

Recall that the mean point of subcloud $\mathrm{M}^{I<c_1>}$ is denoted G^{c_1} and weighted by n_{c_1}, that of subcloud $\mathrm{M}^{I<c_2>}$ is denoted G^{c_2} and weighted by n_{c_2}. The weighted mean point of the two points G^{c_1} and G^{c_2} is denoted G' and weighted by $n' = n_{c_1} + n_{c_2}$ (sum of the weights of the two mean points). Let us denote $I<c_r> = \{i \notin I<c_1, c_2>\}$ and n_{c_r} its cardinality ($n_{c_r} = n - n'$).

Definition 5.3. *The reference cloud $\mathrm{R}^{I<C>}$ associated with cloud $\mathrm{M}^{I<C>}$ is defined by:*

$$\forall i \in I<c_1>, \ \mathrm{R}^i = \mathrm{M}^i + \overrightarrow{\mathrm{G}^{c_1}\mathrm{G}'}$$
$$\forall i \in I<c_2>, \ \mathrm{R}^i = \mathrm{M}^i + \overrightarrow{\mathrm{G}^{c_2}\mathrm{G}'}$$
$$\forall i \in I<c_r>, \ \mathrm{R}^i = \mathrm{M}^i$$

Since $\overrightarrow{\mathrm{G}'\mathrm{G}^{c_1}} = -n_{c_2}\overrightarrow{d}_{obs}/n'$ and $\overrightarrow{\mathrm{G}'\mathrm{G}^{c_2}} = n_{c_1}\overrightarrow{d}_{obs}/n'$ (Chasles's identity, p. 237), the following relations hold:

$$\forall i \in I<c_1>, \ \mathrm{R}^i = \mathrm{M}^i + \frac{n_{c_2}}{n'}\overrightarrow{d}_{obs}$$
$$\forall i \in I<c_2>, \ \mathrm{R}^i = \mathrm{M}^i - \frac{n_{c_1}}{n'}\overrightarrow{d}_{obs}$$
$$\forall i \in I<c_r>, \ \mathrm{R}^i = \mathrm{M}^i$$

Geometrically, the subcloud $R^{I<c_1>}$ is the translation of the subcloud $M^{I<c_1>}$ by vector $\overrightarrow{G^{c_1}G'}$ and the subcloud $R^{I<c_2>}$ is the translation of the subcloud $M^{I<c_2>}$ by vector $\overrightarrow{G^{c_2}G'}$ (see Figure 5.8). Both have the same mean point G', therefore the effect of interest $R^{c_2} - R^{c_1}$ is null. The subcloud $R^{I<c_r>}$ is $M^{I<c_r>}$. Therefore, the mean point of the reference cloud R^I is G.

Figure 5.8
Target example. On the left, cloud $M^{I<C>}$ with its mean point G ($\circledast$), subclouds $M^{I<c_1>}$ (grey points) and $M^{I<c_2>}$ (black points) with the mean point G' ($\odot$) and $M^{I<c_r>}$ (empty circles). On the right, reference cloud $R^{I<C>}$ with its mean point G and point G'.

By construction (translated subclouds), the covariance endomorphisms of subclouds $R^{I<c_1>}$, $R^{I<c_2>}$ and $R^{I<c_r>}$ are equal to those of subclouds $M^{I<c_1>}$, $M^{I<c_2>}$ and $M^{I<c_r>}$. Therefore, the within-C covariance endomorphism of the reference cloud (weighted mean of the covariance endomorphisms of subclouds, Equation 2.12, p. 27), is equal to that of cloud $M^{I<C>}$.

$$Rcov_{I(C)} = Mcov_{I(C)}$$

Proposition 5.6. *The covariance endomorphism Rcov of the reference cloud $R^{I<C>}$ satisfies the following property:*

$$\forall \overrightarrow{u} \in \mathcal{L}, \ Rcov(\overrightarrow{u}) = Rcov_{I(C)}(\overrightarrow{u}) + \frac{n'}{n-n'}\langle\overrightarrow{GG'}|\overrightarrow{u}\rangle\overrightarrow{GG'}$$

Proof. The between-C cloud R^C consists of the two weighted points (G', n') and $(G^{c_r}, (n - n'))$. Then, applying Formula 2.3 (p. 16), its covariance endomorphism is such that $\forall \overrightarrow{u} \in \mathcal{L}, \ Rcov_C(\overrightarrow{u}) = \frac{n'(n-n')}{n^2}\langle\overrightarrow{G'G^{c_r}}|\overrightarrow{u}\rangle\overrightarrow{G'G^r}$. By replacing $\overrightarrow{G'G^{c_r}}$ by $-\frac{n}{n-n'}\overrightarrow{GG'}$, we obtain $Rcov_C(\overrightarrow{u}) = \frac{n'}{n-n'}\langle\overrightarrow{GG'}|\overrightarrow{u}\rangle\overrightarrow{GG'}$. From Theorem 2.3 (p. 28), $Rcov = Rcov_{I(C)} + Rcov_C$, hence the property. $\square$

Remark. In the particular case $C' = C = 2$, the cloud $R^{I<C>}$ is nothing else than the within-C cloud $M^{I(C)}$.

Cloud of deviation-points D_R^J

Consider now the nesting space $\mathcal{J}_R = (R^{I<C>_j})_{j \in J}$. Given a nesting j, we are interested in the deviation $\overrightarrow{d_R^j} = R^{c_2j} - R^{c_1j}$ between the mean points of the two subclouds $R^{I<c_1>_j}$ and $R^{I<c_2>_j}$.

Following the geometric approach (see p. 118), let O be the point that represents the null deviation, the cloud of the deviation-points $D_R^J = (D_R^j)_{j \in J}$ is defined by:

$$\forall j \in J, \, D_R^j = O + \overrightarrow{d}_R^j$$

The mean point of cloud D_R^J is O and, by Proposition 5.5 (p. 119), its covariance endomorphism is equal to $\frac{1}{n-1} \times \frac{nn'}{n_{c_1} n_{c_2}} \times \text{R}cov$.

 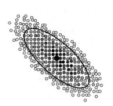

Figure 5.9
Target example. On the left, the reference cloud $R^{I<C>}$ with its mean point G (⊛) and its concentration ellipse. On the right, the deviation cloud D_R^J with its mean point O (●) and its concentration ellipse (the grey levels of points depends on the overlapping).

As we will see later on, all deviations may be expressed in terms of deviations $(\overrightarrow{d}_R^j)_{j \in J}$. In particular, Proposition 5.9 (p. 126) plays an important role in the construction of the compatibility region since it permits to delimit the compatibility interval for each line of the space.

Mahalanobis distance related to the reference cloud

The reference cloud and its structure of covariance are fundamental to determine the compatibility zone. That is why clouds will be studied in the subspace $\mathcal{L}$ equipped with the scalar product induced by the covariance endomorphism of the reference cloud[7].

This scalar product is called R-scalar product and denoted $[\cdot|\cdot]_R$; the associated norm is denoted $|\cdot|_R$ (see definitions in Chapter 2, p. 20); they are such that:

$$\forall \overrightarrow{u}, \overrightarrow{v} \in \mathcal{L}, \, [\overrightarrow{u}|\overrightarrow{v}]_R = \langle \overrightarrow{u}|\text{R}cov^{-1}(\overrightarrow{v}) \rangle \tag{5.1}$$

$$|\overrightarrow{u}|_R^2 = \langle \overrightarrow{u}|\text{R}cov^{-1}(\overrightarrow{u}) \rangle = [\overrightarrow{u}|\overrightarrow{u}]_R \tag{5.2}$$

Step B. Cloud $E^{I<C>}$ associated with $\overrightarrow{e}$

Starting from the reference cloud $R^{I<C>}$ and given $\overrightarrow{e} \in \mathcal{L}$, we construct the cloud $E^{I<C>}$ whose within-C covariance structure is the same as that of cloud $R^{I<C>}$ (hence of that of $M^{I<C>}$) and whose deviation of interest is $\overrightarrow{d}_{obs} - \overrightarrow{e}$.

---

[7] We suppose that the affine support of the reference cloud is $\mathcal{L}$, therefore its covariance endomorphism Rcov is invertible. If not, it is advisable to study in further detail the structure of the initial cloud as well as the group factor under study.

The cloud $E^{I<C>}$ is such that:

$$\forall i \in I<c_1>,\ E^i = R^i - \frac{n_{c_2}}{n'}\left(\overrightarrow{d}_{obs} - \overrightarrow{e}\right)$$

$$\forall i \in I<c_2>,\ E^i = R^i + \frac{n_{c_1}}{n'}\left(\overrightarrow{d}_{obs} - \overrightarrow{e}\right)$$

$$\forall i \in I<c_r>,\ E^i = R^i$$

The mean point E^{c_1} of subcloud $E^{I<c_1>}$ is equal to $G' - \frac{n_{c_2}}{n'}\left(\overrightarrow{d}_{obs} - \overrightarrow{e}\right)$ and that of the subcloud $E^{I<c_2>}$ is $E^{c_2} = G' + \frac{n_{c_1}}{n'}\left(\overrightarrow{d}_{obs} - \overrightarrow{e}\right)$. The deviation $E^{c_2} - E^{c_1}$ is denoted $\overrightarrow{d}_E$ and equal to $\overrightarrow{GG'} + \frac{n_{c_1}}{n'}\overrightarrow{d}_E - (\overrightarrow{GG'} - \frac{n_{c_2}}{n'}\overrightarrow{d}_E) = \overrightarrow{d}_{obs} - \overrightarrow{e}$.

Geometrically, the subcloud $E^{I<c_1>}$ is the translation of the subcloud $R^{I<c_1>}$ by vector $-\frac{n_{c_2}}{n'}\overrightarrow{d}_E$ and the subcloud $E^{I<c_2>}$ is the translation of the subcloud $R^{I<c_2>}$ by vector $\frac{n_{c_1}}{n'}\overrightarrow{d}_E$ (Figure 5.10).

Figure 5.10

Target example. On the left, vectors $\overrightarrow{e}$, $\overrightarrow{d}_{obs}$ and $\overrightarrow{d}_E$. In the middle, the two subclouds $R^{I<c_1>}$ (grey points) and $R^{I<c_2>}$ (black points) with their common mean point G'. On the right, the two subclouds $E^{I<c_1>}$ (grey points) and $E^{I<c_2>}$ (black points) with their mean points E^{c_1} and E^{c_2}.

Equivalently, the subcloud $E^{I<c_1>}$ is the translation of subcloud $M^{I<c_1>}$ by vector $(n_{c_2}/n')\overrightarrow{e}$ and the subcloud $E^{I<c_2>}$ is the translation of subcloud $M^{I<c_2>}$ by vector $-(n_{c_1}/n')\overrightarrow{e}$ (see Figure 5.11).

Figure 5.11

Target example. On the left, subclouds $M^{I<c_1>}$ (grey points) and $M^{I<c_2>}$ (black points) with their mean points G^{c_1} and G^{c_2}. On the right, translated subclouds $E^{I<c_1>}$ and $E^{I<c_2>}$ with their mean points E^{c_1} and E^{c_2} and the deviation vectors $\overrightarrow{d}_{obs} = G^{c_2} - G^{c_1}$ and $\overrightarrow{d}_E = E^{c_2} - E^{c_1}$.

Lemma 5.2. *The within-C covariance endomorphism of cloud* $E^{I<C>}$, *denoted* $\mathrm{Ecov}_{I(C)}$, *is equal to that of cloud* $R^{I<C>}$.

$$\mathrm{Ecov}_{I(C)} = \mathrm{Rcov}_{I(C)}$$

Proof. Since subclouds $E^{I<c>}$ and $R^{I<c>}$ can be deduced from each other thanks to a translation, their covariance endomorphisms are equal. Hence the within-C covariance endomorphisms of clouds $E^{I<C>}$ and $R^{I<C>}$ (weighted average of the covariance endomorphisms of subclouds) are equal. $\square$

Proposition 5.7. *The covariance endomorphism of cloud* $E^{I<C>}$, *denoted* Ecov, *is such that:*

$$\forall \vec{u} \in \mathcal{L}, \ \text{Ecov}(\vec{u}) = \text{Rcov}(\vec{u}) + \tfrac{n_{c_1} n_{c_2}}{nn'} \langle \vec{d_E} | \vec{u} \rangle \vec{d_E}$$

Proof. By Theorem 2.3 (p. 28): $\text{Ecov} = \text{Ecov}_C + \text{Ecov}_{I(C)}$.

— By the preceding lemma and Proposition 5.6 (p. 122), we deduce that $\forall \vec{u} \in \mathcal{L}$, $\text{Ecov}_{I(C)}(\vec{u}) = \text{Rcov}_{I(C)}(\vec{u}) = \text{Rcov}(\vec{u}) - \tfrac{n'}{n-n'} \langle \overrightarrow{GG'} | \vec{u} \rangle \overrightarrow{GG'}$.

— $\text{Ecov}_C(\vec{u}) = \sum\limits_{c \in C} \tfrac{n_c}{n} \langle \overrightarrow{GE^c} | \vec{u} \rangle \overrightarrow{GE^c} = \tfrac{n'}{n} \sum\limits_{c \in C'} \tfrac{n_c}{n'} \langle \overrightarrow{GE^c} | \vec{u} \rangle \overrightarrow{GE^c} + \tfrac{n-n'}{n} \langle \overrightarrow{GG^r} | \vec{u} \rangle \overrightarrow{GG^r}$.

The mean point of (E^{c_1}, n_{c_1}) and (E^{c_2}, n_{c_1}) being G', we deduce from Proposition 2.5 (p. 16): $\sum\limits_{c \in C'} \tfrac{n_c}{n'} \langle \overrightarrow{GE^c} | \vec{u} \rangle \overrightarrow{GE^c} = \langle \overrightarrow{GG'} | \vec{u} \rangle \overrightarrow{GG'} + \tfrac{n_{c_1} n_{c_2}}{n'^2} \langle \overrightarrow{E^{c_1} E^{c_2}} | \vec{u} \rangle \overrightarrow{E^{c_1} E^{c_2}}$.

Yet $\overrightarrow{GG^r} = -\tfrac{n'}{n-n'} \overrightarrow{GG'}$, then $\text{Ecov}_C(\vec{u}) = \tfrac{n}{n'} \left(\langle \overrightarrow{GG'} | \vec{u} \rangle \overrightarrow{GG'} + \tfrac{n_{c_1} n_{c_2}}{n'^2} \langle \vec{d_E} | \vec{u} \rangle \vec{d_E} \right) + \tfrac{n-n'}{n} \left(\tfrac{n'}{n-n'} \right)^2 \langle \overrightarrow{GG'} | \vec{u} \rangle \overrightarrow{GG'} = \tfrac{n_{c_1} n_{c_2}}{n'^2} \langle \vec{d_E} | \vec{u} \rangle \vec{d_E} + \tfrac{n'}{n-n'} \langle \overrightarrow{GG'} | \vec{u} \rangle \overrightarrow{GG'}$.

— $\text{Ecov}(\vec{u}) = \text{Rcov}(\vec{u}) - \tfrac{n'}{n-n'} \langle \overrightarrow{GG'} | \vec{u} \rangle \overrightarrow{GG'} + \tfrac{n_{c_1} n_{c_2}}{n'^2} \langle \vec{d_E} | \vec{u} \rangle \vec{d_E} + \tfrac{n'}{n-n'} \langle \overrightarrow{GG'} | \vec{u} \rangle \overrightarrow{GG'}$, and the proposition. $\square$

Test statistic and level of homogeneity of the two groups

Let us denote $|\cdot|_E$ the Mahalanobis norm attached to cloud E^I that is defined by $\forall \vec{u} \in \mathcal{L}$, $|\vec{u}|_E^2 = \langle \text{Ecov}^{-1}(\vec{u}) | \vec{u} \rangle$. Consider the nesting space $\mathcal{J}_E = (E^{I<C>_j})_{j \in J}$ and the deviations $(\vec{d_E^j} = E^{c_2 j} - E^{c_1 j})_{j \in J}$. The *test statistic*, denoted D_E^2, is the map that associates the squared E-norm of the deviation $\vec{d_E^j}$ with each element of the nesting space $\mathcal{J}_E$:

$$D_E^2 : \mathcal{J}_E \rightarrow \mathbb{R}_{\geq 0}$$
$$E^{I<C>_j} \mapsto |\vec{d_E^j}|_E^2$$

The observed deviation is $\vec{d_E} = \vec{d}_{obs} - \vec{e}$, hence the observed value of D_E^2 is $|\vec{d_E}|_E^2 = |\vec{d}_{obs} - \vec{e}|_E^2$. The *level of homogeneity of the two groups* c_1 and c_2 (*p*-value) is the proportion of nestings $j \in J$ for which the value of the test statistic is greater than or equal to the squared E-norm of the deviation $\vec{d_E}$, that is, $p(D_E^2 \geq |\vec{d_E}|_E^2)$.

Relationship between nesting spaces $\mathcal{J}_E$ and $\mathcal{J}_R$

We now study the relationship between the nesting space $\mathcal{J}_E$ associated with cloud $E^{I<C>}$ and the nesting space $\mathcal{J}_R$ associated with the reference cloud $R^{I<C>}$. Firstly, we express the E-norm as a function of the R-norm. Secondly, we show that the deviation $\overrightarrow{d_{\mathrm{E}}^{j}}$ can be written as a function of the deviation $\overrightarrow{d_{\mathrm{R}}^{j}}$ associated with the reference cloud.

Proposition 5.8. *Letting $\gamma = n_{c_1} n_{c_2}/(nn')$, we have:*

$$\forall \overrightarrow{u} \in \mathcal{L},\ |\overrightarrow{u}|_{\mathrm{E}}^{2} = |\overrightarrow{u}|_{\mathrm{R}}^{2} - \frac{\gamma[\overrightarrow{u}\,|\overrightarrow{d_{\mathrm{E}}}]_{\mathrm{R}}^{2}}{1 + \gamma|\overrightarrow{d_{\mathrm{E}}}|_{\mathrm{R}}^{2}} \tag{5.3}$$

$$|\overrightarrow{d_{\mathrm{E}}}|_{\mathrm{E}}^{2} = \frac{|\overrightarrow{d_{\mathrm{E}}}|_{\mathrm{R}}^{2}}{1 + \gamma|\overrightarrow{d_{\mathrm{E}}}|_{\mathrm{R}}^{2}} \tag{5.4}$$

Proof. E*cov* is the sum of the endomorphism R*cov* and of an endomorphism of rank one corresponding to $\overrightarrow{d_{\mathrm{E}}} = \overrightarrow{d_{obs}} - \overrightarrow{e}$ (Proposition 5.7, p. 125). By Corollary 7.1 (p. 234), we deduce the relations between the E-norm and the R-norm. $\quad\square$

Proposition 5.9 (Relationship between deviations). *Given a nesting j, let n_{11_j}, n_{21_j}, n_{12_j}, n_{22_j} respectively denote the cardinalities of the subsets $I{<}c_1{>}\cap I{<}c_1{>}_j$, $I{<}c_2{>}\cap I{<}c_1{>}_j$, $I{<}c_1{>}\cap I{<}c_2{>}_j$, $I{<}c_2{>}\cap I{<}c_2{>}_j$.*

	$I{<}c_1{>}$	$I{<}c_2{>}$
$I{<}c_1{>}_j$	n_{11_j}	n_{21_j}
$I{<}c_2{>}_j$	n_{12_j}	n_{22_j}

Letting $e^j = (n_{11_j}/n_{c_1}) + (n_{22_j}/n_{c_2}) - (n_{11_j} + n_{21_j} + n_{12_j} + n_{22_j})/n'$, the following relationship between $\overrightarrow{d_{\mathrm{E}}^{j}}$ and $\overrightarrow{d_{\mathrm{R}}^{j}}$ holds.

$$\forall j \in J,\ \overrightarrow{d_{\mathrm{E}}^{j}} = \overrightarrow{d_{\mathrm{R}}^{j}} + e^j \overrightarrow{d_{\mathrm{E}}} \tag{5.5}$$

Proof. $\mathrm{E}^{c1j} = \sum\limits_{i \in I{<}c_1{>}_j} \mathrm{E}^i/n_{c_1} = \sum\limits_{i \in I{<}c_1{>}_j} \mathrm{R}^i/n_{c_1} - n_{11_j}\frac{n_{c_2}}{n'}\overrightarrow{d_{\mathrm{E}}}/n_{c1} + n_{21_j}\frac{n_{c_1}}{n'}\overrightarrow{d_{\mathrm{E}}}/n_{c1}$, thus

$\mathrm{E}^{c1j} = \mathrm{R}^{c1j} - \frac{1}{n'}(n_{11_j}\frac{n_{c_2}}{n_{c_1}} - n_{21_j})\overrightarrow{d_{\mathrm{E}}}$. Similarly, $\mathrm{E}^{c2j} = \mathrm{R}^{c2j} + \frac{1}{n'}(n_{22_j}\frac{n_{c_1}}{n_{c_2}} - n_{12_j})\overrightarrow{d_{\mathrm{E}}}$.

Hence, $\mathrm{E}^{c2j} - \mathrm{E}^{c1j} = \overrightarrow{d_{\mathrm{E}}^{j}} = \overrightarrow{d_{\mathrm{R}}^{j}} + \frac{1}{n'}\left(n_{22_j}(\frac{n'}{n_{c_2}} - 1) - n_{12_j} + n_{11_j}(\frac{n'}{n_{c_1}} - 1) - n_{21_j}\right)\overrightarrow{d_{\mathrm{E}}}$, and the property. $\quad\square$

Step C. Characterization of the compatibility region

The approach we develop in this chapter is similar to that we outlined for the geometric typicality test in Chapter 4 (see §4.2.4, p. 80).

Given $\overrightarrow{\beta} \in \mathcal{L}$, we consider the set of vectors $\overrightarrow{e}$ such that $\overrightarrow{d_{obs}} - \overrightarrow{e}$ is proportional to $\overrightarrow{\beta}$ and among these vectors those that are compatible with $\overrightarrow{d_{obs}}$. Then, we define an "adjusted" compatibility region in the space.

Geometrically, the set of points $D_{\mathrm{E}} = O + \overrightarrow{e}$ such that $\overrightarrow{d_{obs}} - \overrightarrow{e}$ is proportional to $\overrightarrow{\beta}$ defines a line that passes through point $D_{obs} = O + \overrightarrow{d_{obs}}$ and whose direction vector is $\overrightarrow{\beta}$. The set of points D_{E} of this line, for which $\overrightarrow{OD_{\mathrm{E}}}$ is compatible with $\overrightarrow{d_{obs}}$, defines a "compatibility interval" of this line.

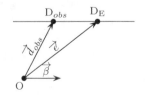

Recall (see Definition 5.1, p. 121) that $\overrightarrow{e}$ is said to be compatible with $\overrightarrow{d}_{obs}$ at level α if $p(D_E^2 \geq |\overrightarrow{d}_{obs} - \overrightarrow{e}|_E^2) > \alpha$.

Since the E-norm can be expressed in terms of the R-norm, we now *equip the space $\mathcal{L}$ with the scalar product related to the reference cloud.*

Compatibility region linked to a vector $\overrightarrow{\beta}$

Let $\mathcal{B}$ be the one-dimensional subspace of $\mathcal{L}$ spanned by the R-unit vector $\overrightarrow{\beta}$. We aim to define the vectors $\overrightarrow{e} = \overrightarrow{d}_{obs} - x\overrightarrow{\beta}$ that are compatible with $\overrightarrow{d}_{obs}$. In other words, we search an *interval of variation* for x.

To this end, we consider the cloud $E^{I<C>}$ associated with vector $\overrightarrow{e}$, the nesting space $\mathcal{J}_E = (E^{I<C>_j})_{j \in J}$ and the family of deviation vectors $(\overrightarrow{d}_E^j)_{j \in J}$. After Proposition 5.9 (p. 126), $\overrightarrow{d}_E^j = \overrightarrow{d}_R^j + \epsilon^j x\overrightarrow{\beta}$ $(j \in J)$. Thus, given j, vectors $\overrightarrow{d}_E^j$, $\overrightarrow{d}_R^j$ and $\overrightarrow{\beta}$ are in a two-dimensional subspace of $\mathcal{L}$ (see Figure 5.12).

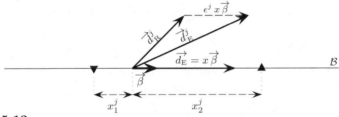

Figure 5.12
Subspace of $\mathcal{L}$ spanned by $\overrightarrow{\beta}$ and $\overrightarrow{d}_R^j$: deviation vectors $\overrightarrow{d}_E^j = x\overrightarrow{\beta}$ and $\overrightarrow{d}_E^j = \overrightarrow{d}_R^j + \epsilon^j x\overrightarrow{\beta}$ associated with nesting j.

Lemma 5.3. *Given an R-unit vector $\overrightarrow{\beta} \in \mathcal{L}$, for each nesting $j \in J$, there exist two numbers $x_1^j < 0$ and $x_2^j > 0$ such that,*

$$\forall x \in [x_1^j; x_2^j] \text{ and } \overrightarrow{e} = \overrightarrow{d}_{obs} - x\overrightarrow{\beta}, \ |\overrightarrow{d}_E^j|_E^2 - |\overrightarrow{d}_{obs} - \overrightarrow{e}|_E^2 \geq 0$$

Proof. Given $j \in J$, we search the values $x \in \mathbb{R}$ such that $|\overrightarrow{d}_E^j|_E^2 - |x\overrightarrow{\beta}|_E^2 \geq 0$. Let $u^j\overrightarrow{\beta} + v^j\overrightarrow{\beta'}$ be the R-orthogonal decomposition of $\overrightarrow{d}_R^j$ onto $\mathcal{B}$ and $\mathcal{B}^\perp$, with $|\overrightarrow{\beta}|_R = |\overrightarrow{\beta'}|_R = 1$ and $[\overrightarrow{\beta}|\overrightarrow{\beta'}]_R = 0$. Hence, we have $\overrightarrow{d}_E^j = u^j\overrightarrow{\beta} + v^j\overrightarrow{\beta'} + \epsilon^j x\overrightarrow{\beta} = (u^j + \epsilon^j x)\overrightarrow{\beta} + v^j\overrightarrow{\beta'}$.

For the sake of simplicity, we will now drop subscript j for u^j, v^j and ϵ^j.

We have $|\overrightarrow{d}_E^j|_E^2 = (u + \epsilon x)^2 + v^2$; $[\overrightarrow{d}_E|x\overrightarrow{\beta}]_R = x(u + \epsilon x)$ and $|\overrightarrow{d}_E^j|_R^2 = |x\overrightarrow{\beta}|_R^2 = x^2$. Applying Formulas 5.3 and 5.4 (p. 126), we obtain:
$|\overrightarrow{d}_E^j|_E^2 = (u + \epsilon x)^2 + v^2 - \frac{\gamma x^2(u + \epsilon x)^2}{1 + \gamma x^2} = \frac{(u + \epsilon x)^2}{1 + \gamma x^2} + v^2$ and $|\overrightarrow{d}_E^j|_E^2 = \frac{|\overrightarrow{d}_E|_R^2}{1 + \gamma|\overrightarrow{d}_E|_R^2} = \frac{x^2}{1 + \gamma x^2}$.
Hence, letting $d^2 = u^2 + v^2$, we have $|\overrightarrow{d}_E^j|_E^2 - |\overrightarrow{d}_E^j|_E^2 = \frac{(\epsilon^2 + \gamma v^2 - 1)x^2 + 2\epsilon ux + d^2}{1 + \gamma x^2}$.
We look for vectors $x\overrightarrow{\beta}$ such that $|\overrightarrow{d}_E^j|_E^2 - |\overrightarrow{d}_E^j|_E^2 \geq 0$, or equivalently, we look for x verifying $(\epsilon^2 + \gamma v^2 - 1)x^2 + 2u\epsilon x + d^2 \geq 0$.

The two roots of the quadratic function $(\epsilon^2 + \gamma v^2 - 1)x^2 + 2u\epsilon x + d^2$ are:

$$x_1 = \frac{-\epsilon u + \sqrt{\epsilon^2 u^2 - d^2(\epsilon^2 + \gamma v^2 - 1)}}{\epsilon^2 + \gamma v^2 - 1} \quad \text{and} \quad x_2 = \frac{-\epsilon u - \sqrt{\epsilon^2 u^2 + d^2(\epsilon^2 + \gamma v^2 - 1)}}{\epsilon^2 + \gamma v^2 - 1}$$

with $x_1 < 0$ and $x_2 > 0$ since the coefficient of x^2 is negative. The function is positive for all x with $x_1 \leq x \leq x_2$, hence $|\overrightarrow{d_E}^j|_E^2 - |\overrightarrow{d_E}|_E^2 \geq 0$ for all vectors $\overrightarrow{e} = \overrightarrow{d}_{obs} - x\overrightarrow{\beta}$.

Particular cases. If $\overrightarrow{d}_R^j = \overrightarrow{0}$, then $\overrightarrow{d}_E^j = \epsilon^j x \overrightarrow{\beta}$ and $|\overrightarrow{d}_E^j|_E^2 - |\overrightarrow{d_E}|_E^2 = \frac{(1-(\epsilon^j)^2)x^2}{1+\gamma x^2} \geq 0$ for all x if $(\epsilon^j)^2 = 1$ and for $x = 0$ if $(\epsilon^j)^2 \neq 1$. Given $j \in J$, if $\epsilon^2 + \gamma v^2 - 1 = 0$ and $d^2 \neq 0$, then $|\overrightarrow{d}_E^j|_E^2 - |\overrightarrow{d_E}|_E^2 = \frac{2\epsilon u x + d^2}{1+\gamma x^2}$. This expression is positive or null for $x > -d^2/(2\epsilon u)$ if $\epsilon u > 0$, for $x < d^2/(2\epsilon u)$ if $\epsilon u < 0$ and for all $x \in \mathbb{R}$ if $\epsilon u = 0$. $\qquad\square$

The following proposition is an immediate consequence of Lemma 5.3.

Proposition 5.10. *Let $\overrightarrow{e} = \overrightarrow{d}_{obs} - x\overrightarrow{\beta}$, the function $x \mapsto p(D_E^2 - \frac{x^2}{1+\gamma x^2} \geq 0)$ is a decreasing step function for $x > 0$ and increasing step function for $x < 0$.*

Given a nesting j and an R-unit vector $\overrightarrow{\beta}$ in $\mathcal{L}$, let $\overrightarrow{e}_1^j = x_1^j \overrightarrow{\beta}$ and $\overrightarrow{e}_2^j = x_2^j \overrightarrow{\beta}$ be the two vectors associated with the values x_1^j and x_2^j defined in Lemma 5.3. The preceding proposition yields the following theorem.

Theorem 5.2. *Given an α-level and an R-unit vector $\overrightarrow{\beta} \in \mathcal{L}$, let us denote $\tilde{x}_1$ (resp. $\tilde{x}_2$) the smaller value $(x_1^j)_{j \in J}$ (resp. the greater value $(x_2^j)_{j \in J}$) such that the proportion of $j \in J$ verifying $x_1^j \leq \tilde{x}_1$ (resp. $x_2^j \geq \tilde{x}_2$) is superior to α. The two points $D_1 = O + (\overrightarrow{d}_{obs} - \tilde{x}_1\overrightarrow{\beta})$ and $D_2 = O + (\overrightarrow{d}_{obs} - \tilde{x}_2\overrightarrow{\beta})$ are the limits of compatibility at level α of the line $(D_{obs}, \overrightarrow{\beta})$.*

Adjusted compatibility region

We deduce from Theorem 5.2 that an interval, which is the set of endpoints of vectors compatible with $\overrightarrow{d}_{obs}$, is associated with each one-dimensional subspace of the space of deviations. But, in the L-dimensional space, we cannot give an analytical form to the compatibility region since this region depends on the distribution of deviation-points in the space. However, it can be shown that an adjusted compatibility region can be defined as the set of deviation-vectors whose endpoints are on or inside a principal ellipsoid of the reference cloud centred at point $D_{obs} = O + \overrightarrow{d}_{obs}$.

The following proposition results from Equation 5.4 (p. 126).

Proposition 5.11. *Let $\overrightarrow{e} \in \mathcal{L}$ and $\overrightarrow{e}' \in \mathcal{L}$ be two vectors such that the R-norms of $\overrightarrow{d}_{obs} - \overrightarrow{e}$ and $\overrightarrow{d}_{obs} - \overrightarrow{e}'$ are both equal to κ, then the squared R-norms of these two vectors are both equal to $\kappa^2/(1 + \gamma\kappa^2)$, with $\gamma = n_{c_1}n_{c_2}/(nn')$.*

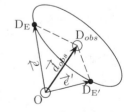

Geometrically, in the space of deviations, the endpoints D_E and $D_{E'}$ of vectors $\overrightarrow{e}$ and $\overrightarrow{e}'$ are on the principal κ-ellipsoid of the reference cloud centred at point D_{obs}.

Consider the cloud E^I defined from a vector $\vec{e} \in \mathcal{L}$ and its R-covariance endomorphism that is expressed in terms of the R-scalar product and is denoted R.E$cov(\vec{u})$ with R.E$cov(\vec{u}) = \sum \frac{1}{n} [\overrightarrow{GE^i} | \vec{u}]_R \overrightarrow{GE^i}$ ($\vec{u} \in \mathcal{L}$) (see §2.3.1, Formula 2.8, p. 21). Since the R-covariance endomorphism of cloud R^I is equal to id_L (Proposition 2.13, p. 21), then R.E$cov(\vec{u}) = \vec{u} + \gamma [\overrightarrow{d_E} | \vec{u}]_R \overrightarrow{d_E}$ (Proposition 5.7, p. 125). From Theorem 7.4 (p. 234), we deduce that $\overrightarrow{d_E}$ is eigenvector of R.Ecov associated with eigenvalue $1 + \gamma |\overrightarrow{d_E}|_R^2$ and that the subspace of $\mathcal{L}$ that is R-orthogonal to $\overrightarrow{d_E}$ is eigenspace of R.Ecov associated with eigenvalue 1 (with multiplicity $L - 1$).

Consider the cloud E'^I defined from a vector $\vec{e}' \in \mathcal{L}$ with $|\overrightarrow{d_{E'}}|_R = \kappa$. The R-covariance endomorphisms of clouds E^I and E'^I have the same eigenvalues, namely 1 with multiplicity $L - 1$ and $1 + \gamma \kappa^2$, therefore, they can be deduced from each other thanks to a plane rotation that maps $\overrightarrow{d_E}$ onto $\overrightarrow{d_{E'}}$.

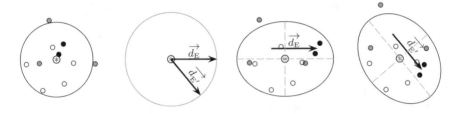

Figure 5.13
Space equipped with the R-norm: (1) reference cloud with its mean point (⊛) and its concentration ellipse (circle with radius 2); (2) two vectors $\overrightarrow{d_E}$ and $\overrightarrow{d_{E'}}$ with the same R-norm ($\kappa = 2.5$); (3) cloud E^I with its concentration ellipse and deviation vector $\overrightarrow{d_E} = E^{c2} - E^{c1}$; (4) cloud E'^I with its concentration ellipse and deviation vector $\overrightarrow{d_{E'}} = E'^{c2} - E'^{c1}$.

From Proposition 5.5 (p. 119), it can easily be proven that the R-covariance endomorphism of the cloud of deviation points $D_E^J = (O + \overrightarrow{d_E^j})_{j \in J}$ is proportional to that of cloud E^I. Similarly for the cloud of deviation points $D_{E'}^J = (O + \overrightarrow{d_{E'}^j})_{j \in J}$ associated with E'^I. Hence, the covariance endomorphisms of the two clouds can be deduced from each other thanks to a plane rotation that maps $\overrightarrow{d_E}$ onto $\overrightarrow{d_{E'}}$. If we fit an L-dimensional normal distribution to the cloud of deviation points, then the proportion of the distribution outside the κ-ellipsoid of cloud is given by $p(\chi_L^2 \geq \kappa^2)$ (see Chapter 2, p. 22). Therefore, if $|\overrightarrow{d_E}|_R = |\overrightarrow{d_{E'}}|_R = \kappa$, then $|\overrightarrow{d_E}|_E^2 = |\overrightarrow{d_{E'}}|_{E'}^2 = \kappa^2/(1 + \gamma \kappa^2)$ (Proposition 5.11, p. 128), and the two proportions $p(D_E^2 \geq |\overrightarrow{d_E}|_E^2)$ and $p(D_{E'}^2 \geq |\overrightarrow{d_{E'}}|_{E'}^2)$ are asymptotically equal. It can be said that the κ-compatibility region is approximately defined by the vectors $\vec{e}$ that are such that $|\overrightarrow{d}_{obs} - \vec{e}|_R^2 = \kappa^2$.

From a *technical standpoint*, we propose an *adjusted compatibility region* defined by the principal ellipsoid of cloud R^I with a scale parameter $\bar{\kappa}$ equal to the mean of the κ values obtained from many (say, 100) direction vectors randomly selected in $\mathcal{L}$.

Geometrically, the adjusted compatibility region is defined by the set of vectors $\vec{e} = \overrightarrow{OE}$ whose endpoints $E = O + \vec{e}$ are inside or on the principal $\overline{\kappa}$-ellipsoid of cloud R^I centred at D_{obs}.

Target example

In Figure 5.14, we present the 95% compatibility region determined from 180 vectors $\vec{\beta}$ (on the left) and that obtained by adjusting the set of the endpoints of compatible vectors by a $\overline{\kappa}$-ellipse (on the right): $\overline{\kappa} = 3.36$; it is the mean of the 360 values calculated for each direction vector whose minimum is 3.288 and maximum 3.445.

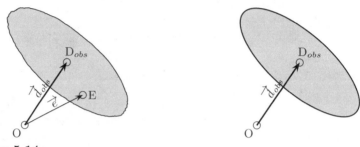

Figure 5.14

Target example. Space of deviations: on the left, 95% compatibility region defined from 180 lines, on the right, adjusted 95% compatibility region defined from the $\overline{\kappa}$-ellipse with $\overline{\kappa} = 3.36$.

Particular case: specific comparison of two groups

In the *specific comparison* of the two groups, the construction of the compatibility region is simpler since the reference cloud is nothing else than the within-groups cloud.

Monte Carlo method

The exact calculations can be performed up to $J = 1,000,000$. But in most cases, the cardinality of the nesting space exceeds this value, so we use the Monte Carlo method in order to approximate the p-value and determine the compatibility region by randomly drawing nestings among the $n!/(n_{c_1}! \, n_{c_2}! \, (n - n')!)$ ones. A useful strategy is to start with 10,000 random permutations and to continue with larger numbers only if p is small enough to make it worthwhile to pursue the analysis, e.g., $p < 0.10$.

We want to underline that simulations are carried out by means of *resampling from the nesting space* and it must be emphasized that the procedure is substantially different from the *bootstrap techniques*.

5.4.4 Approximate Test and Compatibility Region

If n_{c_1} and n_{c_2} are large enough, the cloud of deviation points can be fitted by an L-dimensional normal distribution centred at point O. Then, the distribution of the statistic $T^2 = (n-1)\gamma D_M^2$, with $\gamma = n_{c_1} n_{c_2}/(nn')$, can be approximated by the χ^2 distribution with L degrees of freedom. The observed value of the test statistic D_M^2 is d_{obs}^2, hence the approximate p-value is

$$\widetilde{p} = p\left(\chi_L^2 \geq (n-1)\gamma\, d_{obs}^2\right)$$

Given α and denoting $\chi_L^2[\alpha]$ the critical value of the chi-squared distribution with L degrees of freedom, the approximate $(1-\alpha)$ compatibility region for $\overrightarrow{d}_{obs}$ is defined by the set of vectors $\overrightarrow{e}$ such that $|\overrightarrow{d}_{obs} - \overrightarrow{e}|_R < \widetilde{\kappa}$ with $\widetilde{\kappa}^2 = \frac{1}{\gamma} \times \frac{\chi_L^2[\alpha]}{n-1}/\left(1 - \frac{\chi_L^2[\alpha]}{n-1}\right)$.

The approximate region holds if n is large enough (at least $n > 1 + \chi_L^2[\alpha]$).

Geometrically, the approximate compatibility region is the set of vectors $\overrightarrow{e}$ such that points $E = O + \overrightarrow{e}$ belong to the interior (or are on the boundary) of the principal $\widetilde{\kappa}$-ellipsoid of cloud R^I centred at D_{obs}.

5.4.5 One-dimensional Case

The foregoing test for the homogeneity of two subclouds can be applied as such to a one-dimensional cloud but it leads to a two-tailed test with a nondirectional conclusion. By choosing the *Difference of Means* as test statistic, we will be able to conclude by taking into account the sign of the difference.

If the cloud is one-dimensional ($L = 1$), a numerical variable $x^I = (x^i)_{i \in I}$ can be associated with the set I of the n individuals that are classified into C groups ($C \geq 2$) of cardinalities $(n_c)_{c \in C}$. The means of the C groups are $(x^c = \sum_{i \in I<c>} x^i/n_c)_{c \in C}$.

The homogeneity test presented here amounts to comparing the means of two groups of observations c_1 and c_2. Hence, we consider the partition of I into three classes $I<c_1>$, $I<c_2>$ and $I<c_r> = \{i \notin I<c1, c2>\}$ whose cardinalities are respectively equal to n_{c_1}, n_{c_2} and $n - n'$ (with $n' = n_{c_1} + n_{c_2}$).

Nesting space and test statistic

Recall that J is the set indexing the $(\sum n_c)!/(\prod n_c!)$ possible nestings of type $n_C = (n_{c_1}, n_{c_2}, n - n')$. Given a nesting $j \in J$, the three subsets $(I<c>_j)_{c \in C}$ determine a partition of I of type n_C, hence the nesting space is $\mathcal{J} = (x^{I<C>_j})_{j \in J}$. For $j \in J$, the three means are $(x^{cj} = \sum_{i \in I<c>_j} x^i/n_c)_{c \in C}$.

The test statistic, denoted D, is defined by

$$D: \quad \begin{array}{ccc} \mathcal{J} & \to & \mathbb{R} \\ x^{I<C>_j} & \mapsto & x^{c_2 j} - x^{c_1 j} \end{array}$$

The value of the test statistic D for nesting j is denoted by d^j and its observed value by d_{obs}.

The combinatorial distribution of statistic D is the map that assigns the proportion of nestings whose mean is equal to d^j to each d^j.

$$D(\mathcal{J}) \to [0,1]$$
$$d^j \mapsto p(D = d^j)$$

From Propositions 5.4 and 5.5 (p. 119), we deduce that

- the *mean of D* is equal to 0;

- letting $\gamma = \frac{n_{c_1} n_{c_2}}{nn'}$ and denoting v the variance of variable x^I, the *variance of D* is equal to $\frac{1}{n-1} v / \gamma$.

Combinatorial p-value

The observed difference d_{obs} is located in the distribution of D. For that, we determine the *upper homogeneity level*, denoted $\overline{p}$, which is equal to the proportion of nestings whose difference of means is greater than or equal to d_{obs}: $\overline{p} = p(D \geq d_{obs})$. The *lower homogeneity level* is defined accordingly and denoted $\underline{p}$, with $\underline{p} = p(D \leq d_{obs})$.

The *combinatorial p-value*, which is denoted p, is the smaller of these two proportions:

$$p = \min\{\overline{p}, \underline{p}\}$$

The *p*-value defines the level of homogeneity of the two groups for the *Mean*. The smaller the value of p, the lower the homogeneity.

For any α-level (see p. 31),
- if $p \leq \alpha/2$, the two groups of observations are said to be *heterogeneous* at level $\alpha/2$, on the side of higher values of means for group c_2 if $p = \overline{p}$ or for group c_1 if $p = \underline{p}$;
- if $p > \alpha/2$, we cannot conclude that the two groups are heterogeneous at level α, with respect to the difference of means.

Compatibility interval

The steps for constructing the compatibility interval are the following ones.

1. Given $u \in \mathbb{R}$, we define the variable $e^{I<C>}$:

$$\forall i \in I<c_1>, \quad e^i = x^i + n_{c_2} u/n'$$
$$\forall i \in I<c_2>, \quad e^i = x^i - n_{c_1} u/n'$$
$$\forall i \in I<c_r>, \quad e^i = x^i$$

The mean of the two means e^{c_1} and e^{c_2} is equal to that of x^{c_1} and x^{c_2}. Let e_{obs} denote the difference $e^{c_2} - e^{c_1}$, we have $e_{obs} = d_{obs} - u$.

2. The test statistic, denoted D_e, is defined by

$$D_e : e^{I<C>_j} \mapsto e^{c_2 j} - e^{c_1 j}$$

Given $j \in J$, the difference $e^{c_2 j} - e^{c_1 j}$, denoted e^j, writes $e^j = d^j - \epsilon^j u$ (particular case of Proposition 5.9, p. 126).

3. Letting $\overline{p}_e = p(D_e \geq e_{obs})$, mbox$\underline{p}_e = p(D_e \leq e_{obs})$ and $p_e = \min\{\underline{p}_e, \overline{p}_e\}$, the value u is compatible with d_{obs} at level α if and only if $p_e > \alpha/2$.

Theorem 5.3. *Consider the variable* $u^J = (d_{obs} - d^j)/(1 - \overline{\epsilon}^j)$. *Let us denote* u_m *(resp.* u_M*) the smaller (resp. the greater) value* $(u^j)_{j \in J}$ *such that the proportion of* $j \in J$ *verifying* $u^j \leq u_m$ *(resp.* $u^j \geq u_M$*) is superior to* $\alpha/2$. *The two values* u_m *and* u_M *define the* $(1 - \alpha)$ *compatibility interval.*

Proof. Given j, the property $(D_e \geq e_{obs})$ writes $e^j \geq e_{obs}$ or $d^j - \epsilon^j u \geq d_{obs} - u$, that is, $(d_{obs} - d^j)/(1 - \epsilon^j) \leq u$. Since the proportion of $j \in J$ verifying $u^j \leq u_m$ is superior to $\alpha/2$, one has $p(D_e \geq e_{obs})$ for all values $u \geq u_m$. Similarly $p(D_e \leq e_{obs}) > \alpha/2$ for all values $u \leq u_M$, hence the interval. $\square$

Approximate test and compatibility interval

If n_{c_1} and n_{c_2} are large enough, the distribution of the test statistic D can be fitted by a normal distribution with mean 0 and variance $\operatorname{Var} D = \frac{1}{n-1} v/\gamma$. Thus, statistic $Z = D/\sqrt{\operatorname{Var} D}$ has a standard normal distribution (note that statistics D and Z are permutationally equivalent).

Let z be the variable normally distributed with mean 0 and variance 1, which we write $z \sim \mathcal{N}(0, 1)$. If $z_{obs} = d_{obs}/\sqrt{v/(\gamma(n-1))}$ denotes the observed value of the statistics Z, the approximate p-value is

$$\widetilde{p} = p(z \geq |z_{obs}|)$$

The approximate $(1 - \alpha)$ compatibility interval is the set of values $d \in \mathbb{R}$ that are compatible with 0 at level α. If z_α denotes the critical value of the standard normal variable at level α (recall that $z_\alpha = 1.96$ for $\alpha = 0.05$) , it is such that $|d|/\sqrt{\operatorname{Var} D} < z_\alpha$, or, equivalently d belongs to the interval

$$] - z_\alpha\sqrt{\operatorname{Var} D} \, ; \, z_\alpha\sqrt{\operatorname{Var} D}[$$

5.5 The Case of a Repeated Measures Design

In this section, we briefly present the permutation homogeneity test in a repeated measures (within-subject) design.

The particular case of two repeated measures is of particular interest and is frequently encountered in real life case studies. We will show that, in this particular case, the homogeneity test is equivalent to the geometric typicality test of comparison of the mean point of effect-points to point O (null effect).

5.5.1   Examples of Repeated Measures Design

We begin by describing two situations for which the homogeneity problem can be raised, then we present a hypothetical data set on which we carry out the various steps of the test.

Time reaction experiment

The data are borrowed from a simple reaction-time experiment (see Rouanet and Lépine, 1970). Each of the nine subjects was run under four conditions. The observations consist of the mean reaction time associated with each subject-condition. The question is:

Is there a difference between conditions?

Testosterone

The level of blood testosterone in 11 women is observed five times during one day (see Pesarin and Salmaso, 2010, p. 61). The purpose of the experiment is to evaluate whether the level of testosterone in the blood is subject to change during the day. The question is:

Is there a time-of-day effect?

Hypothetical data

To illustrate the method, we will use the following hypothetical example. It involves four subjects. Each of them is tested under three treatments (or conditions or trials), denoted t_1, t_2 and t_3. The $4 \times 3 = 12$ observations are points in a plane that are depicted in Figure 5.15.

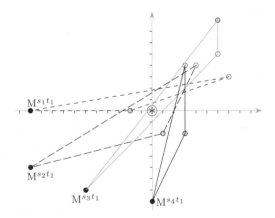

Figure 5.15
Cloud of the $4 \times 3 = 12$ points with orthonormal frame going through the mean point (⊛). The three points of each subject are the vertices of a triangle (● for t_1, ◐ for t_2, ○ for t_3).

Table 5.1 gives the coordinates of points with respect to an orthonormal Cartesian frame of the plane whose origin point is the mean point of the cloud.

Table 5.1
Coordinates of the 12 points with respect to the orthonormal Cartesian frame, and of mean points associated with the three treatments and the four subjects.

	t_1	t_2	t_3	Means
s_1	$(-11,\ 0)$	$(-2,\ 0)$	$(\ 7,3)$	$(-2,\ 1)$
s_2	$(-11,-5)$	$(\ 1,-2)$	$(\ 4,4)$	$(-2,-1)$
s_3	$(\ -6,-7)$	$(\ 6,\ 8)$	$(\ 6,5)$	$(\ 2,\ 2)$
s_4	$(\ \ \ 0,-8)$	$(\ 3,-2)$	$(\ 3,4)$	$(\ 2,-2)$
Means	$(\ -7,-5)$	$(\ 2,\ 1)$	$(\ 5,4)$	$(\ 0,\ 0)$

5.5.2 Crossed Design

Let S denote the set of subjects and T the set of measures (treatments or experimental conditions). The cardinality of S is denoted by n, that of T by T (as the set itself). In repeated measures experiments, each of the n subjects takes all treatments. The experimental factors S and T are said to be crossed, namely, for each pair $(s,t) \in S \times T$, there is a single observed value. Hence, there are nT observations.

Let us denote I the Cartesian product of factors S and T. A cloud of nT points in a geometric space is associated with the set $S \times T = I$. It is denoted M^{ST} (or M^I), and its mean point is denoted G, with $G = \sum \sum M^{st}/nT$.

Besides, the factor T defines a partition of $I = S \times T$ into T classes $(S \times \{t\})_{t \in T}$. The between-$T$ cloud (also called *main cloud*) is denoted M^T (see Figure 5.16). Point M^t is the barycentre of the n points $(M^{st})_{s \in S}$; it is defined by $M^t = \sum_{s \in S} M^{st}/n$. The test concerns the study of the "effect" of factor T, then it pertains to the between-T cloud. Geometrically, it consists in studying the deviations between points of cloud M^T.

Similarly, the factor S defines a partition of $I = S \times T$ into n classes $(\{s\} \times T)_{s \in S}$, hence the between-$S$ and the within-S clouds.
The between-S cloud is denoted M^S with $M^s = \sum_{t \in T} M^{st}/T$.

The within-S cloud, or *residual cloud*, is denoted R^{ST} (Figure 5.16 on the right); it is such that (see Chapter 2, Equation 2.11, p. 27):

$$\forall s \in S, \forall t \in T, \ \overrightarrow{GR}^{st} = \overrightarrow{GM}^{st} - \overrightarrow{GM}^s \quad \text{or} \quad R^{st} = G + \overrightarrow{M^s M}^{st} \quad (5.6)$$

Each experimental unit is "confounded" with the crossing of factors S and T. Therefore, disregarding factor T, the derived baseline data set is characterized by the sole structure of the nesting of units within factor S (restricted exchangeability): it corresponds to the residual cloud R^{ST}.

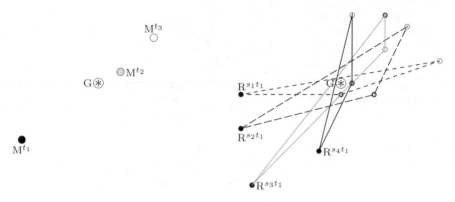

Figure 5.16
On the left, the between-T cloud with mean point G ($\circledast$), on the right, the residual cloud R^{ST} with mean point G ($\circledast$).

Note that the main cloud M^T can be obtained from the initial cloud M^{ST} as well as from the within-S cloud R^{ST} since

$$\overrightarrow{\mathrm{GM}^t} = \sum_{s \in S} \overrightarrow{\mathrm{GM}^{st}}/n = \sum_{s \in S} (\overrightarrow{\mathrm{GM}^{st}} - \overrightarrow{\mathrm{GM}^s})/n = \sum_{s \in S} \overrightarrow{\mathrm{M}^s\mathrm{M}^{st}}/n = \sum_{s \in S} \overrightarrow{\mathrm{GR}^{st}}/n$$

5.5.3   Permutation Crossing Set

In a crossed design, the observations are dependent since they are recorded on the same subjects at different times. Moreover, the problem of testing for possible differences among subjects is not of interest. Then, we can consider that data within each unit are exchangeable with respect to conditions, and that they are independent with respect to units. Therefore we permute observations within each subject independently. Thus, there are $(T!)^n$ possible assignments (crossings) since the $T!$ possible assignments for each subject can be associated with different assignments for every other subject.

In what follows, the set of possible crossings of factors S and T is indexed by J. The crossing space, denoted $\mathcal{J}$, is the set of the $(T!)^n$ clouds $(\mathrm{M}^{I_j})_{j \in J}$.

$$\mathcal{J} = (\mathrm{M}^{I_j})_{j \in J} = (\mathrm{M}^{(ST)_j})_{j \in J}$$

Given $j \in J$ and $t \in T$, the mean point of the n points $(\mathrm{M}^{(st)_j})_{s \in S}$ is denoted H^{tj}, hence the between-T cloud $(\mathrm{H}^{tj})_{t \in T}$ that is denoted H^{Tj} (main cloud for crossing j). This cloud is equally weighted, and the weight of each point is equal to n.

The observed cloud and a "possible cloud" are depicted in Figure 5.17 (p. 137) with the between-T clouds.

	t_1	$t2$	$t3$
$s1$	$\mathrm{M}^{s_1 t_1}$	$\mathrm{M}^{s_1 t_2}$	$\mathrm{M}^{s_1 t_3}$
$s2$	$\mathrm{M}^{s_2 t_1}$	$\mathrm{M}^{s_2 t_2}$	$\mathrm{M}^{s_2 t_3}$
$s3$	$\mathrm{M}^{s_3 t_1}$	$\mathrm{M}^{s_3 t_2}$	$\mathrm{M}^{s_3 t_3}$
$s4$	$\mathrm{M}^{s_4 t_1}$	$\mathrm{M}^{s_4 t_2}$	$\mathrm{M}^{s_4 t_3}$

	t_1	$t2$	$t3$
$s1$	$\mathrm{M}^{s_1 t_1}$	$\mathrm{M}^{s_1 t_2}$	$\mathrm{M}^{s_1 t_3}$
$s2$	$\mathrm{M}^{s_2 t_2} \leftrightarrow \mathrm{M}^{s_2 t_1}$		$\mathrm{M}^{s_2 t_3}$
$s3$	$\mathrm{M}^{s_3 t_1}$	$\mathrm{M}^{s_3 t_2}$	$\mathrm{M}^{s_3 t_3}$
$s4$	$\mathrm{M}^{s_4 t_1}$	$\mathrm{M}^{s_4 t_3} \leftrightarrow \mathrm{M}^{s_4 t_2}$	

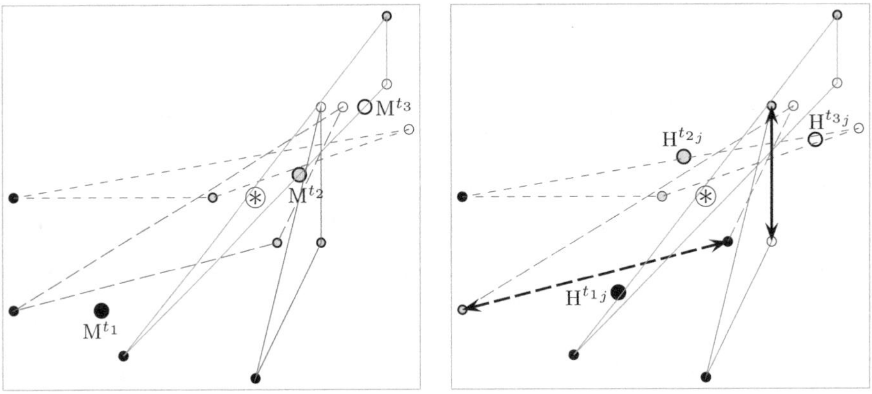

Figure 5.17
On the left, the observed cloud M^{ST} and the between-T cloud M^T; on the right, a "possible cloud" $\mathrm{M}^{(ST)_j}$ associated with crossing j and the between-T cloud H^{T_j}.

5.5.4 Test Statistic

With each $j \in J$, the relevant cloud is the main cloud H^{T_j}. It can be proven that the covariance endomorphism of cloud H^{TJ} is proportional to the covariance endomorphism of the residual (within-S) cloud R^{ST}. This is why, we will work in the affine support of the cloud R^{ST} equipped with the Mahalanobis distance related to the residual cloud R^{ST}.

The covariance endomorphism of cloud R^{ST} is denoted Rcov:

$$\forall \overrightarrow{u} \in \mathcal{L}, \ \mathrm{R}cov(\overrightarrow{u}) = \sum_{s \in S} \sum_{t \in T} \frac{1}{nT} \langle \overrightarrow{\mathrm{GR}}^{st} | \overrightarrow{u} \rangle \overrightarrow{\mathrm{GR}}^{st} \tag{5.7}$$

The Mahalanobis distance, related to cloud R^{ST}, between points A and B is called R-distance and denoted $|AB|_\mathrm{R}$: $|AB|_\mathrm{R} = \langle \mathrm{R}cov^{-1}(\overrightarrow{AB}) | \overrightarrow{AB} \rangle^{1/2}$.

The map, which associates the R-variance of the between-T cloud with each element of the crossing space, is denoted V_R and defines V_R as a statistic. By expressing the variance in terms of the R-distances between couples of points (see Proposition 2.6, p. 16), the statistic writes

$$\begin{aligned} V_\mathrm{R} : \mathcal{J} \ &\to \ \mathbb{R}_{\geq 0} \\ \mathrm{M}^{(ST)_j} &\mapsto \frac{1}{2} \sum_{t \in T} \sum_{t' \in T} \frac{1}{T^2} |\mathrm{H}^{tj} \mathrm{H}^{t'j}|_\mathrm{R}^2 \end{aligned} \tag{5.8}$$

The value of the test statistic for crossing j is denoted $v_\mathrm{R}[j]$.

Considering the observed cloud M^{ST}, the R-variance of the between-T cloud M^T is called *observed value* of the test statistic and is denoted v_{Robs}.

$$v_{Robs} = \tfrac{1}{2}\sum\sum \tfrac{1}{T^2}|\mathrm{M}^t\mathrm{M}^{t'}|_\mathrm{R}^2$$

Combinatorial distribution of the test statistic

Given $j \in J$, let $p(V_\mathrm{R} = v_\mathrm{R}[j])$ denote the proportion of possible crossings for which the value of the test statistic is equal to $v_\mathrm{R}[j]$. The *combinatorial frequency distribution* of statistic V_R is defined as follows.

$$V_\mathrm{R}(\mathcal{J}) \rightarrow [0,1]$$
$$v_\mathrm{R}[j] \mapsto p(V_\mathrm{R} = v_\mathrm{R}[j])$$

This distribution is discrete and will most often be represented by a histogram (see, for example, Figure 5.18, p. 139).

Combinatorial p-value

The proportion of possible crossings that satisfy the property $(V_\mathrm{R} \geq v_{Robs})$ is the *combinatorial p-value*; it defines the level of homogeneity of the T conditions.

$$p = p(V_\mathrm{R} \geq v_{Robs})$$

Hypothetical data

The Cartesian frame of the plane being orthonormal (see Figure 5.15), the matrix of the covariance endomorphism of cloud M^{ST} is the covariance matrix $\mathbf{V} = \begin{bmatrix} 36.5 & 18 \\ 18 & 23 \end{bmatrix}$; the between-$S$ covariance matrix is $\mathbf{V}_S = \begin{bmatrix} 4 & 0 \\ 0 & 2.5 \end{bmatrix}$.

The matrix associated with the endomorphism R*cov* is $\mathbf{V}_\mathrm{R} = \mathbf{V} - \mathbf{V}_S = \begin{bmatrix} 32.5 & 18 \\ 18 & 20.5 \end{bmatrix}$ and its inverse is $\mathbf{V}_\mathrm{R}^{-1} = \frac{1}{342.25}\begin{bmatrix} 20.5 & -18 \\ -18 & 32.5 \end{bmatrix}$.

The coordinates of the mean points $\mathrm{M}^{t_1}, \mathrm{M}^{t_2}, \mathrm{M}^{t_3}$ are provided in Table 5.1 (p. 135). The squared R-distances between the mean points and point G are equal to $|\mathrm{GM}^{t_1}|_\mathrm{R}^2 = \frac{1}{342.25}\begin{bmatrix} -7 & -5 \end{bmatrix}\begin{bmatrix} 20.5 & -18 \\ -18 & 32.5 \end{bmatrix}\begin{bmatrix} -7 \\ -5 \end{bmatrix} = 1.627, |\mathrm{GM}^{t_2}|_\mathrm{R}^2 = 0.124$ and $|\mathrm{GM}^{t_3}|_\mathrm{R}^2 = 0.913$. The observed value of the test statistic is $v_{Robs} = (1.627 + 0.124 + 0.913)/3 = 0.888$.

The number of possible crossings is equal to $(3!)^4 = 1296$.
The distribution of the test statistic is depicted in Figure 5.18 (p. 139).
The p-value is $p = p(V_\mathrm{R} \geq 0.888) = 126/1296 = 0.097$.

5.5.5 Particular Case: Design with Two Repeated Measures

Now, in the case of two repeated measures, we prove the equivalence between the homogeneity test developed in this section and the geometric typicality test presented in Chapter 4 (see §4.4, p. 90).

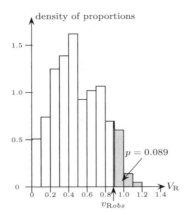

Figure 5.18
Hypothetical data. Histogram of the combinatorial distribution of the test statistic V_R for the global comparison on T. The surface area under the curve past the observed value v_{Robs} (shaded grey area) is equal to the (interpolated) p-value.

Theorem 5.4 (Equivalence property). *For a design with two repeated measures, the homogeneity test is equivalent to the geometric typicality test comparing the mean point of the cloud of individual effect-points to the origin.*

Proof. For $T = \{t_1, t_2\}$, there are 2^n possible crossings (or permutations). The mean point M^s of the two points M^{st_1} and M^{st_2} is the midpoint $(M^{st_1} + M^{st_2})/2$. Then, if $\overrightarrow{d^s} = \overrightarrow{M^{st_1}M^{st_2}}$ denotes the individual effect of subject s, we have $\overrightarrow{M^sM^{st_1}} = \frac{1}{2}\overrightarrow{M^{st_2}M^{st_1}} = -\frac{1}{2}\overrightarrow{d^s}$. In this section, the distance between points is the R-distance defined from the covariance endomorphism $Rcov$ associated with the residual cloud R^{ST}. It is such that $Rcov(\overrightarrow{u}) = \sum_{s \in S} \frac{1}{2n} \langle \overrightarrow{GR}^{st_1} | \overrightarrow{u} \rangle \overrightarrow{GR}^{st_1} + \sum_{s \in S} \frac{1}{2n} \langle \overrightarrow{GR}^{st_2} | \overrightarrow{u} \rangle \overrightarrow{GR}^{st_2}$ $(\overrightarrow{u} \in \mathcal{L})$. From $\overrightarrow{GR}^{st_1} = \overrightarrow{M^sM^{st_1}} = -\frac{1}{2}\overrightarrow{d^s}$ and similarly $\overrightarrow{GR}^{st_2} = \overrightarrow{M^sM^{st_2}} = \frac{1}{2}\overrightarrow{d^s}$, we deduce $Rcov(\overrightarrow{u}) = \frac{1}{4} \sum_{s \in S} \frac{1}{n} \langle \overrightarrow{d^s} | \overrightarrow{u} \rangle \overrightarrow{d^s}$. According to crossing j, the individual effect of subject s is denoted $\overrightarrow{d^{sj}}$ with $\overrightarrow{d^{sj}} = M^{(st_2)_j} - M^{(st_1)_j}$ ($\overrightarrow{d^{sj}}$ is equal to $\overrightarrow{d^s}$ or to $-\overrightarrow{d^s}$). Letting $\overrightarrow{d^j} = \frac{1}{n} \sum_{s \in S} \overrightarrow{d^{sj}}$, we have $\overrightarrow{H^{t_1j}H^{t_2j}} = (\sum_{s \in S} M^{(st_2)_j} - \sum_{s \in S} M^{(st_1)_j})/n = \frac{1}{n} \sum_{s \in S} \overrightarrow{d^{sj}} = \overrightarrow{d^j}$. For a crossing j, the *value of the test statistic* is equal to the R-variance of the cloud of the two points H^{t_1j} and H^{t_2j} (see Formula 2.1, p. 12): $v_R[j] = \frac{1}{4}|\overrightarrow{H^{t_1j}H^{t_2j}}|_R^2 = \frac{1}{4}|\overrightarrow{d^j}|_R^2$.

Now, we follow the approach developed in Chapter 4 for the geometric typicality test, adapting the notations. In the case where the geometric typicality test is applied to a design with two repeated measures (see §4.4, p. 90), the relevant data are the individual effects $(\overrightarrow{d^s})_{s \in S}$ and the reference point is the point O corresponding to the null effect. Considering the cloud of effect-points D^S with $D^s = O + \overrightarrow{M^{st_1}M^{st_2}}$, the test consists in comparing the mean point of the cloud $D^S = (D^s)_{s \in S}$ to the reference point O.

The distances used in the geometric typicality test depend on the endomorphism $\mathrm{D}cov^{\mathrm{O}} = \sum_{s \in S} \frac{1}{n} \langle \overrightarrow{\mathrm{OD}^s} | \vec{u} \rangle \overrightarrow{\mathrm{OD}^s} = \frac{1}{n} \langle \vec{d^s} | \vec{u} \rangle \vec{d^s}$. We have $\mathrm{D}cov^{\mathrm{O}} = 4\,\mathrm{R}cov$ and $|\cdot|_{\mathrm{O}} = \frac{1}{2}|\cdot|_{\mathrm{R}}$, therefore, for a crossing j, the *value of the test statistic* D_{O}^2 (see Formula 4.2, p. 77) is $d_{\mathrm{O}}^2[j] = |\mathrm{OD}^j|_{\mathrm{O}}^2 = |\vec{d}^{\,j}|_{\mathrm{O}}^2 = \frac{1}{4}|\vec{d}^{\,j}|_{\mathrm{R}}^2 = v_{\mathrm{R}}[j]$. $\square$

Since, for a design with two repeated measures, the geometric typicality test and the homogeneity permutation test are equivalent, the geometric typicality test (with its compatibility region) should be preferably used.

We have already studied a design with two repeated measures in the one-dimensional case in the *Student's example* (p. 90) and we will study the multidimensional case in Chapter 6 (see the *Parkinson study*, p. 156).

5.6   Other Methods

Recall that, in this book, we confine ourselves to the permutation approach in a geometric scope. However, the well-known bootstrap techniques and the nonparametric combination of univariate tests are frequently used nonparametric approaches to the analysis of multivariate data.

The principle of *bootstrap procedure* is presented in Chapter 4 (§4.5.1, p. 93). The bootstrap solution for the two-sample problem is outlined in Efron and Tibshirani (1993, Chapter 8, §8.3); see also, e.g., Davison and Hinkley (1997, Chapter 4). The bootstrap procedure can be used for global or specific comparisons of several groups.

The method of *nonparametric combination* (NPC) of dependent permutation tests, which is evoked in Chapter 4 (§4.5.2, p. 94), can be used for all types of comparisons as soon as one-dimensional analyses are performed (see Pesarin and Salmaso, 2010, Chapter 4).

Now we will outline the classical *parametric procedure* for two independent groups: Hotelling's test (see, e.g., Anderson, 2003, pp. 179–180).

Hotelling's Test for Two Independent Groups

The two-sample Hotelling's test is the multivariate extension of the common Student's t-test to two groups.

We consider two independent groups c_1 and c_2 with sizes n_{c_1} and n_{c_2} and the associated subclouds $\mathrm{M}^{I<c1>}$ and $\mathrm{M}^{I<c2>}$ of cloud M^I. The dimensionality of the affine support $\mathcal{M}$ of the cloud $\mathrm{M}^{I<c_1,c_2>}$ is denoted L and $\mathcal{M}$ is equipped with an orthonormal frame $(\mathrm{O}, (\vec{\epsilon_\ell})_{\ell \in L})$. The mean points of the two subclouds are denoted by G^{c_1} and G^{c_2}, the deviation $\mathrm{G}^{c_2} - \mathrm{G}^{c_1}$ by $\vec{d}_{obs}$ and the column vector of its coordinates by $\mathbf{d}_{obs}$. The matrices of the covariance endomorphisms of the two subclouds are denoted $\mathbf{V}_{I<c_1>}$ and $\mathbf{V}_{I<c_2>}$.

The deviation $\vec{\delta}$ between the true mean points is estimated by the deviation $\overrightarrow{d}_{obs}$ between the observed mean points. We denote $\boldsymbol{\delta}$ (italics) the vector-variable associated with $\overrightarrow{\delta}$, $\boldsymbol{d}$ (italics) the vector-variable whose realisation is $\mathbf{d}_{obs}$, and $\boldsymbol{S}$ (italics) the matrix-variable whose realisation is $\mathbf{S} = \left(n_{c_1} \mathbf{V}_{I<c_1>} + n_{c_2} \mathbf{V}_{I<c_2>}\right)/(n_{c_1} + n_{c_2} - 2)$.

Suppose that subclouds $\mathrm{M}^{I<c1>}$ and $\mathrm{M}^{I<c2>}$ are two samples of two L-dimensional normal distributions whose mean points are G^{c_1} and G^{c_2} and with the same covariance matrix $\boldsymbol{\Sigma}$. Then, letting $\tilde{n} = n_{c_1} n_{c_2}/n$, $\sqrt{\tilde{n}}\, \boldsymbol{d}$ is normally distributed $\mathcal{N}_L(\boldsymbol{\delta}, \boldsymbol{\Sigma})$ and $(n-2)\boldsymbol{S}$ is distributed $W_L(n-2, \boldsymbol{\Sigma})$ ($\boldsymbol{\Sigma}$-scaled L-dimensional Wishart with L and $(n-2)$ d.f.) (Anderson, 2003, p. 252).

We wish to test the null hypothesis $\mathcal{H}_0 : \boldsymbol{\delta} = 0$. Under the usual parametric normal-χ^2 sampling model, the null hypothesis $\mathcal{H}_0$ that the true deviation is null can be tested by the classical Hotelling's test.

Under $\mathcal{H}_0$, the statistic $T^2 = \tilde{n}\,(\boldsymbol{d} - \boldsymbol{\delta})^\top \boldsymbol{S}^{-1}(\boldsymbol{d} - \boldsymbol{\delta})$ is distributed as $L\frac{n-2}{n-1-L} F_{L,n-L-1}$ where $F_{L,n-L-1}$ is the central F (Fisher–Snedecor) distribution with L and $(n - L - 1)$ d.f.

Then, the *p-value* of Hotelling's test, that is, the sampling probability that, under $\mathcal{H}_0$, the statistic T^2 exceeds its observed value $t^2_{obs} = \tilde{n}|\overrightarrow{d}_{obs}|^2$, is:

$$\tilde{p} = p(F_{L,(n-L-1)} \geq \tilde{n} \times \tfrac{n-L-1}{L(n-2)}\, |\mathrm{G}^{c_1}\mathrm{G}^{c_2}|^2_S)$$

Denoting $F_{L,n-L-1}[\alpha]$ the critical value of the F distribution with L and $n - L - 1$ degrees of freedom at superior level α, a $(1 - \alpha)$-*confidence region* for $\boldsymbol{\delta}$ is the set of vectors $\overrightarrow{e}$ such that

$$|\overrightarrow{e} - \overrightarrow{d}|^2_S = \tfrac{1}{n} \times \tfrac{(n-2)L}{n-L-1} \times F_{L,n-L-1}[\alpha]$$

5.7 Computations with R and Coheris SPAD Software

As already stated in the preceding chapters, we carry out computations using coordinates of points of cloud over any *orthonormal Cartesian frame* of the geometric space (see p. 55).

We first describe the part of the R script that allows one to perform the testing procedure and the compatibility region in the case of two independent groups (§5.7.1), that is, the methods developed in §5.4 (p. 116). Then, we comment on the use of the full program carrying out the procedures in the general case of independent groups and in the particular case of two groups[8] using Coheris SPAD, a driven-menu software (§5.7.2).

---

[8]For a design with two repeated measures, we used the geometric typicality test presented in Chapter 4.

5.7.1   R Script

The R code, we present now, computes the p-value and the compatibility region for the partial or specific comparison of two groups in the multidimensional case. The data management and the display of results are reduced to the minimum.

Step 1. The data

The beginning of the program deals with reading the data file and organizing data.

The *database* consists in a table with one row per individual. The columns are, in this order, the identifier of the individual, the coordinates (columns X and Y), then the group (column C) the individual belongs to. Names of groups are numerical in the range 1 to C (number of groups). The program proceeds to comparing the first two groups (partial or specific comparison). The opposite data table is that of the *Target example*.

; X ;	Y	;C
$i1$; 0;	-6;	1
$i2$; 3;	-5;	1
$i3$; 7;	-3;	3
$i4$; 3;	-1;	3
$i5$; 6;	0;	3
$i6$; -4;	1;	1
$i7$; 1;	2;	4
$i8$; 3;	2;	4
$i9$; 5;	5;	2
$i10$; 6;	5;	2

Firstly, we specify the working directory (say `"D:/path"`), and read the data file `Target4.txt`, which is supposed to be in the working directory.

Then, we gather the groups other than 1 or 2 into group 3, and choose the type of comparison: partial or specific. If the comparison is specific (`partial <- FALSE`), we select the individuals of groups 1 and 2 (`select`).

```
setwd("D:/path")
base      <- read.table(file = "Target4.txt", header = TRUE,
                 sep = ";", dec = ".", row.names = 1)
base[(which(base[, dim(base)[2]] > 3)), dim(base)[2]] <- 3
# partial <- FALSE                        # if specific comparison
partial <- TRUE                           # if partial comparison
select  <- 1:dim(base)[1]
if(!partial) {select <- which(base[, dim(base)[2]] < 3)}
```

Secondly, we define the $I \times K$-matrix (`X_OM.IK`) of coordinates of individuals and the factor on individuals (`C`) defining the groups; if the comparison is partial, the coordinates of all individuals are kept, otherwise only the coordinates of the first two groups are kept. Then, we determine the cardinality of the set of individuals (`n`), the dimensionality of the space (`K`), the cardinality of each group (`n.C`) and the number of groups (`cardC`). Then, we create the list of the groups to be compared (`lc`) and the number of individuals involved in the comparison (`nprim`).

```
X_OM.IK <- as.matrix(base[select,-dim(base)[2]])   # coordinates
C       <- as.factor(base[select,dim(base)[2]])     # factor
n       <- dim(X_OM.IK)[1]                 # number of individuals
K       <- dim(X_OM.IK)[2]                 # dimensionality of space
n.C     <- as.vector(table(C))             # sizes of groups
```

```
cardC    <- length(n.C)                    # number of groups
lc       <- 1:2                            # groups to be compared
nprim    <- n.C[1] + n.C[2]
if (K < 2) stop("Dimension < 2")
```

Step 2. Parameters of the method

The *parameters* are the limit for notable proportion of variance of the comparison (see p. 28), a reference α-level, the maximum number of possible nestings to be used in calculations (see p. 130), the number of lines for determining the adjusted compatibility region (see p. 128), and possibly an integer (**seed**) in order to initialize the random number generator for MC method (see Footnote 13, p. 56).

```
notable_PV <- 0.04      # notable limit for proportion of variance
alpha      <- 0.05      # alpha level
max_number <- 100000    # maximum number of nestings
n_dir      <- 500       # number of lines for compatibility region
seed       <- NULL      # seed for random number generator
```

Step 3. Descriptive analysis

In this step, we compute descriptive statistics. Firstly, for cloud M^I, we calculate the centred coordinates of points (**X_GM.IK**), then the covariance matrix (**Mcov.KK**), and the variance of the cloud (**Vcloud**). Secondly, we compute, the centred coordinates of the C mean points $(G^c)_{c \in C}$ (**X_GGc.CK**). Finally, we determine the proportion of variance taken into account by the comparison (PV, see p. 28).

```
X_GM.IK   <- sweep(X_OM.IK, 2, colMeans(X_OM.IK), "-")
Mcov.KK   <- t(X_GM.IK) %*% X_GM.IK /n
Vcloud    <- sum(diag(Mcov.KK))
X_GGc.CK  <- as.matrix(aggregate(X_GM.IK, by=list(C), FUN=mean)[,-1])
Var_Cprim <- sum(diag(cov.wt(X_GGc.CK[lc, ],
                        method = "ML", wt = n.C[lc])$cov))
PV        <- (nprim/n) * Var_Cprim/Vcloud

cat("   partial eta-squared ", PV, "\n")
if (PV < notable_PV) {
   cat("   PV is small: there is no point in doing the test.\n")
} else {cat("   descriptively, mean points differ.\n")}
```

Step 4. The testing procedure

Firstly, we construct the nesting table (**Epsilon.JI**), which determines for each nesting $j \in J$ the partition of I into C groups (see §5.3.2, p. 111). If the maximum number of nestings chosen by the user (**max_number**) is greater than the number of possible nestings $(\sum n_c)!/(\prod n_c!)$, then the exhaustive method is performed; if not, the MC method is used based on **max_number** nestings and the associated nesting table is constructed.

```
cardJ <- choose(n, n.C[1])
if(cardC >2){
  for (i in 2:(cardC - 1))
  {cardJ <- cardJ * choose(n - sum(n.C[1:(i-1)]), n.C[i])}
}
if (cardJ < max_number) {                          # exhaustive method
  s.C <- cumsum(n.C); J0 <- choose(n, n.C[1])
  Ep1 <- rbind(combn(n, n.C[1]), combn(n, n - s.C[1])[ , J0:1])
  if(cardC > 2){
    for (c in 2:(cardC - 1)) {
      s0 <- s.C[c - 1]; s1 <- s0 + 1; nr <- n - s.C[c]
      Ep0<- Ep1; J1 <- choose(n - s0, n.C[c])
      Ep1<- rbind(
      matrix(apply(matrix(Ep0[1:s0,],nrow=s0),2,rep,times=J1),nrow=s0),
      matrix(apply(Ep0[(s1):n,],2,combn,m=n.C[c]), nrow=n.C[c]),
    matrix(apply(Ep0[(s1):n,],2,combn,m=nr)[(nr*J1):1,],nrow=nr)[nr:1,])
      J0 <- J0 * J1
    }
  }
  Ep1 <- t(Ep1);  Epsilon.JI <- matrix(1, nrow = J0, ncol = n)
  for (c in 2:cardC) {
    for (j in 1:J0) {Epsilon.JI[j,Ep1[j, (s.C[c-1]+1):(s.C[c])]] <- c}
  }
  rm(Ep0, Ep1)
} else {                                           # MC method
  cardJ <- max_number; set.seed(seed)
  Epsilon.JI <- matrix(0L, nrow = cardJ, ncol = n)
  for (j in 1:cardJ) {Epsilon.JI[j, ] <- sample(C)}
}
```

Secondly, in order to implement the testing procedure, we have to work in the affine support $\mathcal{M}$ of cloud M^I. In addition, the test statistic is based upon the M-norm, which depends on the inverse of the covariance endomorphism of cloud M^I. Therefore, we proceed to the passage from $\mathcal{V}$ (direction of $\mathcal{U}$), equipped with the initial orthonormal basis, to $\mathcal{L}$ (direction of $\mathcal{M}$), equipped with the orthocalibrated principal basis of cloud M^I (see p. 17). The change of basis is done from eigenvectors and eigenvalues of covariance matrix (Mcov.KK) of cloud M^I.

If L is the number of non-null eigenvalues (dimension of the cloud), the change of basis matrix is a $K \times L$-matrix that is called BasisChange.KL. The matrix of standardized principal coordinates of points $(\mathrm{M}^i)_{i \in I}$ is Z_GM.IL (see Equation 2.6, p. 19); that of mean points $(\mathrm{G}^c)_{c \in C}$ is Z_GGc.CL.

Then, we compute the observed value of the test statistic (D2M_obs), namely, the between-C' M-variance (see p. 113), which is equal to the sum of the squared standardized coordinates of the deviation vector $\mathrm{G}^{c_2} - \mathrm{G}^{c_1}$.

```
eig <- eigen(Mcov.KK, symmetric = TRUE)
L   <- sum(eig$values > 1.5e-8); lambda.L <- eig$values[1:L]
```

```
if (L < 2) {warning("One-dimensional cloud")
} else {
  BasisChange.KL <- eig$vectors[ , 1:L] %*%
                    diag(1/sqrt(lambda.L), nrow= L)
  Z_GM.IL   <- X_GM.IK %*% BasisChange.KL
  Z_GGc.CL <- X_GGc.CK %*% BasisChange.KL
  D2M_obs   <- sum( (Z_GGc.CL[1, ] - Z_GGc.CL[2, ])^2 )
```

Thirdly, for each group, we construct the indicator matrix (One_c.JI) of group c and deduce the sum of squared coordinates of points G^{jc} (with respect to the orthocalibrated basis) multiplied by its weight n_c (Zd2_G.JC), then we deduce the value of the test statistic D2M.J. Finally, we determine the p-value.

```
  Zd2_G.JC <- matrix(0, nrow= cardJ, ncol= cardC)
  for (c in (1:cardC)) {
    one.C    <- rep(0L, cardC); one.C[c] <- 1L
    One_c.JI <- matrix(one.C[t(Epsilon.JI)],
                       nrow= cardJ, ncol= n, byrow= TRUE)
    Zd2_G.JC[, c] <- as.matrix( rowSums((One_c.JI %*% Z_GM.IL)^2),
                       nrow= cardJ, ncol= cardC) / n.C[c]
  }
  D2M.J <- { (nprim*rowSums(Zd2_G.JC[,1c])) -
              (n-nprim)*Zd2_G.JC[,cardC]) / (n.C[1]*n.C[2]) }
  D2M_obs <- sum((Z_GGc.CL[2, ] - Z_GGc.CL[1, ])^2)
  n_sup <- sum(D2M.J >= D2M_obs * (1 - 1e-12))
  p_value <- n_sup / cardJ
  cat(" p-value = ", n_sup, "/", cardJ, " = ",
    format(round(p_value, digits= 3), nsmal= 3),"\n", sep= "")
}
```

Step 5. *The compatibility region*

The first step consists in deducing the coordinates of points of the reference cloud (X_GR.IK) from that of the initial cloud (X_GM.IK) and of the observed deviation (Xd_obs.K) using Equation 5.4.3 (p. 121). Then, we proceed to a change of basis from initial to R-orthocalibrated basis. The change of basis matrix (still called BasisChange.KL) is obtained from the eigenvectors and eigenvalues of the covariance matrix of the reference cloud (Rcov.KK). Then, the new coordinates of points of the reference cloud (U_GR.IL) and that of the deviation-points (U_OD.JL) (see p. 122) are calculated. And then, the squared R-norms of $(\overrightarrow{OD}_R^j)_{j \in J}$ are computed (column-matrix C.J); the set of indexes for which this coefficient is null is established (Cnul).

```
Xd_obs.K <- t(t( X_GGc.CK[2, ] - X_GGc.CK[1, ] ))
d.C      <- c(-n.C[2], n.C[1], 0)
X_GR.IK  <- X_GM.IK - t(t(d.C[C])) %*% t(Xd_obs.K)/nprim

Rcov.KK  <- t(X_GR.IK) %*% X_GR.IK /n  # covariance matrix of reference
eig      <- eigen(Rcov.KK, symmetric= TRUE)
```

```
L          <- sum(eig$values > 1.5e-8) ; lambda.L <- eig$values[1:L]
if (L < 2){warning("One-dimensional cloud")
} else {
  BasisChange.KL <- eig$vectors[,1:L] %*% diag(1/sqrt(lambda.L),nrow=L)
  U_GR.IL <- X_GR.IK %*% BasisChange.KL
  U_OD.JL <- matrix(d.C[Epsilon.JI], nrow= cardJ, ncol= n)  %*%
             U_GR.IL/(n.C[1]*n.C[2])
  C.J <- t(t(rowSums(U_OD.JL^2)))      # squared R-norms of deviations
  Cnul <- which(C.J < 1e-12)
```

Then we determine the scale parameter κ of the ellipsoid that defines the adjusted compatibility region from n_dir random lines with R-unit direction vectors $\overrightarrow{\beta}$ (Beta.L).

Firstly, in order to use the relationship between the deviation-vectors (see Proposition 5.9, p. 126), we calculate the coefficients $(\epsilon^j)_{j \in J}$ (epsilon.J). Secondly, for each line $\mathcal{D}$ (loop on d), the coordinates of $(\overrightarrow{OD}_R^j)_{j \in J}$ along this line are computed (U.J). Then, by using Lemma 5.3 (p. 127), for each $j \in J$, we determine the interval $[x_1^j, x_2^j]$ (column-matrices X1.J and X2.J). Finally, applying Theorem 5.2 (p. 128), we obtain two values of κ for each line, hence the $2 * $ n_dir values of κ (kappa.D) whose mean is the adjusted κ value.

```
n11.J <- rowSums(Epsilon.JI[ , which(C == 1)] == 1)
n22.J <- rowSums(Epsilon.JI[ , which(C == 2)] == 2)
nn.J <-  rowSums(Epsilon.JI[ , which(C  !=3)] != 3)
epsilon.J <-  n11.J/n.C[1]  + n22.J/n.C[2]- nn.J/nprim
rm(Epsilon.JI)


rank_inf <- trunc(alpha * cardJ) + 1
kappa.D <- rep(0, 2 * n_dir); eff.D <- rep(0L, 2 * n_dir)
set.seed(seed)
for (d in 1:n_dir){
  xy.L <- runif(L, -1, 1); Beta.L <- t(t(xy.L / sqrt(sum(xy.L^2))))
  U.J <-  U_OD.JL %*% Beta.L
  A.J <- epsilon.J^2 -1 + abs(C.J - U.J^2) * n.C[1]*n.C[2] /n/nprim
  B.J <- epsilon.J * U.J
  Delta.J <- abs( B.J^2 - A.J * C.J)
  x1.J <- (-B.J + sqrt(Delta.J))/A.J   # negative root
  x2.J <- (-B.J - sqrt(Delta.J))/A.J   # positive root
  AC <- setdiff( which(abs (A.J) < 1e-12), Cnul)   #A = 0 & C>0
  if(length(AC) > 0){                              #case A.J=0 and C.J>0
    for (j in AC){
      x1.J[j] = -Inf; x2.J[j] = Inf
      if (B.J[j] >  1e-12) {x1.J[j] <- -C.J[j] / abs(2*B.J[j])}
      if (B.J[j] < -1e-12) {x2.J[j] <- +C.J[j] / abs(2*B.J[j])}
    }
  }
  for (j in Cnul){                        # particular case C.J=0
    if(epsilon.J[j]^2 == 1) { x1.J[j] <- -Inf; x2.J[j] <- Inf
    } else {x1.J[j] <- x2.J[j] <- 0 }
```

```
}
  x1_sorted  <- sort(x1.J); x2_sorted  <- sort(x2.J)
  kappa.D[2*d - 1] <- abs(x1_sorted[rank_inf])
  kappa.D[2*d] <- x2_sorted[cardJ + 1 - rank_inf]
}
if (max(kappa.D) == Inf) {
    warning("Finite compatibility region is not accessible")
} else {
  cat(" mean of ", 2*max(L,n_dir)," kappa values = ",
      round(mean(kappa.D), digits= 2), " in the range ",
      floor(min(kappa.D)*1000)/1000, " to ",
      ceiling(max(kappa.D)*1000)/1000, " \n", sep= "")
}
}
```

5.7.2 R Script Interfaced with Coheris SPAD

For users, it is easier to perform the analyses using the R script interfaced with Coheris SPAD software[9], which is a menu-driven solution.

We will present here the analyses of the *Target example* that are developed in this chapter. The sequence of analyses is quite simple. It comprises two steps (see SPAD diagram in Figure 5.19).

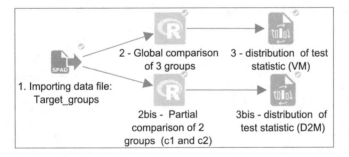

Figure 5.19
Diagram of the SPAD project and data table.

Step 1 is the *import of the data set*. Here the file is a "SPAD data archive" file, which is stored in the project for the sake of the autonomy of the project. Data table is shown in Figure 5.19.

Steps 2 and *2bis* are that of the homogeneity test through the R script for global and partial comparison. *Steps 3* and *3bis* consist in depicting the distribution of test statistic.

Then, the settings are defined using the four tabs: Variables, Analyses, Parameters and Outputs (see Figure 5.20).

---

[9]For a general presentation of Coheris SPAD software, see Chapter 1, p. 6.

- Variables (Figure 5.20)

  Firstly, you have to select the "Homogeneity test for independent groups" R script that was copied in the library of scripts (folder defined by the user). Then, you have to select the *numerical variables* on which the analysis is performed (coordinates of points in a geometric space equipped with orthonormal basis) and the factor defining groups.

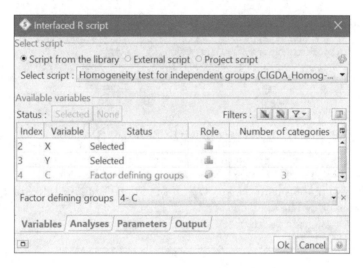

Figure 5.20
Variables dialog box.

- Analyses (Figure 5.21)

  You must choose the type of comparison, the type of analysis (one-dimensional, multidimensional or both) and if you want to perform the test and/or determine the (adjusted) compatibility region. If comparison is partial or specific, the field "comparison" must be completed.

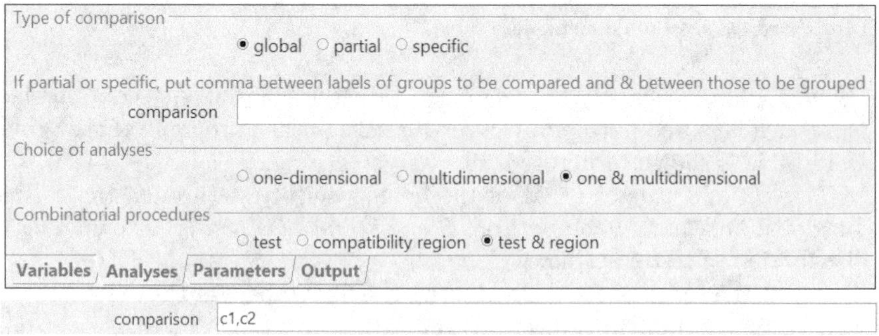

Figure 5.21
Analyses dialog box.

- `Parameters` (Figure 5.22)

  All parameters have a default value corresponding to the more conventional choices. These values can be modified by the user.

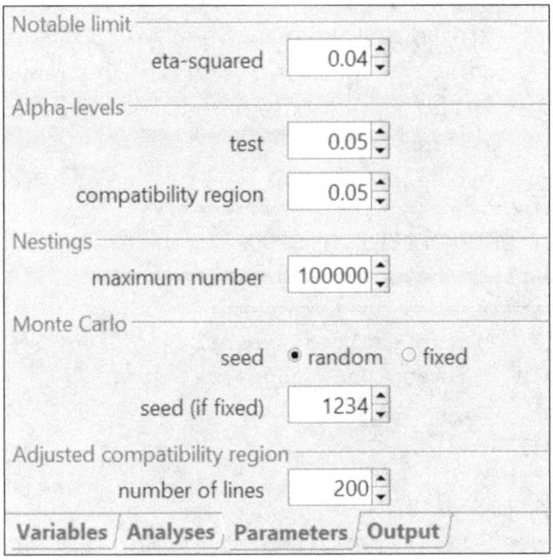

Figure 5.22
`Parameters` dialog box.

- `Outputs` (Figure 5.23)

  The outputs dialog box allows the user to specify the format of the results (number of decimals), to calculate or not the results of approximate methods, and to export or not the test statistic distributions to SPAD.

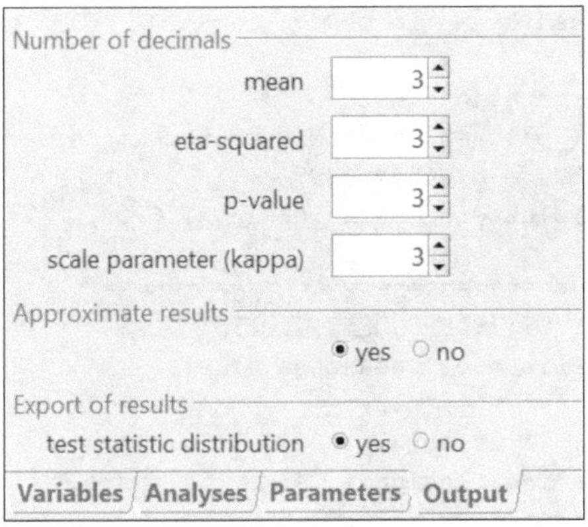

Figure 5.23
`Outputs` dialog box.

Results of analyses

The results of the R script consist of a text file and, if requested, of the values of the distributions of test statistics. The results (text file) can be accessed from the context menu of the method (Right-click – Results – Edition). The distributions can be stored in an external file by using "data export" methods.

The program provides the following results of the homogeneity test applied to the *Target example*, for the global comparison and the partial comparison of two groups.

```
=================================================
    HOMOGENEITY OF INDEPENDENT GROUPS
    global  comparison between mean points
=================================================
  2 variables (X, Y) are analyzed jointly;
  clouds are in a 2-dimensional geometric space.

Descriptive analysis
--------------------
  size of overall cloud: 10
  global comparison: c1, c2, c3 (sum of sizes = 10)
        size
    c1     3
    c2     2
    c3     5
  eta-squared = 13.417/23.000 = 0.583 > 0.04:
  descriptively, mean points differ.

Combinatorial inference
-----------------------
  number of possible nestings: 2520;
  exhaustive method is performed.

~ homogeneity permutation test
  test statistic: between-group M-variance (observed value: 1.001)
  p-value = 37/2520 = 0.015 < 0.05
  The 3 groups are heterogeneous at level 0.05.


=================================================
    HOMOGENEITY OF INDEPENDENT GROUPS
    partial  comparison between mean points
=================================================
  2 variables (X, Y) are analyzed jointly;
  clouds are in a 2-dimensional geometric space.

Descriptive analysis
--------------------
  size of overall cloud: 10
  partial comparison: c1, c2 (sum of sizes = 5)
```

```
       size
  c1    3
  c2    2
  partial eta-squared = 12.417/23.000 = 0.540 > 0.04:
  descriptively, mean points differ.


Combinatorial inference
-----------------------

  number of possible nestings: 2520;
  exhaustive method is performed.


~ homogeneity permutation test
  test statistic: squared M-distance between mean points (observed value:
  7.479)
  p-value = 1/2520 = 0.000 < 0.05
  The 2 groups are heterogeneous at level 0.05.


~ adjusted 95% compatibility region:
  principal kappa-ellipse of reference cloud, with kappa = 3.358
  (mean of 400 values in the range 3.293 to 3.444)
```

Distributions of test statistics

The R script computes the values of test statistics and, if requested, exports them to SPAD. From these results, it is possible to construct the distributions using graphical interface of Coheris SPAD software.

Concluding Remarks

When comparing C' groups (with $C' > 2$), if the result of the homogeneity test is statistically significant (groups are heterogeneous), one might wonder where relevant differences lie. A classical procedure consists in doing $C' - 1$ comparisons with one degree of freedom associated with $C' - 1$ orthogonal contrasts.

For instance, for the nested design of type $n_C = (n_{c_1}, n_{c_2}, n_{c_3})$, the two n_C-orthogonal contrasts $(1, -1, 0)$ and $\left(n_{c_1}/(n_{c_1} + n_{c_2}), n_{c_2}/(n_{c_1} + n_{c_2}), -1 \right)$ can be considered.

Then, according to the descriptive conclusions, we proceed to the partial comparison of two groups (here c_1 and c_2), for the nesting of type $(n_{c_1}, n_{c_2}, n_{c_3})$. Then, we perform the global comparison between group c_3 and the grouping of groups c_1 and c_2, for the nesting of type $(n_{c_1} + n_{c_2}, n_{c_3})$.

Furthermore, the compatibility region, defined for the comparison of two groups, can be constructed.

A study of designs that are more complex than independent groups or repeated measures designs should be explored according to the geometric approach. We will present in Chapter 6 an example in which the group factor is the crossing of two factors, hence the problem of testing the interaction between two factors.

In experimental designs with more than one factor, different strategies for permutations have been proposed in the one-dimensional case; see, e.g., Anderson and Ter Braak (2003); Edgington (2007); Pesarin and Salmaso (2010); Good (2011).

6

Research Case Studies

In this chapter, we apply the methods of combinatorial inference, presented in the previous chapters, to research case studies chosen to exemplify the variety of real situations that can be handled in the present state of the art. Recall that our methods deal with *clouds of points in a geometric space.*

The same strategy is used for all case studies we present in this chapter. Each analysis comprises two phases. The first phase is *descriptive*: the structure of the cloud is studied and descriptive summaries are obtained. The second phase is *inductive*: combinatorial inference procedures are carried out in order to substantiate descriptive findings.

In each study, the analyzed data consist of an Individuals×Variables table[1], and there are structuring factors on the set of individuals.

The steps of analyses are as follows.

1. *Descriptive analysis*

   - Comment elementary statistical analyses and outline the *research questions.*

   - Perform a GDA *method*[2], that is, (1) choose a distance between individuals, (2) determine principal axes and calculate principal coordinates of points, (3) decide how many axes need to be interpreted, then interpret these axes, (4) investigate the cloud of individuals, in connection with the questions of interest, especially the subclouds determined by structuring factors.

   - Looking at the importance of effects, then conclude with the descriptive findings.

2. *Inductive analysis* Attempt to extend the descriptive conclusions of notable effects by inductive analyses using *permutation modelling.*

The case studies differ according to the kinds of questions of interest and each of the foregoing steps may be more or less elaborate.

---

[1] In the Parkinson data, some individuals (healthy subjects) are people, others are defined as pairs (subject, treatment).

[2] Recall that methods we developed in the previous chapters can be applied to any Euclidean cloud whether it comes from GDA method or not.

The Parkinson Study (§6.1)

The data are part of a medical research on Parkinson's disease. They involve genuine experimental factors: *Treatment* factor (before and after drug intake) and *Group* factor (patients vs. healthy subjects).

The dependent variables are measurements of gait performance.

The variables are on different scales; accordingly, the GDA method used is a standard Principal Component Analysis (PCA). Due to the experimental design, PCA is performed on the reference group data (healthy subjects), putting the patient data as supplementary elements; it leads to a two-dimensional geometric model of data, with two axes that are interpreted as a performance axis and a style axis.

Taking the overall cloud of subjects as a "geometric dependent variable", we proceed to two specific analyses and extend inductively the main descriptive conclusions using combinatorial inference. Firstly, we study the main effect of drug intake (geometric typicality test). Secondly, we compare patients after drug intake with healthy subjects (combinatorial typicality test).

The Members of French Parliament and Globalisation in 2006 (§6.2)

One year after the referendum on the European Constitution in France, a research team of *Sciences Po Paris* drew up a questionnaire about globalisation that was sent to the members of Parliament (MPs).

The MPs are divided into parliamentary groups (structuring factor). The question is raised about differences between parliamentary groups, more specifically, between the right-wing and the left-wing.

Using a specific MCA, the structure of the representations of globalisation among the respondents to the questionnaire is studied. The MCA leads to a two-dimensional geometric model of data; the first axis is interpreted as an axis that opposes favourable to unfavourable attitudes, and the second axis opposes protecting agriculture to preserving employment and the environment. Descriptively, we observe differences between the main parliamentary groups of respondents.

Here, the problem is *not* to generalize from the sample of respondents to the population of all members of Parliament as, clearly, the random sampling framework is not applicable. The questions pertain to the set of respondents itself, considered as the object of interest. Therefore, from the cloud of individuals, we proceed to combinatorial inductive analyses (combinatorial typicality tests for Mean and Variance, and homogeneity test) in order to extend descriptive conclusions.

The European Central Bankers Study (§6.3)

In a study on "central bankers in the contemporary global field of power", F. Lebaron analyses the structure of the social space of the European Central Bank (ECB) and, in particular, the collective dynamics inside the ECB. The study population is composed of 63 members of the Governing Council of the ECB. Biographical data were collected and, using specific MCA, clouds of points are constructed.

For these actors, their "economic and monetary approaches" are identified: some members are, for example, supposed to be "hawks" or "doves" (which means more or less concerned with inflation and close to economic-monetary orthodoxy). Differences between "hawks" and "doves" are descriptively observed. Combinatorial inferences (geometric typicality test and homogeneity test) are conducted to be sure that the observed differences are *not due to chance.*

Note that, in this case, we have all members of the ECB; the classical problem of inferring on characteristics of a population on the basis of random sample is out of order.

Cognitive Tests and Education (§6.4)

In a study on metacognitive factors in scientific problem-solving strategies, Rozencwajg (1994) studied a group of 42 seventh graders (ages 12–13). The students, boys and girls, belong to two middle schools that are very different in terms of socio-economic environment.

For our study, we retain six cognitive tests. Then we construct a cloud of individuals with a structuring factor defined by the crossing of gender and type of school.

In such a case, it is classical to study canonical decompositions of the crossing of the two factors and, in particular, their interaction.

A study of complex design goes beyond the scope of the book, but concerning this example, we try to briefly present solutions based on unrestricted permutation of raw data.

About software

The analyses of the case studies presented hereafter are made using Coheris SPAD software and R scripts we wrote ourselves and that are interfaced with SPAD.

Then, data management and GDA analyses are performed using SPAD and inductive analyses are conducted using our R scripts interfaced with SPAD. Thus, we present a menu-driven solution that is very easy to handle.

The steps of the computations are described at the end of each case study.

The data files of the case studies, the Spad projects and the R scripts are available from the authors and from the website of the first author.

6.1   The Parkinson Study

In extensive research on Parkinsonian patients' gait, Ferrandez and Blin (1991) studied an *experimental group* of patients together with a *reference group* of healthy elderly subjects.

There were 15 *patients* (women and men) with ages between 50 and 90 (mean age 72) and heights between 1.50 and 1.85 m (mean height 1.62) and 45 *healthy subjects* (women and men) with ages in the range of 60–92 (mean age 74) and heights in the range of 1.48–1.80 m (mean height 1.61)[3].

In each situation, for every subject, six numerical variables pertaining to gait were recorded, namely *Velocity*, *Stride length*, and durations of *Swing*, *Cycle*, *Stance* and *Double-Support* (see Figure 6.1.1).

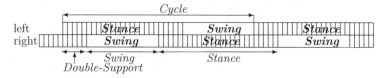

Figure 6.1.1
Description of the four duration variables for two strides.

The design is as follows. There are two groups of subjects, namely an experimental group of 15 Parkinsonian patients, observed twice, before and after drug intake (L-Dopa) and a reference group of 45 healthy subjects, observed once. As a result, the overall data set consists of three groups of six numerical observations, namely one group for the 45 healthy subjects and two matched groups for the 15 Parkinsonian patients; hence $45 + (2 \times 15) = 75$ (6-dimensional) observations.

The *Parkinson study* thus involves multivariate numerical data, with a *genuine design* and a *treatment factor*. To analyze such data, we will combine on the one hand, the approach of Geometric Data Analysis (GDA) by representing data as clouds of points and proceeding to the exploration of these clouds[4], and on the other hand, inductive analyses about the effect of treatment and the differences between patients and healthy subjects.

The main *questions* that will provide a guide throughout the study can be informally stated as follows:

Question A. Does the drug have an effect?

Question B. Do patients after drug intake become more similar to healthy subjects?

---

[3]The description of the tasks can be found in Ferrandez and Blin (1991), together with conventional statistical analyses. Our thanks go to the authors who communicated their data to us.

[4]For an extensive geometric descriptive analysis, see Le Roux (2014a, pp. 347–361).

6.1.1 Overall Cloud of Points

In GDA, stage one consists in representing the data set as a cloud of points in a multidimensional Euclidean space. There are six variables, therefore, at the start, the space is 6-dimensional. In order to construct the cloud of points, we use Principal Component Analysis (PCA)[5] and we begin by studying the structure of data and interpreting axes.

Principal component analysis

Since the six variables are not on a common scale, the chosen GDA method is a PCA based on correlations. It is performed on healthy subjects' data; patients' data are put as supplementary elements (numerical results can be found in Table 6.1.1, p. 158).

PCA reveals that clouds are approximately two-dimensional.

- The first two axes account for 97% of the variance of the cloud (Table 6.1.1-a) and the multiple correlations of all variables with the first two principal variables are very strong ($R_{1-2} \geq 0.973$, see Table 6.1.1-b).
- The qualities of representation of healthy subjects' points in plane 1-2 (see Formula 2.4, p. 17) are quite good since they exceed .90 for 35 of them, and go below .50 for only three of them (see column $\cos^2$ in Table 6.1.1-c). Thus, the cloud of healthy subjects is nearly entirely contained in a plane.
- The two clouds of the 15 patients, whose data are put as supplementary elements, are roughly in this plane since the qualities of representation ($\cos^2$ in Table 6.1.1-d) are very high (all are above 0.84).

Therefore, the clouds projected onto the first principal plane will make up the *basic data set*.

Interpretation of axes

The study of the *space of variables* is summarized by the circle of correlations (see Figure 6.1.2, p. 159). It leads to interpreting the first principal axis as a *performance axis*, that is, performances increase from left (poor) to right (fair), and the second axis as a *style axis*, that is, for equal performance, *Stride length* is longer above and shorter below.

Cloud of individuals

In the *space of individuals*, the *overall cloud* of the two-dimensional representation is shown in Figure 6.1.3 (p. 159). The healthy subjects define a subcloud of 45 points (grey points) of the overall cloud; the reference axes are the principal axes of this subcloud and the origin point is its mean point. Patients are represented by two matched subclouds of 15 points each; the matching

---

[5]For a presentation of PCA in the geometric framework, see Le Roux and Rouanet (2004, Chapter 4) or Le Roux (2014a, Chapter 6).

Table 6.1.1
Results of Principal Component Analysis.

Table 6.1.1-a
Variances of axes (eigenvalues) and variance rates.

$\lambda_1 = 3.9927$	$\lambda_2 = 1.8224$	$\lambda_3 = 0.1711$	$\lambda_4 = 0.0073$	$\lambda_5 = 0.0059$	$\lambda_6 = 0.0007$
$\tau_1 = .665$	$\tau_2 = .304$	$\tau_3 = .029$	$\tau_4 = .001$	$\tau_5 = .000$	$\tau_6 = .000$

Table 6.1.1-b
Correlations of initial variables with the first two principal variables ($r_{\ell 1}$ and $r_{\ell 2}$), and the coefficient of correlation multiple $R_{1-2} = \sqrt{(r_{\ell_1}^2 + r_{\ell_2}^2)}$).

	Veloc.	*Length*	*Swing*	*Cycle*	*Stance*	*Db_Supp.*
$r_{\ell 1}$	0.902	0.665	−0.403	−0.868	−0.958	−0.950
$r_{\ell 2}$	0.389	0.725	0.889	0.494	0.256	−0.210
R_{1-2}	0.983	0.984	0.976	0.999	0.991	0.973

Table 6.1.1-c
Coordinates (y_1, y_2) and qualities of representation ($\cos^2$) of the 45 healthy subjects.

	$\cos^2$	y_1	y_2		$\cos^2$	y_1	y_2		$\cos^2$	y_1	y_2
s17	.959	+3.091	−0.828	s32	.993	+2.887	2.015	s47	.499	+0.464	0.027
s18	.992	−2.420	1.632	s33	.486	−0.233	−0.185	s48	.987	−4.165	2.373
s19	.992	+0.501	1.492	s34	.992	+2.437	−0.923	s49	.905	+0.546	−2.253
s20	.886	−0.153	0.651	s35	.994	+1.961	0.951	s50	.976	+0.751	−3.800
s21	.924	+1.040	0.911	s36	.619	+0.475	0.632	s51	.999	−1.144	−1.261
s22	.999	+2.847	0.756	s37	.991	−2.080	−1.215	s52	.892	+0.198	0.981
s23	.990	+1.350	−3.758	s38	.909	−1.703	0.985	s53	.965	−2.319	0.186
s24	.989	+1.899	0.362	s39	.937	+1.802	0.593	s54	.902	+0.643	−0.751
s25	.783	+0.180	0.840	s40	.996	−3.105	−2.882	s55	.319	−0.169	0.343
s26	.964	−2.319	−0.501	s41	.995	−2.412	−1.535	s56	.843	−0.225	1.165
s27	.776	−0.354	0.750	s42	.972	+2.553	0.786	s57	.767	+0.545	0.182
s28	.948	+0.575	0.515	s43	.930	−3.967	−1.042	s58	.976	−1.142	1.243
s29	.999	+2.568	0.774	s44	.904	−1.599	0.983	s59	.958	+3.669	−0.943
s30	.998	−1.828	−0.136	s45	.988	−1.757	−0.690	s60	.998	−2.300	−0.290
s31	.998	−2.105	1.425	s46	.994	+1.667	0.258	s61	.995	+2.853	−0.815

TABLE 6.1.1-d.
Coordinates (y_1, y_2) and qualities of representation ($\cos^2$) of the 15 patients (supplementary elements).

	before drug intake			after drug intake		
	$\cos^2$	y_1	y_2	$\cos^2$	y_1	y_2
s2	.843	−2.212	0.928	.976	−1.286	1.854
s3	.994	−1.655	−1.288	.938	−0.193	−1.562
s4	.979	−0.948	−5.996	.983	−1.546	−4.651
s5	.971	−0.986	−3.174	.876	+0.158	−0.912
s6	.997	−1.073	−5.265	.991	−0.006	−2.335
s7	.998	−1.217	−4.800	.976	+1.535	−4.765
s8	.953	−1.747	−1.167	.914	−3.465	−1.430
s9	.993	−3.544	−1.048	.977	+0.074	−1.737
s10	.964	−5.217	−1.061	.997	−0.076	−1.814
s11	.935	−5.098	−2.461	.978	−1.777	−1.188
s12	.900	−3.877	−3.426	.976	−1.576	−3.270
s13	.816	−1.152	−0.151	.907	−0.420	1.012
s14	.967	−3.993	−1.038	.968	−5.907	1.014
s15	.997	−0.759	−4.111	.997	+0.434	−4.094
s16	.964	−5.820	−0.794	.929	−1.800	−1.069
Means		−2.620	−2.323		−1.057	−1.663

Coordinates of effect-vector: $\overline{d} = (1.563, 0.660)$

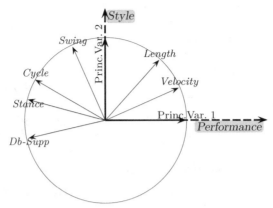

Figure 6.1.2
Space of variables: "circle of correlations".

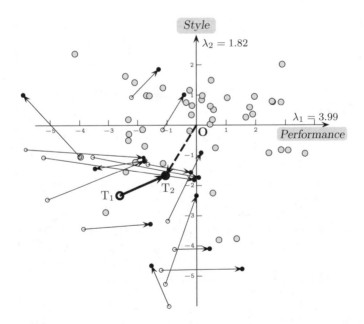

Figure 6.1.3
Overall cloud in plane 1-2: The cloud of 45 healthy subjects (grey circles) with its mean point O; the two subclouds of 15 patients whose points before (○) and after (●) drug intake are joined by arrows and the two mean points T_1 and T_2.

of points is depicted by arrows whose initial points (white points) refer to patients before drug intake and terminal ones (black points) to patients after drug intake. The two mean points for patients (points T_1 and T_2) are joined by an arrow. The mean point of healthy subjects (point O) is joined by an arrow to the one of patients after drug intake.

All the procedures presented in the sequel can be visualized in Figure 6.1.3, which will provide an intuitive guide throughout the study. Geometric intuition, with the subtle interplay between points and vectors, will enable one to follow the strategy without being concerned by the algebraic expressions (e.g., matrix formulations) of the procedures.

Descriptive findings

The geometric analysis (PCA) allows us to give a descriptive answer to questions A and B enunciated on page 156.

1. On the whole, performances are poorer for patients, that is to say, most patient points lie on the left side of Figure 6.1.3.

2. Drug intake appears on the whole to bring patients closer to healthy subjects, especially as regards performance.

3. The mean point of patients after drug intake (point T_2) still lies on the lower left quadrant of Figure 6.1.3 and remains quite distant from the mean point of healthy subjects (point O): roughly speaking, patients have lower performances with shorter strides.

Inductive approach

In MANOVA terms, Question A concerns a *within-subject effect* (with one degree of freedom) and question B a *between-subject effect* (with one degree of freedom). When *Performance* and *Style* are considered jointly, effects are two-dimensional and become geometric vectors that we call *effect-vectors*. In order to try to validate the descriptive results and to ensure that they are not due to chance, we will use the methods of combinatorial inference.

The rest of this section will be devoted to the detailed inductive investigation of Question A (§6.1.2) and Question B (§6.1.3, p. 166). We will carry out *specific analyses*, that is, for each question, we will derive a relevant data set and perform all procedures on that specific data set.

6.1.2 Question A: Effect of Drug

The relevant data set is composed of *two matched groups* of observations (crossed design), that is, of two subclouds of 15 points associated with patients "before" and "after" drug intake.

An individual effect-vector is associated with each patient. It is defined as the deviation from the point "before drug" to the point "after drug" (see

Figure 6.1.3, p. 159). The effect of interest is the geometric vector between the mean point "before drug" (T_1) and the mean point "after drug" (T_2).

The specific cloud is constructed as follows.

1. Restrict the data to the two subclouds of patients (points ○ and •). Join by an arrow the two points of each patient yielding 15 individual effect-vectors together with the mean effect-vector; see Figure 6.1.4 (already shown in Figure 6.1.3 representing the overall cloud).

2. Translate effect-vectors so that their initial points coincide in order to obtain a "field of vectors" as shown in Figure 6.1.5 (point O represents the null effect).

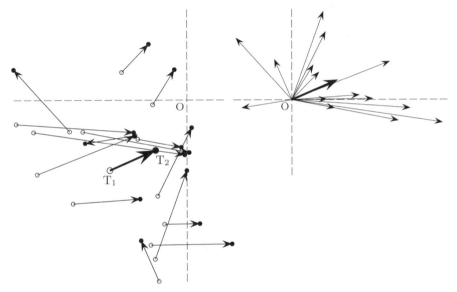

Figure 6.1.4
The two matched subclouds before (○) and after (•) drug intake and the mean effect (in bold).

Figure 6.1.5
The field of effect-vectors with the mean effect-vector (in bold).

3. Replace the vectors by their endpoints, hence individual effects are now conceptualized as points, called effect-points. The endpoint of the mean effect-vector is the mean point of individual effect-points.

Now the *specific cloud* consists of the cloud of 15 individual effect-points, together with its mean point (point G) and the origin point (reference point O), as shown in Figure 6.1.6 (p. 162).

All specific analyses (descriptive and inductive) regarding Question A will be based on this specific cloud.

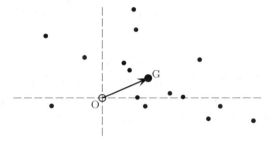

Figure 6.1.6
Question A: specific cloud of effect-points with its mean point and the mean effect $\overrightarrow{OG}$.

First phase: Descriptive analysis

The covariance structure of the specific cloud is shown in Figure 6.1.7 through its indicator ellipse[6].

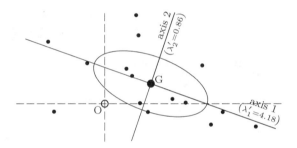

Figure 6.1.7
Question A: Cloud of effect-points with its indicator ellipse.

Magnitude of the mean effect

The orientation of the mean effect-vector shows that, on average, the drug intake tends to improve the performance (Figure 6.1.4, p. 161). The *magnitude of the effect* (effect size) will be assessed with respect to a reference effect. As an index of magnitude, we will take the Mahalanobis norm with respect to the covariance structure of the cloud of effect-points (see §2.3.2, p. 22). The value of this index for the mean drug effect-vector is equal to 1.37. By any standard, such a value can be deemed as large.

Thus, the *descriptive conclusion*:

The mean effect of drug is large in magnitude.

---

[6]The variances along principal axes (eigenvalues) are equal to $\lambda_1' = 4.182$ and $\lambda_2' = 0.859$, the orientation of the cloud is given by the angle of its first principal axis with the performance axis ($\theta_1 = -19.25$ degrees) (detailed calculations are given on page 163).

Recall that the indicator ellipse of a cloud is the principal ellipse whose semi-axes are equal to the standard deviations along principal axes; see §2.3.2 (p. 21).

Numerical results

We will give here some numerical results. They are not necessary to understand the text but can be of some help.

The coordinates of effect-points (differences between "before" and "after" coordinates) can be deduced from Table 6.1.1-d (p. 158). They are given in Table 6.1.2.

Table 6.1.2
Coordinates of effect-points, of the mean effect vector; covariance matrix; magnitude index.

	y_1	y_2
$s2$	0.926	0.925
$s3$	1.462	−0.274
$s4$	−0.597	1.345
$s5$	1.144	2.262
$s6$	1.066	2.930
$s7$	2.752	0.034
$s8$	−1.718	−0.263
$s9$	3.618	−0.688
$s10$	5.142	−0.754
$s11$	3.321	1.273
$s12$	2.302	0.156
$s13$	0.732	1.163
$s14$	−1.914	2.053
$s15$	1.193	0.017
$s16$	4.020	−0.275

Coordinates of the mean effect-vector:
$$\mathbf{d} = \begin{bmatrix} 1.563 \ 0.660 \end{bmatrix}$$

Matrix of the covariance endomorphism Mcov of the cloud of effect-points:
$$\mathbf{V} = \begin{bmatrix} 3.820 & -1.034 \\ -1.034 & 1.220 \end{bmatrix} \quad \begin{array}{l} \lambda'_1 = 4.1816 \\ \lambda'_2 = 0.8587 \end{array}$$

The square of the magnitude index is $\mathbf{d}^\top \mathbf{V}^{-1} \mathbf{d}$:
$$\begin{bmatrix} 1.563 \ 0.660 \end{bmatrix} \begin{bmatrix} 0.3397 & 0.2881 \\ 0.2881 & 1.0639 \end{bmatrix} \begin{bmatrix} 1.563 \\ 0.660 \end{bmatrix}$$
$$= 1.889 \ (= 1.37^2)$$

Second phase: Inductive analysis

At this point, we have established that the drug effect is descriptively large. We now try to corroborate the descriptive conclusion by using inductive analysis in order to answer the natural query:

Is the observed drug effect a genuine one, or might it be due to chance?

In other words, we may wonder whether the data are in favour of a non-null effect of drug or not.

This case is a privileged application of *geometric typicality test*, as described at the beginning of Chapter 4 (see p. 65).

If the treatment has no effect, then for each patient, the "before" point can be permuted with the "after" point: a *permutation* within any pair of patient's points is as likely as the reverse; geometrically, this means that each effect-point may be exchanged with its symmetrical with respect to the reference point O. So we will use the *geometric typicality test* in order to answer the question.

By applying the geometric typicality test to the cloud of the 15 effect-points and taking point O (representing the null effect) as a reference point, we obtain the following results.

Permutation cloud

The permutation cloud consists of $2^{15} = 32\,768$ points (see Figure 6.1.8, the angle of the first principal axis of the permutation cloud with the horizontal one is equal to 0.03 degree).

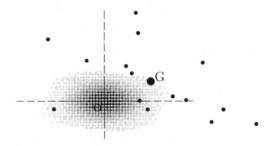

Figure 6.1.8
Cloud of effect-points (black points) and permutation cloud (the grey levels of points depend on the density).

Permutation distribution of test statistic and combinatorial p-value

The histogram of the permutation distribution of the test statistic D_O^2 (see Equation 4.2, p. 77) is depicted in Figure 6.1.9.

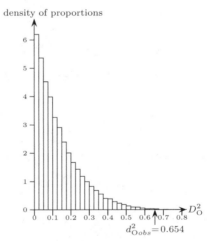

Figure 6.1.9
Question A: permutation distribution of test statistic D_O^2 (based on $2^{15} = 32768$ possible samples); the observed value is $d_{Oobs}^2 = 0.654$.

The observed value of the test statistic depends on the index of magnitude of effect. From the reciprocity formula (see Equation 4.4, p. 77), we obtain $d_{Oobs}^2 = 1.889/(1 + 1.889) = 0.654$.

The *combinatorial p-value* is equal to $36/2^{15} = 0.001$.

Figure 6.1.10 shows the geometric interpretation of the *p*-value.

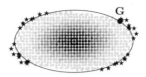

Figure 6.1.10
Sample Cloud. Interpretation of the combinatorial *p*-value: 36 points are on or outside the ellipse of the sample cloud (centred at O) passing through point G.

Thus the conclusion is:

The mean effect of drug is statistically significant at level 0.05: the data are in favour of an effect of drug.

Compatibility region

The 95% *compatibility region*, that is, points compatible with the observed mean point G at level $\alpha = 0.05$ (see §4.2.4, p. 80), is determined from κ values obtained from 250 random lines passing through point G.

The mean of the 2×250 values is equal to 0.767, the minimum is 0.752, the maximum is 0.776; hence the adjusted 95% compatibility region is determined by the principal ellipse of cloud of effect-points with scale parameter $\kappa = 0.767$ (see Figure 6.1.11).

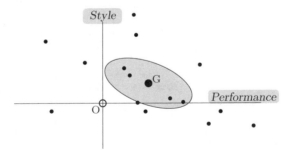

Figure 6.1.11
Question A: adjusted 95% compatibility region ($\kappa = 0.767$).

As shown in Figure 6.1.11, the point O is outside the ellipse: points G and O differ significantly. Moreover, the compatibility region is almost entirely located in the right upper quadrant, which supports a positive effect of drug for both performance and style.

6.1.3　Question B: Comparing Patients after Drug Intake with Healthy Subjects

We will now compare the group of patients after drug intake with the one of healthy subjects taken as a reference (Bienaise and Le Roux, 2017). The steps are similar to those of Question A.

Constructing the specific cloud

The specific cloud is obtained by restricting the overall cloud to the subcloud of healthy subjects (reference cloud) and to the subcloud of patients after drug intake, as shown in Figure 6.1.12.

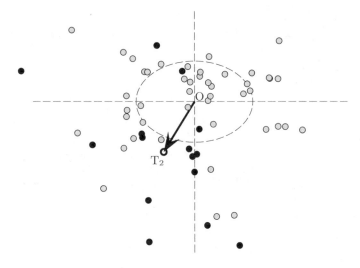

Figure 6.1.12
Reference cloud of the 45 healthy subjects with its mean point O and its indicator ellipse and cloud of the 15 patients after drug intake with its mean point T_2, together with the deviation vector $\overrightarrow{OT_2}$, in plane 1-2.

The *effect of interest* is the deviation between the two mean points O and C, namely the geometric vector $T_2 - O$.

First phase: Descriptive analysis

The orientation of the deviation-vector $\overrightarrow{OT_2}$ shows that the mean point of patients after drug intake is still at the bottom left of the diagram.

Magnitude of the observed deviation

As a magnitude index of the deviation between mean points, we take the Mahalanobis norm associated with the covariance structure of the reference cloud (see p. 22). By Proposition 2.12 (p. 20), we obtain $|\overrightarrow{d_{obs}}|^2 =$

$(-1.057)^2/3.9927 + (-1.663)^2/1.82237 = 1.798 = 1.341^2$. By regarding this value as large, we conclude:

> Descriptively, *the mean difference between patients after drug intake and healthy subjects is large.*

Second phase: Inductive analysis

The observed deviation is of large magnitude, therefore we now attempt to evaluate the atypicality of the group of Parkinsonians by addressing the following query:

> *Is it possible to assimilate patients after drug intake to healthy subjects?*

The answer is provided by performing the *combinatorial typicality test* presented in Chapter 3.

Sample cloud

The number of *possible samples of size 15* that can be extracted from the reference group of healthy persons is the binomial coefficient $\binom{45}{15} = 3.44 \times 10^{11}$. This number is too large to construct the whole set of possible samples. So a subset of 1,000,000 possible samples is generated with the Monte Carlo method (see Figure 6.1.14, p. 168).

Distribution of test statistic and combinatorial p-value

The histogram of the distribution of the test statistic (D^2) based on 1,000,000 samples is shown in Figure 6.1.13.

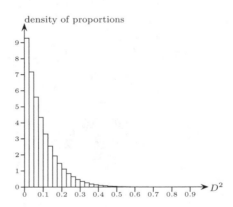

Figure 6.1.13
Question B: distribution of test statistic D^2 based on 1,000,000 samples among the $\binom{45}{15}$ possible samples; the observed value is $D^2_{obs} = 1.798$.

Among the 1,000,000 samples, no mean point of possible sample clouds is found outside of or on the principal ellipse of the cloud of the healthy subjects passing through point C, hence $p = 0/1000000$ (see Figure 6.1.14). Hence, the group of patients after drug intake is *atypical* of the reference set.

> *The data are in favour of a difference between patients after drug intake and healthy subjects (p < 0.001).*

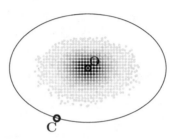

Figure 6.1.14
Sample Cloud. Interpretation of the observed level: there is no point on or outside the ellipse of the sample cloud (centred at O) passing through point C ($p = 0.000$).

Compatibility region

We now determine the 95% compatibility region, that is, the set of points that are compatible with point T_2 at level $\alpha = 0.05$. This region is depicted in Figure 6.1.15: it is delineated by a principal ellipse of the cloud of the healthy subjects with $\widehat{\kappa} = 0.514$, which is translated so that its centre is point T_2.

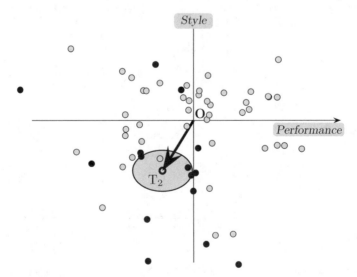

Figure 6.1.15
Clouds of the 45 healthy subjects (reference cloud, grey points); cloud of the 15 patients after drug intake (black points) with its mean point T_2; 95% compatibility region ($\kappa = 0.514$).

The mean point O of the group of healthy subjects is outside the ellipse: the group of patients after drug intake is atypical of the reference set, at level .05. Moreover, the compatibility ellipse being located in the lower left quadrant of Figure 6.1.15, all points in the other three quadrants lead to significant results: the patients cannot be assimilated to healthy subjects, as they are less performing and make smaller steps.

6.1.4 Concluding Comments

As a conclusion to this study, we can assert that the data are in favour of an effect of drug on the Parkinsonians' gait, but the difference between patients after drug and healthy subjects remains significantly large.

In this case study, for question A, the preferred method is the permutation test. This is partly due to the smallness of the data size, which makes the assumptions required in the parametric framework (such as normality) unverifiable. For question B, the issue is to know if the patients after drug intake can be likened to samples of healthy subjects. In this case, the combinatorial typicality test seems to be the most appropriate.

6.1.5 Steps of the Study Using R and Coheris SPAD Software

We now present the analysis carried out using Coheris SPAD with interfaced R scripts (see Figure 6.1.16). At the very beginning, there is the *import of the*

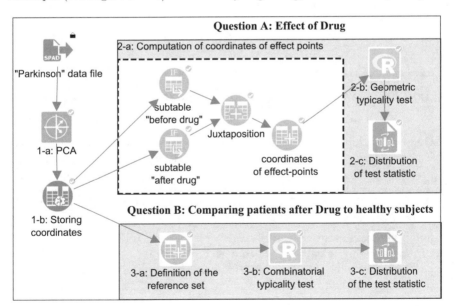

Figure 6.1.16
The SPAD project.

data set, a "SPAD data archive" file called `Parkinson Data`, which is stored in the SPAD project.

The *first step* is the construction of geometric clouds. For that, we perform the PCA of the six variables with standardization of the six variables (parameters: `Normed (correlations)`) with the healthy subjects as active elements and Parkinsonians as supplementary ones (Figure 6.1.16, p. 169, icona 1-a). Then we store the first two principal coordinates of all subjects (healthy subjects, Parkinsonians before and after drug intake) (icona 1-b).

The *second step* consists in answering Question A.

2-a We compute the coordinates of the effect-points on axes 1 and 2 (selection of subtables "before" and "after", then juxtaposition of the tables and calculation of the differences).

2-b We perform the geometric typicality test using the interfaced R script;

2-c From the values of the test statistic exported to SPAD, we draw the histogram of the statistic distribution.

The *third step* consists in answering Question B.

3-a We construct an indicator variable in order to define the reference set (healthy subjects).

3-b We perform the geometric typicality test using the interfaced R script; the group is defined by choosing the supplementary data "after"" drug intake.

3-c From the values of the test statistic exported to SPAD, we draw the histogram of the statistic distribution.

6.2 The Members of French Parliament and Globalisation in 2006

On May 29, 2005, a referendum on the European Constitution[7] took place, in France. There was a majority of "no" votes. One year later, a research team of *SciencesPo Paris*[8] decided to study the *representations of globalisation* among the members of the French Parliament. For this purpose, they developed a questionnaire, which was sent to the 577 Members of Parliament (MPs).

The respondents to the questionnaire are 163 in number. They are divided into parliamentary groups as follows: 92 respondents (out of 364) belong to the UMP group (the right-wing), 48 (out of 150) belong to the PS group (socialist party, the left-wing), 15 (out of 30) belong to UDF group (the centre-wing), 5 (out of 22) belong to Communist party, 3 are not affiliated with any group.

---

[7]The question was: "Do you approve the bill authorising the ratification of the treaty establishing a Constitution for Europe?"

[8]This study was led by Zaki Laïdi, Cynthia Fleury and Brigitte Le Roux; it was a part of a master's dissertation for a group of 12 students.

The questionnaire comprises a general question on the positivity or negativity of the globalisation process. The other questions fall under four main headings, namely: the winners and losers of globalisation; unemployment and relocations; the role of the World Trade Organization; the problems of France with regard to globalisation.

In the study presented here, 12 questions, which are the most characteristic of each theme, have been retained.

The main research questions can be stated as follows:

1. What are the main "representations" of globalisation among the MPs?

2. Are there atypical parliamentary groups?

3. Can we say that the PS group (left) and the UMP group (right) are different with regard to their representations of globalisation?

6.2.1 Elementary Descriptive Analysis

To begin with, we provide a few comments on the answers to the twelve questions[9] used for the construction of the geometric cloud of MPs.

- *Globalisation process* (1 question)
  Among the 92 respondents belonging to the UMP group (right), 39 of them say that globalisation is a decisively positive *process* while, in the socialist group (left), they are only 2 out of 48. On the contrary, more than one third (18) of the members of the PS group say that globalisation is an overall negative process, while they are only 3 out of 92 in the UMP group.

- *Winners and losers of globalisation* (2 questions)
  All the 48 members of the PS group think that the *beneficiaries* of globalisation are multinationals and the financial markets. The respondents belonging to the right party (UMP) are more divided: 42 choose this response and 45 answer that those who are benefiting are the consumers or everyone. As for the *losers*, the PS respondents cite the workers, while a majority of the UMP ones cite the farmers.

- *Employment* (3 questions)
  The cleavage between the UMP and PS groups is still strengthening regarding the *employment* problems at the level of electoral division: 34 PS members (out of 48) think that they are the consequence of the global competition, whereas 66 UMP members (out of 92) answer that the problems are more linked to the structure of the French labour market than to globalisation.
  As far as the labour market is concerned, three quarters of the members

---

[9]The Globalisation questionnaire and its main findings have been commented upon by political scientists in the newspaper "Le Figaro" (May 29, 2006), and Telos Website.

of the UMP group consider that the *job flexibility* ought to be increased, which is rejected by all but one of the PS members.

Nearly all UMP members think that *job insecurity* comes from the existing system—overly protective for those who have a job and too closed for those who are looking for a job—while nearly all PS members think that it is due to the ultraliberal policy.

- *Business lines and priority sectors* (2 questions)

  Seventy-four UMP members (out of 92) think that it is normal that certain *business lines* are cut out for the benefit of other countries in order to redeploy industries with a high added value. This view is shared by 26 PS respondents (out of 48) but rejected by 19 others who think that these business lines must be maintained at all costs.

  The respondents of the two groups are also opposed on the definition of the sectors that Europe ought to protect as a *priority*: the public service for the PS group (77%), agriculture for the UMP group (43%).

- *The role of the World Trade Organization (*WTO*)* (2 questions)

  Twenty-nine members of the PS group (out of 48) say that the *WTO* is going too far in opening up markets, whereas it is the opinion of only 13 out of 92 UMP members, and 75 say that the WTO does its best to set up a framework for the liberalisation.

  Among *priority* interests that France must defend at WTO, 35 of the PS group say the recognition of a social clause, while 36 (out of 92) of the UMP group advocate the protection of agriculture and 28 the access of the markets of the developing countries to French companies.

- *France and globalisation* (2 questions)

  Fifty-nine percent in the UMP group think that *France faces globalisation* with more difficulty than the other European countries while 73% in the PS group say that it is with equal difficulties.

  Almost all UMP respondents (86 out of 92) think that the *problems of France* are more the consequence of its internal blockages than of globalisation itself, while three-quarters of the members of the PS group (34 out of 45) think exactly the opposite.

6.2.2   The GDA of the Questionnaire

The representation of data as *clouds of points in a geometric space* is made through a Multiple Correspondence Analysis (MCA). We use specific MCA, a variant of MCA that permits to discard the categories "no-answer" and "other" from the determination of principal axes[10]. More precisely, the analysis is performed from the answers of the 163 respondents to 12 questions of the questionnaire, with 29 active response categories.

---

[10]See Le Roux (1999); Le Roux and Rouanet (2004, Chapter 5, pp. 203–210). For an elementary presentation, see Le Roux and Rouanet (2010, Chapter 3).

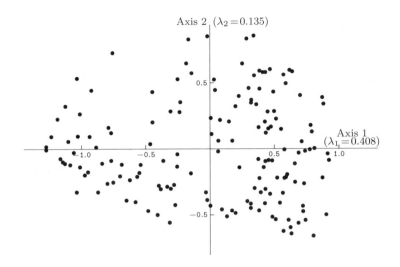

Figure 6.2.1
Cloud of the 163 MPs in the first principal plane.

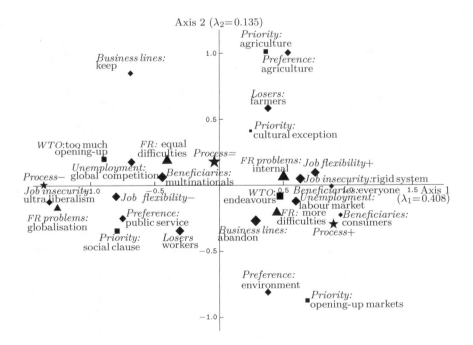

Figure 6.2.2
Cloud of 29 categories in the first principal plane (sizes of markers are proportional to response frequencies; the names of questions are written in italics).

Dimension of the cloud

The first axis is important ($\lambda_1 = 0.408$, variance rate: 28%) and well separated from the second one ($\lambda_2 = 0.135$, variance rate: 9%). After Axis 3, the variances of axes decrease regularly. The first two axes take into account 37 percent of the variance of the cloud and their cumulated modified rate reaches 95 percent.

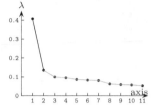

Then[11], we retain the first principal plane for our study.

The cloud of the 163 MPs[12] who answered the questionnaire and the cloud of the 29 categories are represented in plane 1-2 and depicted in Figures 6.2.1 and 6.2.2 (p. 173).

Interpretation of axes

Here, we use the cloud of categories. The interpretation of an axis is based on all categories whose contributions to axis exceed 2.5%.

For the *interpretation of Axis 1*, we select 17 categories contributing together to 85 percent of the variance of the axis. Hence, they provide a good summary of Axis 1. These 17 categories are depicted in Figure 6.2.3.

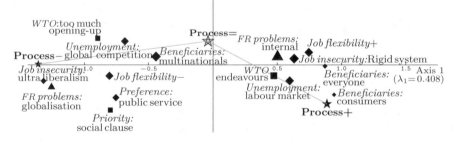

Figure 6.2.3
Interpretation of Axis 1: 17 categories most contributing to Axis 1 (we add "*Process=*"), depicted in Plane 1-2.

The following categories are found on the right of Axis 1: globalisation profits to everyone and to consumers; the problems of unemployment are more linked to the structure of the labour market than to globalisation; the flexibility of employment is not too large; job insecurity is a consequence of ultraliberal policy; the WTO does its best to manage liberalisation and exchanges.

---

[11] For modified rates in specific MCA, see Le Roux and Rouanet (2010, p. 63).
[12] The cloud of the MPs with their names can be found on the website:
www.telos-eu.com/fr/vie-politique/deputes-et-mondialisation/
que-pense-votre-depute-de-la-mondialisation.html.

By contrast, on the left of Axis 1, one finds : globalisation profits to multi-nationals; the problems of unemployment are the direct consequence of globalisation; the flexibility of employment is large enough; job insecurity is a consequence of the rigid system; the WTO does its best to manage liberalisation and exchanges; the WTO goes too far in the opening up of markets.

Moreover, the three categories about the process of globalisation (written in bold in Figure 6.2.3) are ranked along the first axis, from the left with a negative judgement, to the right with a positive judgement.

The first axis opposes a favourable to an unfavourable overall attitude toward globalisation.

Similarly, we *interpret Axis 2*. The nine categories whose contributions to Axis 2 exceed 2.5% are depicted in Figure 6.2.4. Together they contribute to 88 per cent of the variance of Axis 2.

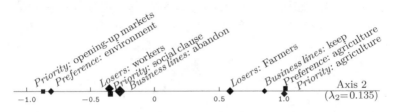

Figure 6.2.4
Interpretation of Axis 2: 9 categories most contributing to Axis 2 (depicted on axis).

The second axis opposes protecting agriculture to preserving employment and the environment.

Cloud of individuals

Furthermore, we consider that belonging to a parliamentary group defines a *structuring factor* of the set of respondents[13]. Each parliamentary group defines a subcloud of the overall cloud. We will now comment on the locations of the mean points and the dispersions of the subclouds associated with the three groups.

The coordinates of the mean points of subclouds and the scaled deviations from mean points to the origin are given in Table 6.2.1 (p. 176). A scaled deviation between two points is a descriptive indicator of the magnitude of deviation between these points. It is regarded as notable if it is greater than 0.4 and small if not (see Chapter 2, p. 22).

Recall the formulas of scaled deviations. Let $\overline{y}_1$ and $\overline{y}_2$ denote the principal coordinates of a mean point on Axis 1 and Axis 2. The scaled deviation from

[13]See Le Roux and Rouanet (2010, Chapter 4) and Le Roux (2014b).

mean point to a reference point (here the origin) is equal to $d_1 = (\bar{y}_1 - 0)/\sqrt{\lambda_1}$ for Axis 1, to $d_2 = (\bar{y}_2 - 0)/\sqrt{\lambda_2}$ for Axis 2, and to $\sqrt{(d_1^2 + d_2^2)}$ for Plane 1-2 (see Chapter 3, §3.5, especially pp. 53 and 54).

Table 6.2.1

For each subcloud: coordinates of the mean point in plane 1-2, scaled deviations on Axis 1, Axis 2 and in plane 1-2 (notable deviations are written in bold).

	size	coordinates	scaled deviations Axis 1	Axis 2	Plane 1-2
UMP	92	(0.425, 0.060)	**0.67**	0.16	**0.68**
UDF	15	(0.187, −0.065)	0.19	−0.18	0.34
PS	48	(−0.741, −0.099)	**−1.16**	−0.27	**1.19**

The cloud of the respondents with the mean points and the concentration ellipses[14] of the subclouds of the main three parliamentary groups, namely UMP (right), UDF (centre) and PS (left), is depicted in Figure 6.2.5.

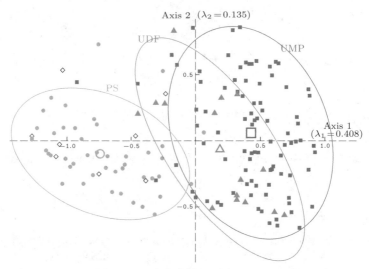

Figure 6.2.5

Cloud of the 163 respondents (MPs) in plane 1-2, marked according to their parliamentary group: 92 UMP (■), 15 UDF (▲), 48 PS (●) and 8 other (◇). The subclouds of the main three groups with their mean points and their concentration ellipses.

Descriptive conclusions *are that, on Axis 1 and in Plane 1-2, deviations are notable for the* UMP *and* PS *groups and small for the* UDF *group, and that, on Axis 2, all deviations are small.*

---

[14]Recall that the concentration ellipse of a subcloud is a geometric summary of a subcloud (see Chapter 2, §2.3.2, p. 21).

From Figure 6.2.5 (p. 176), we see that the dispersions of subclouds are very different. The *variance ratio*, that is, the variance of the subcloud divided by the overall variance, can be taken as a descriptive indicator of importance. All variance ratios are greater than 1.2 or less than 0.53 (see Table 6.2.2): *descriptively, they are notable.* In particular, on Axis 2, the variances of the UMP and UDF subclouds are much greater than that of the reference one, whereas the variance of the PS subcloud is much smaller.

Table 6.2.2
For each subcloud, variances and variance ratios on Axis 1, Axis 2 and in Plane 1-2 (notable deviations are written in bold).

	size	variances			variance ratios		
		Axis 1	Axis 2	plane 1-2	Axis 1	Axis 2	plane 1-2
UMP	92	0.103	0.162	0.265	**0.25**	**1.20**	**0.49**
UDF	15	0.108	0.179	0.287	**0.26**	**1.33**	**0.53**
PS	48	0.122	0.062	0.185	**0.30**	**0.46**	**0.34**
overall	163	0.408	0.135	0.543			

Descriptive findings

It seems clear that the structure of the representations of globalisation for the MPs cannot be characterized as one-dimensional, but, at least, two-dimensional. Moreover, the geometric analysis reveals a differentiation between parliamentary groups on Axis 1 and in plane 1-2 (Figure 6.2.5, p. 176):

With respect to globalisation, PS *members tend to lie on the left side of the plane (unfavourable), and* UMP *members on the right side (favourable).*

The UMP *and* UDF *groups are very dispersed on the second axis, whereas the* PS *group is rather compact.*

Intuitively, the PS *and* UMP *groups appear to be quite distinct.*

Inductive approach

Following the descriptive conclusions, we address questions such as:

Question A. Is the subcloud of each group atypical from the overall cloud according to mean point? (typicality problem for location parameter)

Question B. Is the subcloud of each group atypical from the overall cloud on Axis 2 according to variance? (typicality problem for scale parameter)

Question C. Are the PS and UMP groups heterogeneous or not? (homogeneity problem)

In what follows, we work on the projection of the cloud of respondents onto the first principal plane, and we study the subclouds corresponding to the three main parliamentary groups. The number of respondents in each group is equal to 92 (UMP), 15 (UDF) and 48 (PS).

6.2.3 Question A. Typicality of Subclouds for Mean

For answering question A, in order to extend the descriptive conclusions, we will proceed to inductive analyses of the cloud projected on the first principal axis and on the first principal plane. We will use the combinatorial methods of typicality, which are presented in Chapter 3.

> Recall the *principle of the combinatorial typicality methods.*
> The basic idea is to take the set of the 163 respondents as a *reference set* and to compare *each group* to the samples of the reference set—namely, the *subsets* of the reference set—that have the same size as the group. Then, a test statistic is chosen, and the proportion of possible samples, whose value of the test statistic is more extreme than (or as extreme as) that of the group, is computed. This proportion defines the typicality level of the group for the test statistic with respect to the reference set.

The cardinality of the sample set (set of possible samples) is equal to $\binom{163}{92}$ for the UMP group and to $\binom{163}{48}$ for the PS group. These numbers are very large (greater than 10^{40}), therefore we will use the MC method by drawing one million random samples from the sample set (see Remark (1), p. 44).

Typicality on the first principal axis (see Chapter 3, §3.5.2, p. 52)

The basic numerical data set consists of the coordinates of points on Axis 1. For each group, we determine the deviation between the mean of coordinates of the group and the mean of the reference set. *Descriptively*, the deviations are notable for the UMP and PS groups.

Then, taking the mean as a statistic of interest, we will assess the *typicality level* of these two groups by performing the combinatorial typicality test for *Mean* (the deviation for the UDF group is small, hence, there is no point in doing the test). Then, the combinatorial distributions of the *Mean* for samples of the same size as that of the groups are determined. For UMP and PS groups ($n_c = 92$ and $n_c = 48$), they are depicted in Figure 6.2.6.

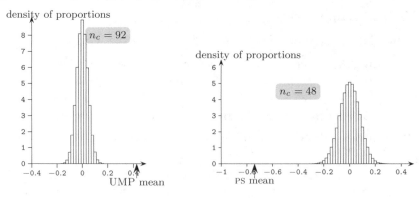

Figure 6.2.6
Combinatorial distributions of the *Mean* for sample sizes $n_c = 92$ and $n_c = 48$.

For the UMP subcloud, whose mean is on the right side of Axis 1, and for the PS subcloud, whose mean is on the left side of Axis 1, the p-values are extremely small (much smaller than conventional α-levels: $p \ll 0.001$).

Hence the conclusions:

On the first axis, the UMP *and* PS *group subclouds are atypical of the overall cloud, on the "favourable" side for* UMP *and on the "un-favourable" side for* PS.

Moreover, we determine the 95% *compatibility interval* for the mean. They are respectively equal to $[+0.338, +0.511]$ for the UMP group, to $[-0.002, +0.377]$ for the UDF group and to $[-0.892, -0.588]$ for the PS group (see Figure 6.2.7).

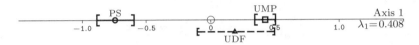

Figure 6.2.7
95% compatibility intervals for PS, UDF, UMP. with observed means and the mean of the reference set ($\bigcirc$).

For the PS group, the interval lies entirely on the left side of the origin, whereas for the UMP group, it lies on the right side.

Remarks. (1) In accordance with the construction of the compatibility region, the reference value 0 is outside the compatibility intervals for the UMP and PS groups, and inside for the UDF group.

(2) Compatibility intervals bring an essential complement to tests because they give the set of values that are compatible with the observed one. To a certain extent, they evaluate the precision of the estimation of the mean.

(3) In our methodology, the typicality level of the UDF group on Axis 1 was not performed since the deviation is descriptively small. Thus, there is no point in inquiring about the typicality testing. It just so happens that the result of the test is not statistically significant ($p = 0.117$): there is no "conflict" between the descriptive conclusion (deviation is small) and the inductive one (result is not statistically significant).

(4) In the same vein, we did not study the typicality level of groups on Axis 2, since all deviations are small (see Table 6.2.1, p. 176). When tests were performed, they were found statistically significant for UMP and PS: *a statistically significant deviation does not mean "important deviation".*

Typicality in plane 1-2 (see Chapter 3, §3.5.3, p. 54)

In the first principal plane, the scaled deviation between the mean point of the subcloud and that of the overall cloud is notable for the UMP subcloud as well as for the PS subcloud (see Table 6.2.2, p. 177), therefore we study the typicality level of these two groups.

The principle of the test is always the same. However the test statistic is the squared M-distance between mean points (see Equation 3.1, p. 36), which depends on the covariance structure of the reference cloud, that is, test statistic takes into account the shape of the reference cloud.

The combinatorial distributions of the test statistic, for sample sizes $n_c = 92$ and $n_c = 48$, are depicted in Figure 6.2.8.

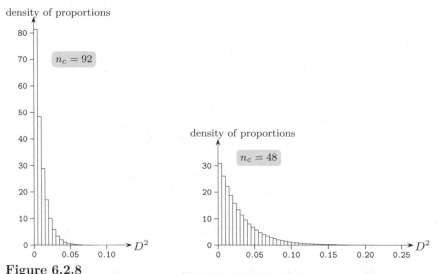

Figure 6.2.8
Combinatorial distribution of test statistic D^2 for sample sizes 92 and 48. The observed values are equal to 0.469 for UMP and 1.42 for PS (too far to be represented in the graphs).

For both groups, the p-values are extremely small (much smaller than conventional α-levels: $p \ll 0.001$). Hence the *conclusion*:

> *In the first principal plane of the overall cloud, the* UMP *and* PS *sub-clouds are atypical of the overall cloud; the deviations between their mean point and the centre of the overall cloud cannot be due to chance—beyond any reasonable doubt—therefore, one is entitled to search for a sociological interpretation.*

In order to specify the essence of the atypicality, we construct the compatibility regions (see p. 55). The 95% compatibility regions are depicted in Figure 6.2.9 (p. 181).

We can see immediately that the PS compatibility region is entirely located in the lower left quadrant of the figure, that is, on the side of unfavourable representations of globalisation in conjunction with the preservation of environment and employment. By contrast, the UMP compatibility region is located in the upper right quadrant, that is, on the side of favourable representations of globalisation in conjunction with agricultural protection. Observe the largeness of the UDF compatibility region, due to the smaller size of the UDF subcloud ($n_c = 15$).

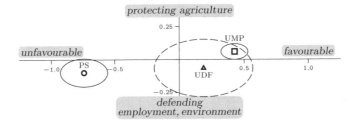

Figure 6.2.9
Combinatorial 95% compatibility ellipses for mean points of UMP ($\kappa = 0.17$), UDF ($\kappa = 0.60$), and PS ($\kappa = 0.30$).

Combinatorial or geometric typicality method? In order to answer question A, we choose the combinatorial typicality methods (Chapter 3). In the combinatorial typicality test, on the one hand, the group is compared to samples of the reference set; on the other hand, the test statistic (the squared M-distance) depends on the *covariance structure of the reference cloud*. As a consequence, the distribution of points of the reference cloud plays an important role in the construction of the test through both sample space and test statistic.

We may have chosen the geometric typicality methods presented in Chapter 4, which are dealing with the typicality of the subcloud mean point with respect to a reference point. The test statistic (see Equation 4.2, p. 77) depends on the *subcloud* and on the *reference point*, namely the centre of the overall cloud. There is no further reference to the overall cloud.

The geometric typicality methods lead to the same conclusions. For instance, in the first principal plane, the p-values of the UMP and PS groups are very small ($p \ll 0.001$), whereas the p-value of the UDF group is equal to 0.096 (non significant). For each group, the adjusted compatibility region (see §4.2.4, p. 80) is defined by a principal ellipse of the group, hence they differ from each other regarding shape. The (adjusted) 95% compatibility regions are depicted in Figure 6.2.10.

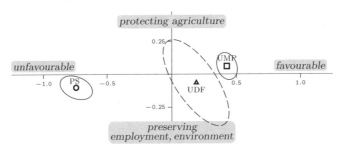

Figure 6.2.10
Adjusted geometric 95% compatibility ellipses for mean points of UMP ($\kappa = 0.26$), UDF ($\kappa = 0.78$), and PS ($\kappa = 0.37$).

6.2.4   Question B. Typicality for Variance on Axis 2

We will now answer question B (see p. 177). Recall that the descriptive conclusion is that, on Axis 2, the UMP and UDF groups are very dispersed, whereas the PS group is rather compact. Then we perform, using MC method with one million samples, the combinatorial typicality test by taking the *Variance* as a test statistic. The distributions are depicted in Figure 6.2.11.

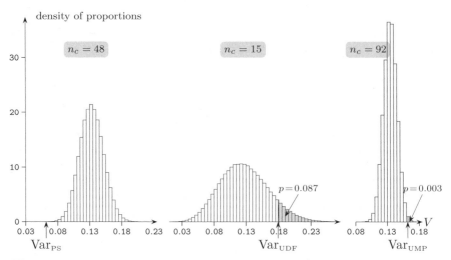

Figure 6.2.11
Histograms of the combinatorial distributions of the test statistic *Variance* for sample sizes 48 (left), 15 (middle) and 92 (right). The observed values of the test statistic are equal to 0.062 for PS (far on the left side of the graph), 0.179 for UDF and 0.162 for UMP.

The p-values are equal to 0.003 for UMP, to 0.087 for UDF and near zero for PS; at usual levels, the result of the test is statistically significant for the UMP and PS groups, and not statistically significant for the UDF group. Thus, the low variance for the PS subcloud and the high variance for the UMP subcloud cannot be due to chance, and we are entitled to search for a sociological interpretation. However, the variance of the UDF subcloud along Axis 2 cannot be said to be atypical of the one of the overall cloud of respondents.

The *compatibility interval* with the upper limit equal to 0.09 for PS and with the lower limit equal to 0.14 for UMP reinforces the *conclusion of compactness of the* PS *group and of scattering of the* UMP *group along Axis 2*.

Figure 6.2.12
95% compatibility intervals for variance with respect to the overall cloud (the variance of the reference set is $\lambda_2 = 0.135$).

6.2.5 Question C. Homogeneity of the UMP and PS groups

We now study the homogeneity of the UMP and PS groups in plane 1-2. We perform a specific comparison, that is, data are restricted to the two groups.

Constructing the specific cloud

The specific cloud is obtained by restricting the overall cloud to the two subclouds associated with the two groups under study, as shown in Figure 6.2.13.

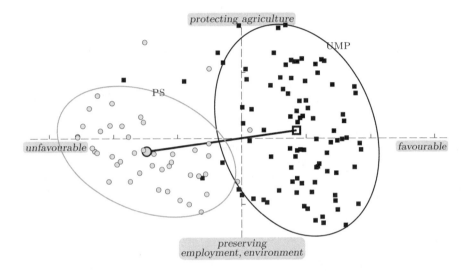

Figure 6.2.13
Specific cloud in plane 1-2: 140 MPs marked according to the two parliamentary groups: 92 UMP (■), 48 PS (●) with the mean point, the concentration ellipse of each group and the deviation between the two mean points.

The number of points of the specific cloud is equal to $92 + 48 = 140$. Its covariance structure is defined by the covariance matrix $\begin{bmatrix} 0.4159 & 0.0130 \\ 0.0130 & 0.1335 \end{bmatrix}$, hence the variance of the cloud is $V_{\text{cloud}} = 0.4159 + 0.1335 = 0.5494$.

The *effect of interest* is the deviation between the two mean points.

Descriptive analysis

From the coordinates of the two mean points (Table 6.2.1, p. 176), we deduce the coordinates of the deviation between these two points $(1.166, 0.1586)$. Applying Formula 2.1 (p. 12), we calculate the variance of the cloud of the two mean points: $\frac{92 \times 48}{140^2}(1.166^2 + 0.1586^2) = 0.3120$. Hence $\eta^2 = 0.3120/0.5494 = 0.57$: this value is notable (see Chapter 2, p. 28).

Descriptively, the deviation between the two mean points is notable.

Inductive data analysis

In order to extend the descriptive conclusion, we use the homogeneity test for two independent groups (see Chapter 5, §5.4, p. 116), more precisely, we perform a specific comparison of the two groups (see p. 109).

> The *principle of the test* is the following one.
>
> If the two groups with sizes n_{c_1} and n_{c_2} are homogeneous, we are free to exchange the observations of the two groups. In permutation theory, a set of $(n_{c_1} + n_{c_2})!/(n_{c_1}! \, n_{c_2}!)$ possible clouds of the same structure is generated by exchanging points between the two subclouds (all arrangements of $n_{c_1} + n_{c_2}$ points into two subclouds of n_{c_1} and n_{c_2} points). With each possible cloud there is attached a difference vector $\overrightarrow{d}$ between the two mean points. Let us consider the squared Mahalanobis distance (D^2) attached to the specific cloud; this squared distance is taken as test statistic (see Chapter 5, §5.4), hence *permutation distribution* of test statistic, which is the analog of a sampling distribution.
>
> The proportion of vectors $\overrightarrow{d}$ that are more extreme than $\overrightarrow{d}_{obs}$, i.e., such that $|\overrightarrow{d}|^2 \geq |\overrightarrow{d}_{obs}|^2$, defines the *p*-value of the permutation test.

The *permutation distribution* of the test statistic is shown in Figure 6.2.14.

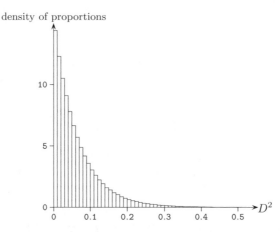

Figure 6.2.14
Histogram of the combinatorial distribution of the test statistic D^2 (the observed value is equal to 3.38).

The observed value of the test statistic is equal to 3.38, hence the *p*-value is extremely small ($p \ll 0.001$). The deviation between the two mean points is highly significant.

> *Beyond any reasonable doubt, in the principal plane 1-2, there is heterogeneity between subclouds* UMP *and* PS*: The deviation between the mean points is not due to chance.*

For the construction of the *compatibility region*, the reference cloud is the within-groups cloud whose deviation between the two mean points is null. The reference cloud is represented in Figure 6.2.15 with its concentration ellipse and the deviation between the mean points of UMP and PS subclouds (observed deviation $\overrightarrow{d}_{obs} = \overrightarrow{\mathrm{OD}_{obs}}$).

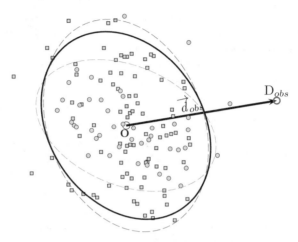

Figure 6.2.15
The reference cloud with its concentration ellipse (in black), the two subclouds with their concentration ellipses (dashed line) and deviation between the two mean points.

The adjusted 95% compatibility region for the observed deviation $\overrightarrow{d}_{obs}$ is defined by the translation of the principal κ-ellipse of reference cloud by vector $\overrightarrow{d}_{obs}$ (its centre is $\mathrm{D}_{obs} = \mathrm{O} + \overrightarrow{d}_{obs}$) and with scale parameter $\kappa = 0.44$. It is depicted in Figure 6.2.16.

Figure 6.2.16
For UMP-PS deviation, adjusted 95% compatibility ellipse (translation of the principal κ-ellipse of the reference cloud by vector $\overrightarrow{d}_{obs}$), with $\kappa = 0.44$.

The deviations that are compatible with $\overrightarrow{d}_{obs}$ are represented by vectors whose initial point is O and terminal point belongs to the interior of the ellipse.

All deviations compatible with $\overrightarrow{d}_{obs}$ are large, which reinforces the conclusion of heterogeneity of the two groups.

6.2.6   Concluding Comments

For the three questions raised on page 177, the problem is *not* to generalize from the sample of respondents to the population of all the members of parliament; the questions pertain to the set of respondents itself, considered as the object of interest. Indeed, had all members answered the questionnaire, and the descriptive conclusions been found similar, the problem would have remained the same. Clearly the random sampling framework (inferring on characteristics of the population on the basis of a random sample) is not applicable. For such nonrandom problems, the methods of *combinatorial inference* provide a well suited solution.

6.2.7   Steps of the Study Using R and Coheris SPAD Software

We now present the SPAD project that performs the statistical analyses that are previously presented. It comprises three folders (Figure 6.2.17). The Process diagram folder includes the four steps of the data analysis: first and foremost, the construction of the cloud of points, then the study of the three questions raised p. 177. The Archived data folder contains the data file of the questionnaire (Dataset), then that with the principal coordinates of the cloud of individuals (Cloud & groups). The Graphics folder contains the graphs of the cloud of MPs and that of the cloud of response categories.

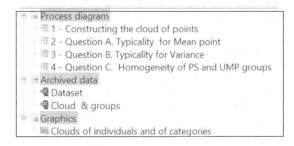

Figure 6.2.17
SPAD project with the three folders.

- The *first diagram* (Figure 6.2.18) sets about constructing the cloud of points and storing the principal coordinates of individual points.

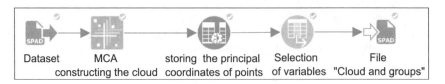

Figure 6.2.18
1 - Constructing the cloud of points.

At the very beginning, there is the *import of the data set*, here a "SPAD data archive" file called `Dataset`, which is stored in the `Archived data` folder. Then, a specific MCA, as described in §6.2.2 (p. 172), is performed so as to construct geometric clouds. After that, the first two principal coordinates of individual points are added to the data file, and (last icona) the new data set is registered into a `Cloud & groups` file, and stored in the `Archived data` folder.

- The *second diagram* (Figure 6.2.19) shows the use of the "combinatorial typicality test" R script for answering question A. Firstly, `Cloud & groups` file is imported from the `Archived data` folder. Secondly, the individuals who belong to the reference are identified through the definition of a "reference" indicator. Thirdly, an analysis is performed for each parliamentary group (for configuring the method, see Chapter 3, §3.6.2, p. 59). The last step is the plotting of the histogram of the statistic distribution.

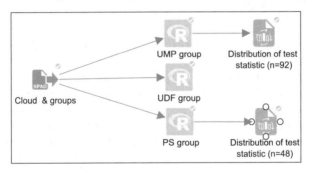

Figure 6.2.19
2 - Question A: Typicality for Mean point.

- The *third diagram* is similar to the second one, but the statistic to be chosen is the *Variance* instead of the *Mean*.
- The *fourth diagram* (Figure 6.2.20) is the specific comparison of two groups (UMP and PS).

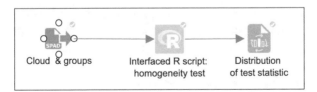

Figure 6.2.20
4 - Question C: Homogeneity test of PS and UMP groups.

The successive steps are (1) importing data, (2) performing the homogeneity test using the R script (parameters are described in Chapter 5, §5.7.2, p. 147), (3) plotting the distribution of the test statistic.

6.3   The European Central Bankers Study

In the context of the financialisation of the global economy and the importance of financial elites (Bourdieu, 2000), the role of central banks has become particularly important. In a study on "central bankers in the contemporary global field of power", Lebaron and Dogan (2016) analyse the structure of the social space of the European Central Bank (ECB) and, in particular, the collective dynamics inside the ECB.

The study population is composed of the 63 members of the Governing Council of the ECB in office since 1999.

The chosen methodology is prosopographical: *biographical data*[15] were collected from various sources on the Internet. Thus, eleven indicators were constructed on the basis of this information; they pertain to educational background (educational field, educational level, location of studies abroad) and professional activity (career path in university, politics, administration, private finance, other private institutions and in the central bank itself) with identification of the "mean career" of each member.

In addition, the researchers studied, on the basis of the investigation of the journalistic literature about "Hawks" or "Doves" at the ECB, the "economic and monetary approach" of the ECB members according to their public and explicit *position-takings*. Being "Hawk" means opposing a radical adaptive strategy that uses a range of non-conventional measures. In contrast, the "Doves" favour a certain degree of flexibility and loosening of the monetary policy. Nineteen individuals are regarded as "Dove", sixteen as "Hawk", eight as "moderate" and, for twenty individuals, it was not possible to identify their position.

The main *research questions* can be stated as follows.

1. What is the structure of the social space of the ECB according to the social characteristics of its members?

2. Do the positions of bankers in the ECB social space ("field effect") explain their position-takings on the "economic and monetary policy" (structuring factor)[16]?

We begin by constructing the *social space of the* ECB and then by investigating *position-takings* as structuring factor.

---

[15]Our thanks go to F. Lebaron who communicated the data and commented upon an earlier version of the present text.

[16]Such a question leads to an *explanatory use of* GDA, enlarged with *structured data* and *inductive* analyses.

6.3.1 Elementary Descriptive Analysis

To begin with, we briefly describe the 11 indicators used for the construction of the space and provide the frequencies of categories.

1. *Education* (3 indicators)

   ◇ Discipline: Economics ($n = 50$), Law ($n = 8$), Other or not known ($n = 5$);
   ◇ Diploma: Bachelor's degree ($n = 4$), Master's degree ($n = 18$), PhD ($n = 38$), not known ($n = 3$);
   ◇ Studies abroad: no ($n = 36$), UK ($n = 8$), USA ($n = 13$), Europe except UK ($n = 4$), not known ($n = 2$).

2. *Professional trajectory* (7 indicators) with exclusive categories: make main career (main), make career (yes), does not make career (no) in the sector:

   ◇ Private finance: main ($n = 20$), yes ($n = 10$), no ($n = 33$);
   ◇ Other private: main ($n = 7$), yes ($n = 10$), no ($n = 44$), not known ($n = 2$);
   ◇ Central bank: main ($n = 19$), yes ($n = 11$), no ($n = 33$);
   ◇ Administration: main ($n = 20$), international ($n = 5$), national ($n = 8$), no ($n = 30$);
   ◇ Politics: main ($n = 5$), yes ($n = 11$), no ($n = 47$);
   ◇ University: main ($n = 11$), yes ($n = 20$), no ($n = 32$);
   ◇ Number of sectors during the career: one ($n = 14$), more than one ($n = 49$).

3. *Public visibility* (1 indicator with three levels)

   low ($n = 32$), medium ($n = 16$), high ($n = 12$), unknown ($n = 3$).

6.3.2 The Social Space of the ECB

Stage one consists in representing the data set as clouds of points in a multidimensional geometric space. The construction of clouds is made through a variant of Multiple Correspondence Analysis (MCA), called specific MCA[17], which allows one to discard the "unknown" and the too infrequent categories (frequency lower than four) from the determination of principal axes.

Dimension of clouds

Together, the first two axes take into account 22% of the variance of the cloud and the cumulated modified rate reaches 59%. From the fourth axis, the variances of axes decrease regularly (see opposite diagram). Three axes can be retained for interpretation. Their variances are $\lambda_1 = 0.235$, $\lambda_2 = 0.214$, $\lambda_3 = 0.181$ (that of Axis 4 is 0.153).

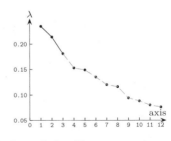

The cloud of the 33 "active" categories and that of the 63 ECB members projected onto the first principal plane are respectively depicted in Figures 6.3.1 and 6.3.2 (p. 190).

---

[17]See Le Roux (1999); Le Roux and Rouanet (2004, Chapter 5, pp. 203–210).

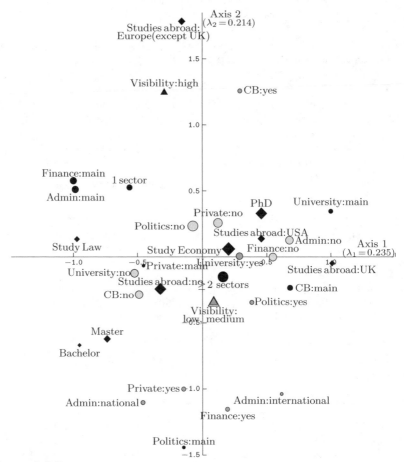

Figure 6.3.1
Cloud of the 33 categories in the first principal plane.

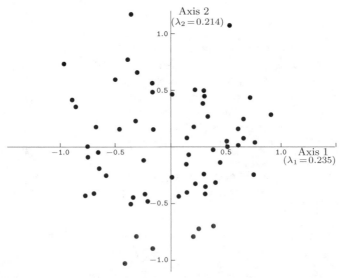

Figure 6.3.2
Cloud of the 63 ECB members in the first principal plane.

Interpretation of axes

The interpretation of axes is based on categories whose contributions to axis exceed the average contribution.

Twelve categories are retained for the *interpretation of Axis 1*. Together, they contribute to 82 percent of the variance of Axis 1. Hence they provide a good summary. Figure 6.3.3 shows the categories most contributing to Axis 1 as projected on Axis 1.

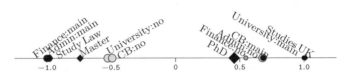

Figure 6.3.3
Interpretation of Axis 1: 12 categories most contributing to Axis 1.

To the left of Axis 1, one finds: main career in Finance, in Administration, no career at University, no career at CB, as well as Master's degree and Education in law. To the right of the axis, one finds opposing categories: no career in Finance, in Administration, main career at university, at CB, PhD and studies in the UK.

The first axis can be interpreted as an axis of academic capital, opposing higher to lower levels of academic capital.

Similarly, we proceed to the *interpretation of Axis 2*. The 11 categories whose contributions to Axis 2 exceed the average contribution are depicted in Figure 6.3.4. Together, they contribute to 78 percent of the variance of axis.

Figure 6.3.4
Interpretation of Axis 2: 11 categories most contributing to Axis 2.

To the left of Axis 2, one finds: main career in Politics, career in Finance, in international or national administration, and Master's degree. On the other side, one finds: main career in Finance, and administration, career in CB, high public visibility, and studies in Europe (except the UK). The second axis opposes political careers with administrative or private sequences to high levels of legitimacy in the field of central banking and finance.

Axis 2 highlights an opposition between dominant "insiders" and political "outsiders".

The *interpretation of Axis 3* is based on the 12 categories that are depicted in Figure 6.3.5 as projected on the axis. Together, they contribute to 82 percent of the variance of axis.

To the left of the axis, one finds: European international educational trajectory, related to the worlds of international organizations, universities and marginally central banks, with a plurisectoral orientation. To the right of the axis, one finds: educational trajectories in the US and an early specialization in central banks with a monosectoral orientation.

Axis 3 is related to the international aspects of the educational and professional trajectories.

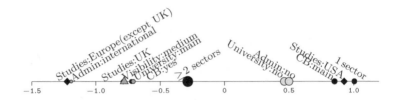

Figure 6.3.5
Interpretation of Axis 3: 12 categories most contributing to Axis 3.

Cloud of individuals

We will now study in detail the cloud of individuals equipped with the structuring factor "position-taking". Figure 6.3.6 shows the two subclouds "Hawk" and "Dove", which are geometrically summarized by their mean points and their concentration ellipses (see Chapter 2, p. 21).

Table 6.3.1 gives the principal coordinates of the mean points and Table 6.3.2 the scaled deviations between the mean points and the origin, for the first three principal axes (and in the first principal plane).

Table 6.3.1
Coordinates of the two points of the two subclouds on the first three principal axes.

	size	Axis 1	Axis 2	Axis 3
Dove	19	0.204	−0.115	−0.008
Hawk	16	−0.079	0.247	0.076

Table 6.3.2
For each subcloud: scaled deviations from mean points to origin on the first three principal axes and in principal plane 1-2 (notable deviations are written in bold).

	Axis 1	Axis 2	Axis 3	plane 1-2
Dove	**0.420**	−0.249	−0.020	**0.488**
Hawk	−0.163	**0.533**	0.179	**0.557**

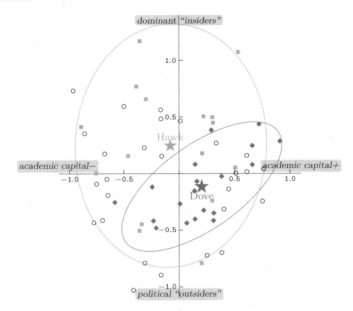

Figure 6.3.6
Cloud of the 63 ECB in the first principal plane:the two subclouds "Hawk" (■) and "Dove"
(◆) with their mean points (★) and their concentration ellipses.

Note about scaled deviations. If $\overline{y}_1$ denotes the principal coordinate of the
mean point of a subcloud on Axis 1, the scaled deviation from the mean point
to the origin on Axis 1 (whose variance is λ_1) is equal to $d_1 = \overline{y}_1/\sqrt{\lambda_1}$. Similarly, the scaled deviation on Axis 2 is $d_2 = \overline{y}_2/\sqrt{\lambda_2}$. The scaled deviation
in plane 1-2 is equal to $\sqrt{(d_1^2 + d_2^2)}$ (see Chapter 3, §3.5, especially pp. 53
and 54). A scaled deviation is regarded as *notable* if it is greater than 0.4
(see Chapter 2, p. 22).

We will now comment on these results.

For "Dove", deviations are notable on Axis 1 and in plane 1-2, but small
on Axes 2 and 3. For "Hawk", deviations are notable on Axis 2 and in plane
1-2, but small on Axes 1 and 3.

These results invite us to study the comparison of the two subclouds in the
first principal plane. Figure 6.3.6 shows that (1) the group "Dove" is mainly
located in the right half-plane with only four members (over 19) on the other
side, (2) the group "Hawk" is very dispersed and mainly located in the upper
half-plane with only four members (over 16) on the other side.

We will now assess the importance of the deviation between the two groups
in the first principal plane. As a descriptive index, we take the partial η^2 that
is equal to $0.065 > 0.04$. Hence, the deviation is of large magnitude.

Calculation of the partial eta-squared coefficient. From the coordinates of
two mean points (see Table 6.3.1, p. 192), we deduce that their squared

distance is equal to $(0.204 - (-0.079))^2 + (-0.115 - 0.247)^2 = 0.2111$. The variance of the two points is equal to $\frac{19 \times 16}{(19+16)^2} \times 0.2111 = 0.0524$ (Formula 2.2, p. 12). The variance of the overall cloud in the first principal plane is equal to $\lambda_1 + \lambda_2 = 0.235 + 0.214 = 0.449$. Hence the partial eta-squared coefficient is $\eta'^2 = \frac{19+16}{63} \times 0.0524/0.449 = 0.0291/0.449 = 0.065$ (Formula 2.14, p. 28).

Descriptive findings

The MCA shows that the structure of the social space of the ECB cannot be characterized as one-dimensional, but is at least two- even three-dimensional.

The position-takings "Dove" and "Hawk" appear to be different. The group "Hawk" is mainly located on the upper half-plane of the graph, while the group "Dove" is more concentrated in the right half-plane.

Inductive approach

Following the descriptive conclusions, we address questions such as:

Question A. Are the "Hawk" and "Dove" subclouds atypical from the overall cloud according to mean point? (typicality problem for location parameter)

Question B. What is the region of the space compatible with the mean point of each subcloud? (geometric compatibility region)

Question C. Are the "Hawk" and "Dove" groups heterogeneous? (homogeneity test)

6.3.3   Question A: Typicality of Subclouds for Mean Point

For answering Question A, in order to extend the descriptive conclusions, we proceed to inductive analyses by studying the cloud projected on the first principal plane. In order to assess the level of typicality of groups "Dove" and "Hawk" with respect to the overall set of ECB members, we will use the combinatorial typicality test outlined in Chapter 3.

> Recall briefly the principle of the test.
>
> It consists in locating the *group* under study among all *possible samples* (subsets) of the *overall set* of individuals, the sizes of samples being equal to that of the group under study.
>
> Considering the cloud of individuals, the *test statistic* depends on the Mahalanobis distance between the mean point of the sample subcloud and that of the overall cloud (this statistic takes into account the shape of the overall cloud). From the combinatorial distribution of the test statistic, the *typicality level* of the group with respect to the overall set of individuals is defined as the proportion of possible samples, for which the value of the test statistic is greater than (or equal to) that of the group.

For the "Dove" group, the cardinality of the sample set is equal to $\binom{63}{19}$; for the "Hawk" group, it is equal to $\binom{63}{16}$. These numbers are very large ($> 3{\times}10^{14}$), hence, the *exact tests of typicality* are carried out through the MC method by drawing one million random samples from the sample set (see Remark (1), p. 44). The *combinatorial distributions* of the test statistic for sample sizes $n_c = 19$ ("Dove") and $n_c = 16$ ("Hawk") are shown in Figure 6.3.7.

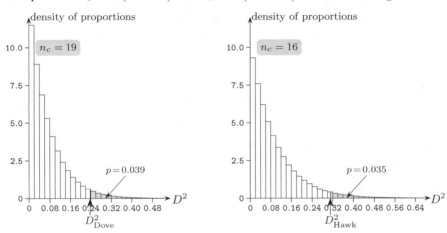

Figure 6.3.7
Combinatorial distributions of statistic D^2 (squared Mahalanobis distance) for sample sizes $n_c = 19$ and $n_c = 16$. The p-values correspond to the area under the curve past the observed value (shaded grey area) with $D^2_{\text{Dove}} = 0.238$ and $D^2_{\text{Hawk}} = 0.311$.

For the "Dove" group, the observed value of the test statistic is equal to $0.488^2 = 0.238$ (see Table 6.3.2, p. 192) and the level of typicality (p-value) is $p = 0.039$. Since $p < 0.05$, we conclude that, in the first principal plane, the "Dove group" is atypical of the overall set of ECB members with respect to *Mean Point*, at level 0.05. For the "Hawk" group, the observed value of the test statistic is equal to $0.557^2 = 0.311$; the level of typicality is $p = 0.035$. Hence the conclusion of atypicality.

> *In the first principal plane, the "Dove" and "Hawk" subclouds are atypical of the cloud of all ECB members; their deviations from the centre cannot be due to chance—beyond any reasonable doubt—, and one is therefore entitled to search for a sociological interpretation.*

6.3.4   Question B: Compatibility Regions

Question B raises the problem of assessment of uncertainty about the location of the mean point of a subcloud. A solution is provided by the compatibility region defined by using the geometric typicality test (see Chapter 4, p. 80).

The method consists in taking each point of the space as a reference point.

A permutation space is associated with each point. It is generated by the symmetry about this point. The test statistic depends on the reference point and on the shape of the cloud. The compatibility region at level 0.05 for the mean point of the group subcloud is the set of points of the space for which the result of the geometric typicality test is nonsignificant at this level.

The solution we propose takes into account two target properties: (1) any point of the compatibility region taken as a reference point provides a non-significant result of the test, (2) the shape of the compatibility region depends on that of the subcloud (it can be adjusted by an ellipsoid that is homothetic with the principal ellipsoid of the subcloud)[18].

The exhaustive method is performed since the numbers of permutations are respectively equal to $2^{16} = 65536$ and $2^{19} = 524288$ (less than one million). The scale parameters of the ellipses are 0.653 for "Dove" and 0.742 for "Hawk". The compatibility ellipses are depicted in Figure 6.3.8.

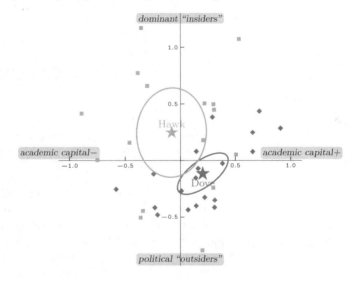

Figure 6.3.8
Compatibility ellipses at level 0.05 for "Dove" and "Hawk" groups in Plane 1-2.

The compatibility region for the mean point of the "Dove" subcloud is clearly in the lower-right quadrant, whereas that for the mean point of the "Hawk" subcloud is quite large and mainly in the upper half-plane. We can conclude that:

The position-takings ("Dove" and "Hawk") can be related to the position of the ECB members in the ECB social space.

---

[18]Some authors (see, e.g., Greenacre, 1984; Lebart, 2007) proposed a solution based on "partial bootstrap". The bootstrap method amounts to determining the mean of the subcloud reweighted with the bootstrap weights (0, 1, 2, ...) that characterize random draws with replacement among the points of the subcloud. Our method is deeply different and geometric in essence.

6.3.5 Question C: Homogeneity of subclouds

We will now assess the level of homogeneity of the two groups, and thus try to corroborate the descriptive conclusion of difference between the two groups, in plane 1-2. The homogeneity test presented in Chapter 5 provides an appropriate response to Question C.

In this particular case, we take into account the cloud of all ECB members, then we proceed to a *partial comparison* between the two groups.

> For a partial comparison of two groups, considering the subsets that have the *same sizes* as the groups under study, the permutation space is made up of *all pairs of disjoint subsets* that can be extracted from the overall set of individuals.
>
> Then, we consider the deviation between the mean points of the two subclouds associated with each possible pair of subsets and take the squared M-distance between mean points (Mahalanobis distance related to the overall cloud) as test statistic. Note that the test statistic depends on the *covariance structure of the overall cloud*.
>
> The *homogeneity level* of the two subsets with respect to the overall set of individuals is defined as the proportion of possible pairs of subclouds, for which the value of the test statistic is greater than (or equal to) that of the pair of subclouds under study.

The number of possible pairs of subsets (with sizes 16 and 19) is equal to $63!/(16!19!(63-35)!)$ ($> 2 \times 10^{27}$), hence we use the MC method with one million rearrangements. The histogram of the permutation distribution of the test statistic is shown in Figure 6.3.9.

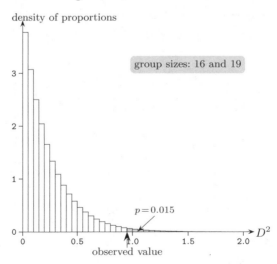

Figure 6.3.9
Histogram of the permutation distribution of the test statistic D^2 (the observed value is equal to 0.952).

Then, the *homogeneity level* of the two subclouds with respect to the overall cloud of ECB members is defined by the proportion of possible pairs of subclouds (that are comprised of 16 and 19 points) whose statistic value is greater than (or equal to) that of the observed pair ("Dove", "Hawk").

The observed value of the test statistic, that is, the squared M-distance between the two mean points is equal to $(0.204-(-0.079))^2/0.235+(-0.115-0.247)^2/0.214 = 0.953$ (see Table 6.3.1, p. 192). The p-value is equal to 0.015 < 0.05. Hence, we can conclude:

The "Dove" and "Hawk" groups are heterogeneous, at level 0.05.

6.3.6   Concluding Comments

In this study we have all members of the ECB; the classical problem of inferring on characteristics of a population on the basis of a random sample is unsuitable.

The typicality test shows that the "Dove" and "Hawk" groups cannot be assimilated to a random sample of the ECB members. Considering the number of individuals for whom we could not obtain the information, we constructed geometric compatibility region for each group. These regions are quite large, however, we can assert the two groups are heterogeneous.

6.3.7   Steps of the Study Using R and Coheris SPAD Software

We now present the SPAD project, which performs the statistical analyses that we previously presented.

It comprises three folders: Process diagram, Archived data and Graphics (Figure 6.3.10).

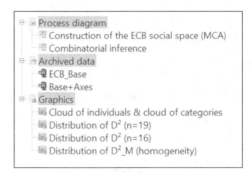

Figure 6.3.10
SPAD project for analysing the ECB data.

The **Process diagram** folder includes two steps: first and foremost, the construction of the social space of the ECB, then the inductive analyses for answering the three questions raised p. 194.

The `Archived data` folder contains the basic data file (`ECB_Base`), and the file (`Base+Axes`) that includes the principal coordinates of the points of the cloud of individuals.

The `Graphics` folder contains the graphs of the cloud of individuals, that of the cloud of categories, and the distributions of the test statistics.

Now, we will briefly comment on the two diagrams.

• The *first diagram* (Figure 6.3.11) sets about constructing the clouds and storing the principal coordinates of individual points. At the very beginning, there is the *import of the data set* (a "SPAD data archive" file called `ECB_Base`, which is stored in the `Archived data` folder). Then, a specific MCA (as described in §6.3.2, p. 189), is performed so as to construct geometric clouds. After that, (1) the selection of the categories whose contributions exceed the average is done for interpreting axes (upper icona), (2) the first two principal coordinates of individual points are added to the data file, and (last icona) the new data set is saved into the `Base+Axes` file.

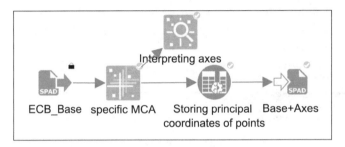

Figure 6.3.11
Construction of the ECB social space (cloud of points).

• The *second diagram* (Figure 6.3.12, p. 200) shows the use of the R scripts interfaced with Spad for answering questions A, B and C. Firstly, the `Base+Axes` file is imported from the `Archived data` folder. Then, the analyses are performed in order to answer the three questions.

Question A (see §6.3.3, p. 194). The R script `CIGDA_Comb-v1.R` is used (see §3.6.2, p. 59). For that, an indicator of the individuals who belong to the reference (here all individuals) must be defined beforehand. Then, the choice of the group under study is done by using the `PositionTaking` indicator and an analysis is performed for each group. The last step consists in the drawing of the histogram of the test statistic distribution.

Question B (see §6.3.4, p. 195). For answering Question B, the R script `CIGDA_Geo-v1.R` is used (see §4.6.2, p. 101)).

Question C (see §6.3.5, p. 197). The R script `CIGDA_Homog-v1.R` is used, as described in §5.7.2 (p. 147), for the partial comparison of the two groups and the histogram of the statistic is drawn.

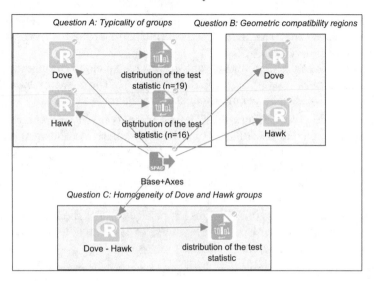

Figure 6.3.12
Combinatorial inference analyses.

6.4   Cognitive Tests and Education

In a study on metacognitive factors in scientific problem–solving strategies, P. Rozencwajg (1994) studied a group of forty-two 12 to 13-year-old seventh graders from two middle schools in the metropolitan Paris area.

The students belong to *two middle schools* that are very different in terms of socioeconomic environment: one school (denoted a_1) is located in a priority education zone (underprivileged environment), and the other (denoted a_2) is located in a medium-level socioeconomic environment. The *gender* of the students was also noted. In school a_1, there are 5 boys and 9 girls; in school a_2, there are 17 boys and 11 girls. For every student, the scores to six *cognitive tests* were recorded.

The six tests are:

General intelligence test (*g*-factor test) is made of several items; each item is a table with diagrams in rows and columns; the subject has to discover the relation that characterizes the change from one row to another and that from one column to another. One diagram is missing, and the subject has to choose, among several diagrams, the one that satisfies both relations on rows and columns.

Numerical test, which aims at evaluating one's ability to handle numbers, consists of numerical operations to be complemented.

Verbal test is a task that consists in completing a sentence by choosing among a set of terms the appropriate one.

Spatial test consists in finding, among a set of diagrams oriented in different directions, the ones that are the same as a target-diagram.

FDI test aims to measure the dimension "field dependence-independence"; the subject has to find a simple figure in a complex drawing. This dimension opposes analytical to global skills.

RI (reflective-impulsive) cognitive style test measures latencies, which were initially measured by J. Kagan on the Matching Familiar Figures Test (MFFT); a high latency indicates a high degree of thinking.

Thus, the study involves multivariate numerical data and two structuring factors, namely *Gender* and *Status* (socioeconomic environment).

The aim of the study is to figure out to what extent Status *and* Gender *explain the position of students in the cognitive space.*

To analyze such data, we combine on the one hand, the approach of Geometric Data Analysis (GDA) by representing data as clouds of points and exploring the clouds, and on the other hand, the structured data analysis supplemented by inductive analyses.

6.4.1 Elementary Descriptive Analysis

The first remark is that the scales of the tests are different. The second remark is that all the correlations between tests are positive, except that between the spatial test and the reflective-impulsive test, which is negative but near zero.

Table 6.4.1
Means, standard deviations (*SD*) and table of correlations.

	Mean	SD		Gen.	Num.	Verbal	Spatial	FDI	RI
General	11.3	3.3	General	1.					
Numerical	10.3	3.4	Numerical	0.63	1.				
Verbal	16.3	4.4	Verbal	0.54	0.62	1.			
Spatial	23.7	14.6	Spatial	0.20	0.42	0.29	1.		
FDI	11.6	4.0	FDI	0.74	0.52	0.51	0.37	1.	
RI	6289.7	3146.2	RI	0.59	0.36	0.26	−0.14	0.36	1.

6.4.2 Geometric Data Analysis: Cognitive Space

First of all, we perform geometric data analysis (see Chapter 1, p. 4) in order to construct the cognitive space and represent students by a cloud of points in a multidimensional Euclidean space. There are six variables, therefore, at the beginning, the space is 6-dimensional. Thus, we begin by studying the structure of the cognitive space and interpreting axes.

Principal component analysis

Since the six variables are not on a common scale (see Table 6.4.1), a Principal Component Analysis (PCA) based on correlations has been performed[19].

PCA reveals that clouds are approximately two-dimensional. As a matter of fact, the first two axes account for 74% of the variance of the cloud (see Table 6.4.2-a, p. 203) and the multiple correlations of all variables with the first two principal variables are high ($R_{1-2} \geq 0.768$, see Table 6.4.2-b).

In addition, the qualities of representation of students' points in plane 1-2 (see the definition on p. 17) are quite good since they are greater than 0.5, except for only eight of them. Thus, the cloud of students is nearly entirely contained in a plane; therefore its projection onto the first principal plane will make up the basic data set for further analyses.

Interpretation of axes

The study of the *space of variables* is summarized by the circle of correlations (see Figure 6.4.1, p. 203).

All cognitive tests are positively correlated with Axis 1 and the *g*-factor is highly correlated (0.88, see Table 6.4.2-b), hence the interpretation:

Axis 1 is an axis of general cognitive abilities.

The spatial test and the RI test are highly correlated with Axis 2, therefore Axis 2 opposes students who have good performance in spatial test—which is done in limited time, and thus requires fast responses—to students who take a longer time to produce their response to the matching test.

Axis 2 is an axis of processing speed.

Cloud of individuals

The cloud of the 42 students projected in the first principal plane is shown in Figure 6.4.2 (p. 203).

6.4.3   Structured Data Analysis

Status and *Gender* are structuring factors of the set of students, each with two levels. We illustrate the approach of structured data analysis (see Chapter 1, page 5) by studying these two structuring factors. Carrying over the experimentally-minded language, we may speak of the "effect of *Status*" (assimilated to an independent variable) on the cognitive space ("geometric dependent variable"). Knowing the socioeconomic environment (status) of a student, we would like to know the position of the student in the cognitive space (language of prediction).

---

[19]For a presentation of PCA in the geometric framework, see Le Roux and Rouanet (2004, Chapter 4) or Le Roux (2014a, Chapter 6). For a detailed study of the PCA of the dataset, see Richard (2004, Chapter 6) and Le Roux (2014a, pp. 173–178).

Table 6.4.2
Results of Principal Component Analysis.

Table 6.4.2-a Variances of axes (eigenvalues λ) and variance rates (τ).

$\lambda_1 = 3.219$	$\lambda_2 = 1.213$	$\lambda_3 = 0.590$	$\lambda_4 = 0.478$	$\lambda_5 = 0.314$	$\lambda_6 = 0.186$
$\tau_1 = .537$	$\tau_2 = .202$	$\tau_3 = .098$	$\tau_4 = .080$	$\tau_5 = .0.052$	$\tau_6 = .031$

Table 6.4.2-b ; Correlations of cognitive tests with the first two principal variables (r_{ℓ_1} and r_{ℓ_2}), and coefficients of multiple correlation $R_{1-2} = \sqrt{(r_{\ell_1}^2 + r_{\ell_2}^2)}$.

	General	*Numerical*	*Verbal*	*Spatial*	*FDI*	*RI*
$r_{\ell 1}$	0.881	0.825	0.757	0.437	0.828	0.560
$r_{\ell 2}$	−0.241	0.150	0.132	0.788	0.045	−0.701
R_{1-2}	0.913	0.838	0.768	0.901	0.829	0.897

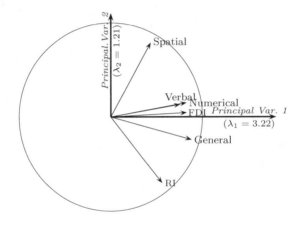

Figure 6.4.1
Space of variables: "circle of correlations".

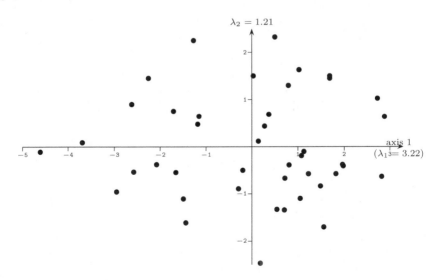

Figure 6.4.2
Cloud of the 42 students in the first principal plane.

The *Status* factor is denoted $A = \{a_1, a_2\}$, and the *Gender* factor is denoted $B = \{b_1, b_2\}$. *Status* and *Gender* are *crossed factors* since there are students of both genders in each school. They define four groups of students: $a_1b_1 =$ (underprivileged,boy), $a_1b_2 =$ (underprivileged,girl), $a_2b_1 =$ (medium,boy) and $a_2b_2 =$ (medium,girl). The cloud of students is divided into four subclouds that are depicted in Figure 6.4.3 with their mean points; each student's point is joined to the mean point of her/his group.

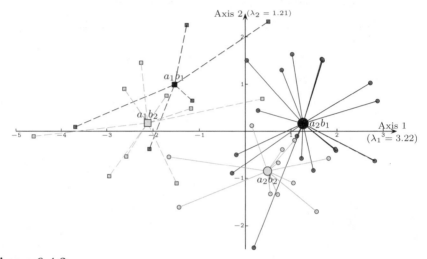

Figure 6.4.3
The four subclouds of students with their mean points (■ for a_1b_1 (underprivileged,boy); ■ for a_1b_2 (underprivileged,girl); ● for a_2b_1 (medium,boy); ● for a_2b_2 (medium,girl)), in the first principal plane.

The weighted cloud of the four category mean points is denoted $\mathrm{G}^{A \times B}$ and called *Status× Gender* cloud. The coordinates of the four points in Plane 1-2 are given in Table 6.4.3, and the cloud is depicted in Figure 6.4.4 (p. 205).

Table 6.4.3
Cloud $\mathrm{G}^{A \times B}$: weights; coordinates of points and variances on Axis 1 and Axis 2.

Coordinates	weights	Axis 1	Axis 2
$\mathrm{G}^{a_1b_1}$	5	−1.538	0.990
$\mathrm{G}^{a_1b_2}$	9	−2.115	0.174
$\mathrm{G}^{a_2b_1}$	17	1.259	0.164
$\mathrm{G}^{a_2b_2}$	11	0.484	−0.847
Var $\mathrm{G}^{A \times B}$		1.943	0.322

The variance of the cloud $\mathrm{G}^{A \times B}$ in the first principal plane being equal to $1.943 + 0.322 = 2.265$ and that of the cloud of students to $\lambda_1 + \lambda_2 = 4.432$ (see Table 6.4.2-a, p. 203), the η^2 coefficient is equal to $2.265/4.432 = 0.51$: by any standard, such a value can be deemed large (see Chapter 2, p. 28).

Descriptively, the global difference between the four groups is large.

Thus, we will now take the weighted cloud $G^{A \times B}$ as the basic data set for the descriptive analyses that follow.

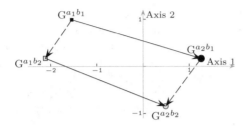

Figure 6.4.4
Weighted cloud $G^{A \times B}$ in the first principal plane, with $G^{a_1 b_1}$, (underprivileged,boy), ■; $G^{a_1 b_2}$, (underprivileged,girl), ▪; $G^{a_2 b_1}$, (medium,boy), ●; $G^{a_2 b_2}$, (medium,girl), ○.

The mean point of the 14 students of school a_1 is denoted G^{a_1} and that of the 28 students of school a_2 is denoted G^{a_2}. These two points define the *Status* cloud, denoted G^A, and shown in Figure 6.4.5. Its variance in Plane 1-2 (between-*Status* variance) is equal to $1.822 + 0.108 = 1.930$. The *Status* cloud takes into account 85% of the variance of the *Status×Gender* cloud; the η^2 coefficient is equal to 0.44 (quite a large value)[20].

Similarly, we get the *Gender* cloud (denoted G^B). Its variance in Plane 1-2 (between-*Gender* variance) is equal to $0.427 + 0.136 = 0.563$ (see Table 6.4.4). The *Gender* cloud takes into account 25% of the variance of the *Status×Gender* cloud; the η^2 coefficient is equal to 0.13 (a large value).

Table 6.4.4
Coordinates of the points of clouds G^A and G^B on Axis 1 and Axis 2 and their variances.

Coordinates	n	Axis 1	Axis 2	Coordinates	n	Axis 1	Axis 2
G^{a_1}	14	−1.909	0.466	G^{b_1}	22	0.623	0.352
G^{a_2}	28	0.954	−0.233	G^{b_2}	20	−0.686	−0.387
Var G^A	42	1.822	0.108	Var G^B	42	0.427	0.136

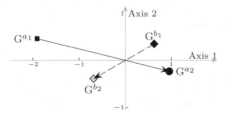

Figure 6.4.5
The mean points of underprivileged status G^{a_1} (■) and of medium-level status G^{a_2} (●); the mean points of boys G^{b_1} (◆) and of girls G^{b_2} (◇), in the first principal plane.

---

[20] Recall that η^2 is regarded as notable if it is greater than 0.04 (see Chapter 2, p. 28).

Descriptively, the difference between the two socioeconomic statuses and that between boys and girls are large.

When two factors A and B are crossed, it is customary to break down the source of variation $A \times B$ into three sources, namely the two main sources A (*Status*) and B (*Gender*) and the interaction source that we denote $A \cdot B$.

There are three usual *decompositions of the source of variation $A \times B$*: (1) the additive-source (denoted $A+B$) and the interaction source $A \cdot B$, (2) the main source A and the B within-A source (denoted Bwithin$A$), (3) and, in a symmetrical way, the main source B and the A within-B source (Awithin$B$).

Main effect of Status and effects of Status within-Gender

For *Status*, one has $2 - 1 = 1$ degree of freedom. The main effect of *Status* is the geometric vector joining the two points G^{a_1} and G^{a_2}; the effect-vector is shown in Figure 6.4.5.

For each level of the factor *Status*, there is an effect of *Gender*. The two effect-vectors $G^{a_2 b_1} - G^{a_1 b_1}$ and $G^{a_2 b_2} - G^{a_2 b_2}$ are called effects of *Status* within-*Gender* and are represented in Figure 6.4.6. They are slightly different.

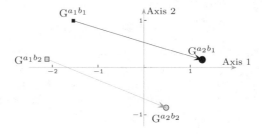

Figure 6.4.6
The effect-vectors of *Status* for boys and girls in plane 1-2.

From Table 6.4.3 (p. 204), one calculates the variances of the subclouds $G^{A/b_1} = (G^{a_1 b_1}, G^{a_2 b_1})$ and $G^{A/a_2} = (G^{a_1 b_2}, G^{a_2 b_2})$; the weighted mean of the two variances is the variance of *Status* within-*Gender*, denoted $\mathrm{Var}\, G^{A\mathrm{within}B}$ (see Table 6.4.5).

Table 6.4.5
Variances of the two subclouds $G^{A/b1}$ and $G^{A/b2}$ and variance of $G^{A\mathit{within}B}$.

	weights	Axis 1	Axis 2	Plane 1-2
Var G^{A/b_1}	14	1.373	0.120	1.493
Var G^{A/b_2}	28	1.671	0.258	1.930
Var $G^{A\mathit{within}B}$	42	1.515	0.186	1.701

Main effect of Gender and effects of Gender within-Status

For *Gender*, one has $2 - 1 = 1$ degree of freedom. The main effect of *Gender* is the geometric vector $G^{b_2} - G^{b_1}$, which is shown in Figure 6.4.5.

The effects of *Gender* within-*Status* are defined by two effect-vectors $G^{a_1 b_2} - G^{a_1 b_1}$ and $G^{a_2 b_2} - G^{a_2 b_1}$ (see Figure 6.4.7). They are slightly different.

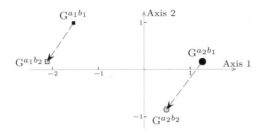

Figure 6.4.7
The effect-vectors of *Gender* for each status in the first principal plane.

From Table 6.4.3 (p. 204), one calculates the variances of the two subclouds $G^{B/a_1} = (G^{a_1 b_1}, G^{a_1 b_2})$ and $G^{B/a_2} = (G^{a_2 b_1}, G^{a_2 b_2})$; the weighted mean of the two variances is the variance of *Gender* within-*Status*, denoted $\operatorname{Var} G^{B\text{within}A}$ (see Table 6.4.6).

Table 6.4.6
Variances of the two subclouds $G^{B/a1}$ and $G^{B/a2}$ and variance of $G^{B\text{within}A}$.

	weights	Axis 1	Axis 2	Plane 1-2
$\operatorname{Var} G^{B/a_1}$	14	0.076	0.153	0.229
$\operatorname{Var} G^{B/a_2}$	28	0.143	0.244	0.387
$\operatorname{Var} G^{B\text{within}A}$	42	0.121	0.213	0.334

Interaction and additive cloud

The *Status* effect-vectors are little different between boys and girls (Figure 6.4.6, p. 206). Equivalently, the *Gender* effect-vectors are little different from one school to another (Figure 6.4.7, p. 207). In such a situation, it is commonly said that there is an *interaction effect*[21] between the two factors. The interaction effect-vector is equal to the difference between the two *Status* effect-vectors within-*Gender* or equivalently between the two *Gender* effect-vectors within-*Status*. Hence it is equal to $G^{a_1 b_1} - G^{a_2 b_1} - G^{a_1 b_2} + G^{a_2 b_2}$, and its coordinates in Plane 1-2 are $(-0.198, -0.196)$. It is clearly small.

The *Status*× *Gender* cloud $G^{A \times B}$ can be fitted by an *additive cloud*, that is, a cloud without interaction and with the same main clouds as for cloud $G^{A \times B}$. Thus, the additive cloud, denoted P^{A+B}, is such that its interaction

---

[21]The term "interaction" is metaphorical and does not imply any substantive interplay between *Status* and *Gender*.

effect $P^{a_1b_1} - P^{a_2b_1} - P^{a_1b_2} + P^{a_2b_2}$ is null and $P^A = G^A$ and $P^B = G^B$ (see Table 6.4.7 and Figure 6.4.8)[22].

Table 6.4.7
Coordinates of the four points of the additive cloud P^{AB} and of deviations from initial point to additive point in plane 1-2 and variances of clouds.

Coordinates	n	Axis 1	Axis 2		Coordinates	n	Axis 1	Axis 2
$P^{a_1b_1}$	5	−1.452	1.075		$G^{a_1b_1} - P^{a_1b_1}$	5	−0.0859	−0.0850
$P^{a_1b_2}$	9	−2.163	0.127		$G^{a_1b_2} - P^{a_1b_2}$	9	0.0477	0.0472
$P^{a_2b_1}$	17	1.234	0.139		$G^{a_2b_1} - P^{a_2b_1}$	17	0.0253	0.0250
$P^{a_2b_2}$	11	0.523	−0.808		$G^{a_2b_2} - P^{a_2b_2}$	11	−0.0390	−0.086
Var P^{A+B}	42	1.941	0.320		Variances	42	0.00202	0.00198

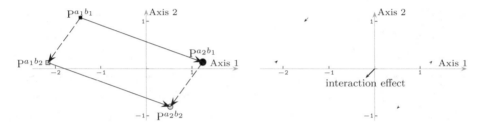

Figure 6.4.8
Additive cloud (left); the four deviations (vectors) from additive cloud to *Status× Gender* cloud with the interaction effect (right), in the first principal plane.

In the first principal plane, the variance of the additive cloud is equal to $1.941 + 0.320 = 2.261$; it takes into account 99.8% of the variance of the *Status× Gender* cloud and 51% of the variance of the overall cloud (η^2 coefficient is equal to $2.261/4.432 = 0.51$, a value that is quite large).

> *Descriptively, in the cognitive space, the fitting of the* Status×Gender *cloud by an additive cloud is fairly good.*

The coordinates of the interaction effect-vector on Axis 1 and Axis 2 are equal to $(-0.197, -0.195)$. It can be easily shown that the interaction variance is equal to the squared norm of the interaction effect-vector divided by $\left(n/n_{a_1b_1} + n/n_{a_1b_2} + n/n_{a_2b_1} + n/n_{a_2b_2}\right)$, that is, $0.077/19.4 = 0.004$.

> *In the cognitive space, the interaction between* Status *and* Gender *is extremely weak.*

> **Structure effect**. By construction, the effects of *Gender* within-*Status* in the additive cloud are equal, hence the corresponding effect-vector can be

---

[22]The coordinates of the additive cloud are obtained by a weighted multiple regression of the coordinates of cloud $G^{A \times B}$ on $A - 1$ and $B - 1$ dummy variables of factors A and B.

taken as the *averaged effect-vector* of *Gender* within-*Status*, and similarly for the effect-vectors of *Status* within-*Gender*.

The averaged effect-vector of *Gender* within-*Status* (as it appears in the additive cloud) is rather different from the main effect of *Gender* (see Figure 6.4.9). This is because *Gender* and *Status* factors are correlated, that is, the *Gender* ratios vary among schools (the ratio of boys is one third in school $a1$ and two thirds in school $a2$). For the same reason, the averaged effect of *Status* within-*Gender* is different from the main effect of *Status*.

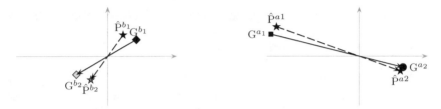

Figure 6.4.9
On the left, main effect of *Gender* (solid line) and averaged effect-vector of *Gender* within-*Status* (dashed line); on the right, main effect of *Status* (solid line) and averaged effect-vector of *Status* within-*Gender* (dashed line).

When factors are strongly correlated, the "within" effects of a factor may greatly differ from the main effects of this factor, or even have opposite signs[23]. In this case, we say that there is a *structure effect*[24].

Structure effect and interaction are two different things. In our example, there is practically no interaction but structure effect. The opposite can also occur, namely interaction and no structure effect (orthogonal design). The procedures presented here hold for the general case involving both structure effect and interaction.

Decompositions of variance

There are three additive decompositions of the variance of the *Status*×*Gender* cloud (Table 6.4.8): (additive)+(interaction), *Status*+(*Gender* within-*Status*); *Gender*+(*Status* within-*Gender*).

Table 6.4.8
Additive decompositions of variances on Axis 1, Axis 2 and in Plane 1-2.

	Status× *Gender*	additive	inter− action	*Gender*	*Status* within-*Gender*	*Status*	*Gender* within-*Status*
Axis 1	1.943	1.941	0.002	0.427	1.515	1.822	0.121
Axis 2	0.322	0.320	0.002	0.136	0.186	0.108	0.213
Plane 1-2	2.265	2.261	0.004	0.563	1.701	1.930	0.334

---

[23] This phenomenon is often stated in regression terms: when regressor variables are correlated, the regression coefficients of a variable in a multiple regression differ from those in a simple regression.

[24] For a detailed study in the one-dimensional case, see Rouanet et al. (2002).

Interaction effect is very small, therefore the variance of the additive cloud (2.261) is almost equal to that of the *Status*× *Gender* cloud (2.265).

> The correlation between factors is revealed by the fact that the variance of the additive cloud is different from the sum of the between-*Status* and between-*Gender* variances: $2.261 < 2.493 (= 0.563 + 1.930)$.

Descriptive findings

The geometric analysis (PCA) shows that the structure of the cognitive space is mainly two-dimensional, and, the structured data analysis of the cloud of students reveals that, in the cognitive space,

1. the four groups are well differentiated;
2. the interaction effect between factors is nearly null, that is, the crossing of the two factors can be adjusted by an additive model;
3. the main effect of *Status* and that of *Gender* are both of large magnitude.

Inductive approach

Following the descriptive conclusions, we address three questions.

In the cognitive space,

Question A: are the four groups heterogeneous or not?

Question B: is there an interaction between *Status* and *Gender*?

Question C: is there an effect of *Status* (socioeconomic environment)?

6.4.4　Inductive Data Analysis

In what follows, we work on the projection of the cloud of students onto the first principal plane.

The statistical tests are homogeneity tests. The group of permutations that are used throughout inductive analyses is the one associated with the *nesting of type* $(5, 17, 9, 11)$, that is, the partition of the set of 42 students into four groups of sizes 5, 17, 9, and 11. The number of possible nestings is equal to $42!/(5!17!9!11!)$ ($> 2 \times 10^{21}$). This number is very large, so we proceed to the exact test by using the Monte Carlo method with one million partitions.

Question A: Homogeneity of the four groups of students

Descriptively, we concluded that the four groups are well differentiated ($\eta^2 = 0.51$). In order to extend the descriptive conclusion, we will proceed to the homogeneity test for independent groups presented in Chapter 5 (§5.3, p. 109).

Recall the *principle of the homogeneity test* for independent groups. Saying that the groups of observations are homogeneous amounts to considering that the subdivision into groups may be ignored, which implies that any observation belonging to a group might be *exchanged* with one of an other group. The *exchangeability principle* invites us to consider the *baseline cloud of points* obtained by disregarding the subdivision into groups, and then to reallocate, in all possible ways, the points of this baseline cloud to subclouds that have the same number of points as the observed subclouds.

Given a partition of the cloud into subclouds (a possible nesting cloud), we take the between-classes M-variance[25] as a test statistic, which is equal to the sum of the between-classes variances along principal axes weighted by the inverses of eigenvalues (variances of axes).

Then, the *permutation distribution* of test statistic (the analogue of a sampling distribution) is established. The proportion of possible nesting clouds for which the test statistic is more extreme than (or as extreme as) the observed one defines the p-value of the permutation test.

The comparison is a global one. The test statistic is the between-groups M-variance and is denoted $V_{M(A \times B)}$. Its distribution is shown in Figure 6.4.10. The observed value of the test statistic is the M-variance of cloud $G^{A \times B}$; it is denoted $V_M G^{A \times B}$ and equal to $\frac{1.943}{3.219} + \frac{0.322}{1.213} = 0.869$.

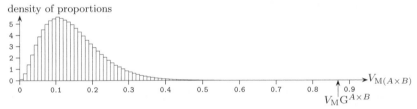

Figure 6.4.10
Histogram of the distribution of the test statistic M-variance associated with the source of variation $A \times B$; the observed value of the statistic test is equal to 0.869.

Among one million possible nesting clouds, no between-$A \times B$ variance is greater than (or equal to) the observed one, hence $p = 0/10^6$. We conclude:

> *The data are in favour of differences between the four groups of students in the cognitive space ($p < 0.001$).*

6.4.5 Question B: Interaction effect

Considering the set of possible nesting clouds (constructed for Question A), we choose the M-variance of interaction as a test statistic, which is denoted $V_{M(A \cdot B)}$. The histogram of the distribution of the interaction M-variance, determined from a random sample of one million possible nestings, is given in Figure 6.4.11.

---

[25]The M-variance (see page 21) is determined from the Mahalanobis distance related to the overall cloud, hence the test statistic depends on the covariance structure of the overall cloud and consequently takes its shape into account.

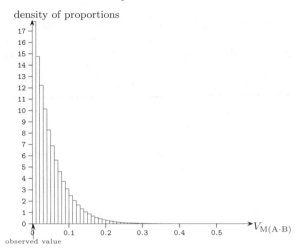

Figure 6.4.11

Histogram of the distribution of the M-variance of interaction; the observed value is equal to 0.00226.

Descriptively, the observed variance of interaction is very small, so there is no point in making the test. It just so happens that the test is not statistically significant (the observed value of the test statistic is $\frac{0.00202}{3.219} + \frac{0.00198}{1.213} = 0.00226$ (see Table 6.4.7, p. 208) and the p-value is equal to $956283/1000000 = 0.96$). Regarding the variance, the interaction effect is very small (and the result of the test is nonsignificant).

> *These results point toward additive effects of factors* Status *and* Gender *in the cognitive space.*

6.4.6   Question C: Main effects

Assuming that interaction effect is not present, we will test the main effects of the factors *Status* and *Gender*, which are both descriptively important.

Again we consider the set of possible nesting clouds (constructed for Question A), and choose the M-variance of main effects as a test statistic. For the source of variation A (*Status*), the distribution of the test statistic, determined from a random sample of one million possible nestings, is depicted in Figure 6.4.12 (p. 213).

The observed value of the test statistic is equal to $\frac{1.822}{3.219} + \frac{0.108}{1.213} = 0.655$ (see Table 6.4.4, p. 205), and the p-value is equal to $0/1000000$. This highly significant result leads us to the following conclusion.

> *In the first principal plane of the cognitive space, there is heterogeneity of subclouds corresponding to "underprivileged" status and "medium-level" status. Beyond any reasonable doubt, the deviation between the mean points cannot be due to chance.*

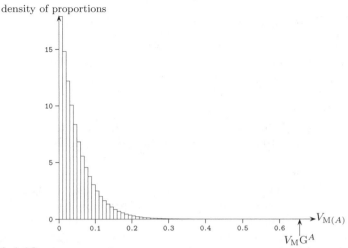

Figure 6.4.12
Distribution of the statistic M-variance for source of variation A (*Status*); the observed value is equal to 0.655.

Similarly, we study the main effect of *Gender*. The distribution of the test statistic is depicted in Figure 6.4.13 (p. 213).

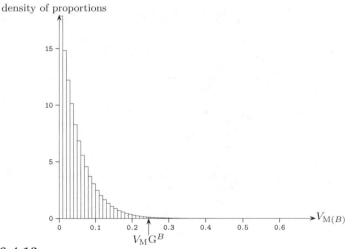

Figure 6.4.13
Distribution of the statistic M-variance for source of variation B (*Gender*); the observed value is equal to 0.245.

The observed value of the test statistic is equal to $\frac{0.427}{3.219} + \frac{0.136}{1.213} = 0.245$ (see Table 6.4.4, p. 205), and the p-value is equal to $4586/1000000 = 0.005$. The result is significant at level 0.05, therefore we state the following conclusion.

In the first principal plane of the cognitive space, there is heterogeneity of subclouds of boys and girls.

6.4.7 Concluding Comments

With this example, we wanted to show that the combinatorial inference can be applied to observational data with a slightly complex design, insofar as students are nested in the crossing of two factors with two levels each.

In experimental data, the relationships between factors take the form of a more or less complex factorial design, involving simple relations such as nesting and crossing, or more complex ones (Latin squares, etc.). The factorial design can itself be constructed from basic factors: a typical example being the case where the basic factor *Subjects* is crossed with a *Treatment* factor, as in the *Parkinson Study*[26] (§6.1, p. 156).

In observational data, even in the absence of prior design, there are also relationships between structuring factors. In many observational data sets, nesting and crossing are sufficient to characterize the whole factorial structure. But in observational data, the crossing relation is usually not orthogonal (as opposed to experimental data, where orthogonality is often ensured by design), that is, structuring factors are typically *correlated*, as in this study.

As shown in this chapter, typicality tests and homogeneity tests afford a solution to numerous problems raised in GDA.

6.4.8 Steps of the Study Using R and Coheris SPAD Software

We now present the SPAD project with interfaced R script and the specific R script (described hereinafter) that performs the statistical analyses we previously presented. It comprises three folders (Figure 6.4.14).

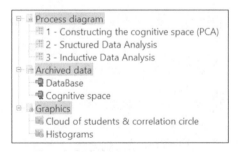

Figure 6.4.14
SPAD project with the four folders.

The `Process diagram` folder includes the three steps of the data analysis: first and foremost, the construction of the cognitive space (§6.4.2, p. 201); then the structured data analysis (§6.4.3, p. 202) and finally the inductive data analysis (§6.4.4, p. 210).

The `Archived data` folder contains two files: the initial data file

---

[26] In experimental data, for two-way ANOVA with balanced design in the one-dimensional case, see, e.g., Anderson and Ter Braak (2003); Pesarin and Salmaso (2010, Chapter 11).

(`DataBase`), and the file including the first two principal coordinates of individual points together with the two structuring factors (`Cognitive space`).

The `Graphics` folder contains the graph of the cloud of students and that of the variables (circle of correlations), as well as the histograms of the test statistics.

Now, let us comment on the three diagrams.

• The *first diagram* (Figure 6.4.15) sets about constructing the cloud of points and storing the principal coordinates of points. At the very beginning, there is the *import of the data set*, a "SPAD data archive" file called `DataBase`, which is stored in the `Archived data` folder. Then a PCA is performed so as to construct geometric clouds (as described in §6.4.2, p. 201). After that, the first two principal coordinates of students' points are added to the data file, and (last two icons) the first two principal coordinates and the two factors are selected and registered into a file called `Cognitive space` that is stored in the `Archived data` folder.

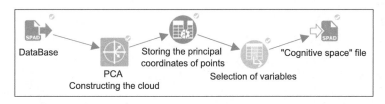

Figure 6.4.15
1 - Constructing the cognitive space.

• The *second diagram* (Figure 6.4.16) shows how to calculate the coordinates of the additive cloud.

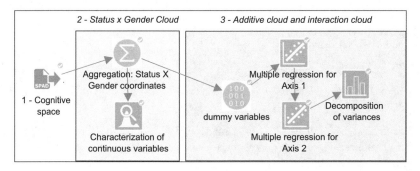

Figure 6.4.16
2 - Structured data analysis.

Firstly, the `Cognitive space` file is imported from the `Archived data` folder. Secondly, the coordinates of the four mean points are computed, and then the between- and within-classes decomposition of variances is given (Table 6.4.8, p. 209). Thirdly, we compute the dummy variables of *Status* and *Gender* and perform the multiple regression of Axis 1 and that of Axis 2 onto two dummy variables in order to obtain the coordinates of the additive cloud and of the interaction deviations (Table 6.4.7, p. 208). Finally we obtain the decomposition of variances into (additive)+(interaction) (Table 6.4.8, p. 209).

- The *third diagram* (Figure 6.4.17) shows the use of the interfaced R script for testing the homogeneity of the four groups and the main effects of *Status* and *Gender*. For the main effects, the R script gives the values of the squared M-distance, and then we deduce that of the M-variance. Then (in the grey square) the interaction and additive distributions of the M-variances are computed by using the R script that is commented hereafter. Lastly, the histograms of the M-variance distributions are depicted.

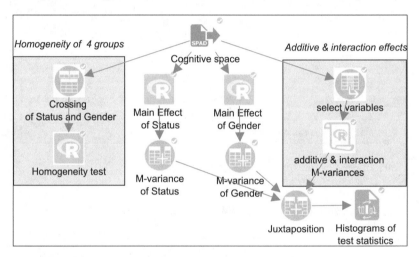

Figure 6.4.17
3 - Inductive data analysis.

R script for calculation of distributions of the additive and interaction M-variances

The *first step* consists in importing the SPAD *data table* into R. Then several elements, which are necessary for the calculations, are determined; these elements are commented in the script.

```
# === Step 1. Data ===
base <- SPAD.readAsDataFrame()
ncol <- dim(base)[2]                       # number of columns of base
Y.IL <- as.matrix(base[ , 1:(ncol-2)])     # principal coordinates
n <- dim(Y.IL)[1]                          # number of cases
L <- dim(Y.IL)[2]                          # number of principal axes
```

```
Lambda.L <- colSums(Y.IL^2)/n          # variances of axes
A    <- as.factor(base[, (ncol-1)])    # A factor (Status)
B    <- as.factor(base[, ncol])        # B factor (Gender)
C    <- factor(paste(as.vector(A),as.vector(B), sep ="")) # crossing AxB
n.C <-as.vector(table(C))              # frequencies of groups
names(n.C)<-levels(C); cardC <- length(n.C)  # number of groups
C_int.C<- matrix(c(1,-1,-1,1),nrow= 1)  # interaction contrast and
w_int <- 1/(n * sum(n.C^-1))           # inverse of its squared norm
```

In order to use the script outside of SPAD, you can import the data in the following way. Copy the file and then read it from the file called "clipboard" (`base <- read.table("clipboard")`). The copying can be done in Excel file (by painting out the data file and then using right-clicking on the mouse, and selecting COPY, or Ctrl+C) or in a text file. Put a blank cell in the upper left corner, which signals to the function that the first row contains the column labels and the first column the row labels.

	A	B	C	D	E
1		Axis1	Axis2	A	B
2	s1	0.37216673	0.68580532	a1	b2
3	s2	-3.6884874	0.09391212	a1	b1
4	s3	2.81685922	-0.6292803	a2	b1
5	s4	0.70828774	-1.3409296	a2	b2

Another way is the following: if the file `Cognitive.txt` file is in your working directory, then the command `base <- read.table("Cognitive.txt")` reads the data.

The *second step* is the *descriptive analysis*.

The coordinates of the mean points of the groups are computed (`Y_G.CL`), then the results are printed. After that, the variances of the $A \times B$ cloud (`V_AxB.L`) and that of interaction (`V_int.L`) on the L axes are calculated by using the formula of page 208, and the variances of the additive cloud (`V_add.L`) are deduced by substraction; then the variances in the space (`tmp2`) are calculated and the results are printed.

```
# === Step 2. Descriptive analysis ===
                              # coordinates of mean points
Y_G.CL <- as.matrix(aggregate(Y.IL, by=list(C), FUN=mean)[,-1])
cat(" Weights and means of  mean points,\n")
print(cbind(n.C,Y_G.CL))
                              # variances of sources of variation
V_AxB.L <- n.C %*% Y_G.CL^2/n;
V_int.L <- w_int * (C_int.C %*% Y_G.CL)^2
V_add.L <- V_AxB.L - V_int.L;
row.names(V_AxB.L) <- "AxB";row.names(V_int.L) <- "A.B"
row.names(V_add.L) <- "A+B"
                              # printing of results
cat("\n Decomposition of variances\n")
tmp1 <- rbind(V_AxB.L, V_add.L, V_int.L)
tmp2 <- t(t(c(sum(V_AxB.L), sum(V_int.L), sum(V_add.L))))
```

```
colnames(tmp2) <- "Plane 1_2"
print(round((cbind(tmp1,tmp2)),digits=5))
```

The *third step* is the *inductive analysis*.

Firstly, the observed values of test statistic in the space (VM_AxB, VM_int, VM_add) are deduced from the variances computed at step 2. Then, in order to easily calculate the distribution of the test statistic, the standardized principal coordinates of the individual points are calculated (Z.IL), then—using a method similar to that described on page 143—for each nesting *j* (sample), the indicator vector (one.C) of group *c* is defined and the coordinates of the mean points (Z_Cj.CL) associated with the nesting are computed, then the M-variance associated with each source of variation is deduced. Finally, the tests are performed and the results printed.

```
# === Step 3. Inductive analysis ===
                              #observed M-variances in plane 1-2
VM_AxB <- sum(V_AxB.L %*% diag(1/Lambda.L))
VM_int <-  sum(V_int.L %*% diag(1/Lambda.L))
VM_add <- VM_AxB - VM_int
                          # Distributions of test statistics
Z.IL <- Y.IL %*% diag(1/sqrt(Lambda.L)) # standardized principal coord.
cardJ <- 1000000                      # number of nestings
VM.AxB.J <- VM_int.J <- VM_add.J <- matrix(0, cardJ)
Z_Cj.CL <- matrix(0,nrow=cardC, ncol=L)
for (j in 1:cardJ) {
  AA <- sample(C)
  for (c in (1:cardC)) {               # coordinates of mean points for j
    one.C <- rep(OL,cardC);  one.C[c]=1
    Z_Cj.CL[c, ] <- t(one.C[AA]) %*% Z.IL/n.C[c]
  }
                          # M-variances of sources
  VM.AxB.J[j] <- sum(n.C %*% Z_Cj.CL^2)/n              # AxB
  VM_int.J[j] <- w_int * sum((C_int.C %*% Z_Cj.CL)^2)  # A.B
}
VM_add.J <- VM.AxB.J - VM_int.J                        # A.B

n_int <- sum(VM_int.J >= VM_int*(1-1e-12))# testing A.B and A+B
n_add <- sum(VM_add.J >= VM_add*(1-1e-12))
cat("\n","p-value\n for A+B ", n_add,"/",cardJ,' = ',
    round(n_add/cardJ, digits = 3),"\n for A.B ", n_int,"/",cardJ,
    ' = ', round(n_int/cardJ, digits = 3), sep="")
```

In order to construct the histograms of distributions, the results are exported to SPAD.

```
# export of distributions to SPAD
SPAD.setDataFrame(cbind(VM.AxB.J,VM_add.J, VM_int.J))
```

The results of the script are the following ones.

```
Weights and means of the four mean points,
           n.C      Axis1        Axis2
mediumboy   17  1.2588449   0.1643558
mediumgirl  11  0.4837496  -0.8466595
underboy     5 -1.5376148   0.9897596
undergirl    9 -2.1148373   0.1744899

Decomposition of variances
      Axis1    Axis2 Plane 1_2
AxB 1.94257  0.32182   2.26439
A+B 1.94055  0.31984   0.00400
A.B 0.00202  0.00198   2.26039

p-value
 for A+B: 0/1e+06 = 0
 for A.B: 956283/1e+06 = 0.956
```

The preceding R script works for the crossing of two factors with two levels each.

7

Mathematical Bases

The importance of invariance properties
in geometric settings cannot be overemphasized.
P. Suppes

In the present chapter, we give a short overview of the background that is necessary to make the book mathematically self-contained. In accordance with the formal-geometric approach of Geometric Data Analysis (GDA) (Le Roux and Rouanet, 2004, pp. 6–9), abstract linear algebra is given a central place.

Let us briefly comment on this point.

In GDA, the clouds of points are constructed from numerical tables[1]. The construction is an elaborate process that comprises two phases:

1. *Formalization*, to express the basic statistical objects in mathematical terms;

2. *Application of mathematical theory*, to characterize statistical procedures and conduct relevant proofs.

To account for this process, abstract linear algebra is ideally suited, because it allows one to clearly distinguish *vectors* (elements of a vector space), *points* (elements of a geometric space) and *sets of numbers* (e.g., coordinates) and to describe the interplay between these entities. The current matrix approach to statistics cannot account for this process, but *matrix formulas* are useful as a shorthand notation, and are thus used in this book.

The review of this chapter, where most proofs are sketched or simply omitted, is no substitute for the study of textbooks (e.g., Halmos (1958), Godement (1966), Mac Lane and Birkhoff (1967), etc.).

The chapter is organized as follows. We first recall (with the shorthand idea in mind) the basic matrix operations (§7.1). Then we review finite-dimensional vector spaces (§7.2), Euclidean properties and spectral decomposition (§7.3). Finally we present the basics of multidimensional geometry (§7.4).

---

[1]This situation is seen to differ from that of geometry in mathematics, where the objects can be studied directly, taking an arbitrary orthonormal reference basis to carry out computations.

7.1   Matrix Calculus

A *matrix* is a rectangular array of real numbers arranged in rows and columns that obeys the rules of matrix calculus.

A matrix of *elements* or *entries* a_{ij} $(a_{ij} \in \mathbb{R})$, with rows indexed by a set I and columns indexed by a set J, is called an $I{\times}J$-matrix. Capital boldface letters are used to denote matrices; their elements are denoted by the corresponding italic lower-case with appropriate indices. We will write $\mathbf{A} = [a_{ij}]$.

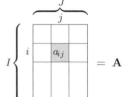

Operations on Matrices

Transposition

The *transpose* of the $I{\times}J$-matrix $\mathbf{A} = [a_{ij}]$ is the $J{\times}I$-matrix obtained by interchanging rows and columns. It is denoted $\mathbf{A}^\top$ with $\mathbf{A}^\top = [a_{ji}]$.

Addition

The sum of two $I{\times}J$-matrices $\mathbf{A} = [a_{ij}]$ and $\mathbf{B} = [b_{ij}]$ is the $I{\times}J$-matrix $\mathbf{A} + \mathbf{B} = [a_{ij} + b_{ij}]$.
— *Properties*: $\mathbf{A} + \mathbf{B} = \mathbf{B} + \mathbf{A}$; $(\mathbf{A} + \mathbf{B}) + \mathbf{C} = \mathbf{A} + (\mathbf{B} + \mathbf{C})$; $(\mathbf{A} + \mathbf{B})^\top = \mathbf{A}^\top + \mathbf{B}^\top$; the product of $\mathbf{A} = [a_{ij}]$ by a scalar λ is $\lambda\mathbf{A} = [\lambda a_{ij}]$.

A *null matrix* or *zero matrix* is the one whose elements are all zeros. It is denoted $\mathbf{0}$ and verifies the following property: for all $\mathbf{A}$, $\mathbf{A} + \mathbf{0} = \mathbf{0} + \mathbf{A} = \mathbf{A}$.

Matrix multiplication

The product of the $I{\times}J$-matrix $\mathbf{A} = [a_{ij}]$ by the $J{\times}K$-matrix $\mathbf{B} = [b_{jk}]$ (the same set J indexing the columns of $\mathbf{A}$ and the rows of $\mathbf{B}$) is defined as the $I{\times}K$-matrix $\mathbf{C} = \mathbf{AB}$, such that $\mathbf{C} = [c_{ik}]$ with $c_{ik} = \sum_j a_{ij}b_{jk}$.

Remark. The product $\mathbf{BA}$ may be meaningless even if $\mathbf{AB}$ exists; the two products can co-exist only if $\mathbf{A}$ is an $I{\times}J$-matrix and $\mathbf{B}$ is a $J{\times}I$-matrix, then $\mathbf{AB}$ is an $I{\times}I$-matrix and $\mathbf{BA}$ is a $J{\times}J$-matrix. Even if $J = I$ it can be easily verified that in general $\mathbf{AB} \neq \mathbf{BA}$. We must therefore always distinguish between premultiplication and postmultiplication.
— *Properties*:
$\mathbf{A}(\mathbf{B} + \mathbf{C}) = \mathbf{AB} + \mathbf{AC}$; $(\mathbf{A} + \mathbf{B})\mathbf{C} = \mathbf{AC} + \mathbf{BC}$; $(\mathbf{AB})^\top = \mathbf{B}^\top\mathbf{A}^\top$.

Row and Column Matrices

Matrices consisting of only one row or column are often termed vectors. We use boldface lower-case letters to denote these.

An $I{\times}1$-matrix, or an I-column vector, is in short termed an I-*column*.

An $1 \times J$-matrix, or *J-row*, is often written as the transpose of a J-column. An 1×1-matrix is a number and is written as an italic lower-case letter. The column, all elements of which are equal to 1, is denoted $\mathbf{e}$.

Let $\mathbf{x} = [x_i]$ and $\mathbf{y} = [y_i]$ be two I-columns then $\mathbf{x}^\top \mathbf{y} = \mathbf{y}^\top \mathbf{x} = \sum_i x_i y_i$ is a number, whereas $\mathbf{xy}^\top = [x_i y_{i'}]$ and $\mathbf{yx}^\top = [y_i x_{i'}]$ are $I \times I$-matrices.

Square Matrices

A matrix is square if its rows and columns are indexed by a same set I. The terms $(c_{ii})_{i \in I}$ of the square matrix $\mathbf{C} = [c_{ii'}]$ are called *diagonal elements*.

For a square matrix $\mathbf{C}$ it is convenient to write $\mathbf{C}^2 = \mathbf{CC}$. A square matrix $\mathbf{C}$ is *idempotent* if $\mathbf{C}^2 = \mathbf{C}$.

Symmetric matrix

A square matrix $\mathbf{S} = [s_{ii'}]$ is said to be *symmetric* if $\mathbf{S}^\top = \mathbf{S}$, that is, $s_{ii'} = s_{i'i}$ for all i and i'.

— *Property*: $\forall \mathbf{A} = [a_{ij}]$, $\mathbf{AA}^\top$ is a symmetric $I \times I$ matrix and $\mathbf{A}^\top \mathbf{A}$ is a symmetric $J \times J$ matrix.

Diagonal matrix

A diagonal $I \times I$-matrix is a square matrix with all non-diagonal elements equal to zero; it is written $\mathbf{D}_I = [d_i]$. A diagonal matrix with all diagonal elements equal to 1 is called *identity matrix* and denoted $\mathbf{I}$ (or more precisely $\mathbf{I}_I$ for the identity $I \times I$-matrix): $\forall \mathbf{C} = [c_{ii'}]$, $\mathbf{IC} = \mathbf{CI} = \mathbf{C}$.

Premultiplying the $I \times J$-matrix $\mathbf{A} = [a_{ij}]$ by the diagonal matrix $\mathbf{D}_I = [d_i]$ multiplies the ith row by d_i, hence the matrix $\mathbf{D}_I \mathbf{A} = [d_i a_{ij}]$.

Postmultiplying the $I \times J$-matrix $\mathbf{A} = [a_{ij}]$ by the diagonal matrix $\mathbf{D}_J = [d_j]$ multiplies the jth column by d_j, hence the matrix $\mathbf{AD}_J = [a_{ij} d_j]$.

Trace

The *trace* of a square matrix $\mathbf{C} = [c_{ii'}]$ is the number $\sum_i c_{ii}$ (sum of the elements on the main diagonal). It is denoted $\mathrm{tr}(\mathbf{C})$.

— *Properties*: If $\mathbf{AB}$ and $\mathbf{BA}$ are both square matrices, $\mathrm{tr}(\mathbf{AB}) = \mathrm{tr}(\mathbf{BA})$. In particular, if $\mathbf{x}$ and $\mathbf{y}$ are I-columns, $\mathrm{tr}(\mathbf{xy}^\top) = \mathrm{tr}(\mathbf{y}^\top \mathbf{x}) = \sum_i x_i y_i$. Observe that $\mathrm{tr}(\mathbf{AB}) \neq \mathrm{tr}(\mathbf{A}) \times \mathrm{tr}(\mathbf{B})$.

Determinant

Let S_I be the set of all $I!$ permutations of the elements of I and let $\left(\sigma(i)\right)_{i \in I}$ denote a permutation. Let s be the number of inversions in the permutation, that is, the total number of times in which an element is followed by elements that precede it in the initial order. The *determinant* of a square matrix $\mathbf{C} = [c_{ii'}]$, denoted $\det(\mathbf{C})$, is equal to $\sum_{\sigma \in S_I} (-1)^s \prod_{i \in I} c_{\sigma(i)\,i}$.

— *Properties*: $\det(\mathbf{C}^\top) = \det(\mathbf{C})$; $\det(\mathbf{AB}) = \det(\mathbf{A}) \times \det(\mathbf{B})$.

The determinant of a diagonal matrix $\mathbf{D}_I = [d_i]$ is merely the product of its diagonal elements: $\det(\mathbf{D}_I) = \prod_i d_i$. By definition, $\mathbf{C}$ is *singular* if $\det(\mathbf{C}) = 0$.

— *Property*: $\mathbf{AB} = \mathbf{AC}$ and $\mathbf{A}$ non-singular $\Rightarrow \mathbf{B} = \mathbf{C}$.

Inverse matrix

The *inverse* of a non-singular square matrix $\mathbf{C}$ is the unique square matrix denoted $\mathbf{C}^{-1}$ such that $\mathbf{CC}^{-1} = \mathbf{C}^{-1}\mathbf{C} = \mathbf{I}$.

— *Properties*: $(\mathbf{C}^\top)^{-1} = (\mathbf{C}^{-1})^\top$; $\det(\mathbf{C}^{-1}) = (\det(\mathbf{C}))^{-1}$; $(\mathbf{AB})^{-1} = \mathbf{B}^{-1}\mathbf{A}^{-1}$; the inverse of the diagonal matrix $\mathbf{D}_I = [d_i]$ is $\mathbf{D}_I^{-1} = [1/d_i]$.

Rank of a Matrix

The product of an I-column by a J-row is an $I \times J$-matrix of rank 1. In a matrix of rank 1, all rows are proportional and all columns are proportional.

Any $I \times J$-matrix can be expressed as a sum of matrices of rank 1; by definition the rank L of the matrix is the minimum number of matrices of rank 1 necessary to express the matrix. Thus, if the rank of $\mathbf{M}$ is equal to L, $\mathbf{M}$ can be expressed as the sum of L matrices of rank 1: $\mathbf{M} = \sum_\ell \mathbf{a}_\ell \mathbf{b}_\ell^\top$. Equivalently, $\mathbf{M}$ is a product of an $I \times L$-matrix $\mathbf{A}$ by a $L \times J$-matrix $\mathbf{B}$: $\mathbf{M} = \mathbf{AB}$.

7.2   Finite-Dimensional Vector Space

The concept of *vectors* (elements of a vector space) applies to entities that may not be reducible to numbers; for example, forces in mechanics are formalized as vectors, in so far as they add up vectorially ("the parallelogram of forces") and are multiplied by numbers ("scalars") to yield other forces. Abstract linear algebra[2], in brief *linear algebra*, is the general theory of *vector spaces*. Linear algebra involves two sorts of notions: the *purely vectorial* ones and the *Euclidean* ones; both are well covered in mathematical textbooks.

Throughout this book, it is assumed that all vector spaces under consideration are defined over the field $\mathbb{R}$ of real numbers and we confine our attention to finite-dimensional vector spaces. Thus, in this book, "vector space" will always mean *finite-dimensional vector space over* $\mathbb{R}$.

In this section, we recall the definition of vector space (§7.2.1), briefly review the vectorial notions namely bases (§7.2.2), homomorphisms, bilinear and quadratic forms (§7.2.3), and conclude with matrix formulas (§7.2.4).

---

[2]"Abstract" in this context means "general" and is not to be opposed to "concrete" but to "numerical".

7.2.1 Vector Space

Definition 7.1. *A vector space (over* $\mathbb{R}$*) consists of a (non-empty) set* $\mathcal{X}$ *of elements called* vectors *together with two binary operations, the vector addition* $+ : \mathcal{X} \times \mathcal{X} \to \mathcal{X}$ *and the multiplication by scalars* $\cdot : \mathbb{R} \times \mathcal{X} \to \mathcal{X}$*, satisfying the following axioms: for all* $x, y, z \in \mathcal{X}$ *and for all* $\lambda, \mu \in \mathbb{R}$,

(1) Vector addition $+ : (x, y) \mapsto x + y$

    1. $(x + y) + z = x + (y + z)$;
    2. $x + y = y + x$;
    3. there exists a null vector*, i.e., $\forall x \in \mathcal{X} : 0 + x = x$;*
    4. to every vector $x \in \mathcal{X}$ there corresponds a unique vector $-x \in \mathcal{X}$ such that $x + (-x) = 0$.

(2) Multiplication by scalars $\cdot : (\lambda, x) \mapsto \lambda \cdot x$

    1. $\lambda \cdot (x + y) = \lambda \cdot x + \lambda \cdot y$;
    2. $(\lambda + \mu) \cdot x = \lambda \cdot x + \mu \cdot x$;
    3. $(\lambda\mu) \cdot x = \lambda \cdot (\mu \cdot x) = \mu \cdot (\lambda \cdot x)$;
    4. $\forall x \in \mathcal{V}$, $1 \cdot x = x$.

For vector spaces in general, we will denote vectors by $(x,\ y,\ \ldots)$ and the null vector by 0; for the multiplication by scalars, we omit the point.

7.2.2 Subspaces, Bases, Coordinates

Given a family[3] of vectors $(x_i)_{i \in I}$ in $\mathcal{X}$ and a family of scalars $(\lambda_i)_{i \in I}$, their *linear combination* is the vector $\sum \lambda_i x_i$.

A family $(x_i)_{i \in I}$ of vectors in $\mathcal{X}$ is said to be *linearly independent* if for every family $(\lambda_i)_{i \in I}$ of scalars,

$$\sum \lambda_i x_i = 0 \text{ implies that } \lambda_i = 0 \text{ for all } i \in I$$

Subspace

Given a vector space $\mathcal{X}$, a non-empty subset $\mathcal{L}$ of $\mathcal{X}$ is a *vector subspace* if $\forall x, y \in \mathcal{L}$, $\forall \lambda, \mu \in \mathbb{R}$, $\lambda x + \mu y \in \mathcal{L}$.

A family of vectors $(x_\ell)_{\ell \in L}$ in $\mathcal{L}$ *spans* $\mathcal{L}$ if, for every $y \in \mathcal{L}$, there is a family of scalars $(\lambda_\ell)_{\ell \in L}$ such that $y = \sum \lambda_\ell x_\ell$; the set $(x_\ell)_{\ell \in L}$ is called a *spanning set* of $\mathcal{L}$. The *sum of two subspaces* $\mathcal{U}$ and $\mathcal{V}$ consists of all sums $u + v$ where $u \in \mathcal{U}$ and $v \in \mathcal{V}$ and is denoted $\mathcal{U} + \mathcal{V}$. If $\mathcal{U} \cap \mathcal{V}$ is the null vector, the sum is called *direct sum* and denoted $\mathcal{U} \oplus \mathcal{V}$.

---

[3]A family of vectors $(x_i)_{i \in I}$ in $\mathcal{X}$ is simply a function $x : I \to \mathcal{X}$. When considering a family indexed by a set I, there is no reason to assume that I is ordered. The crucial point is that every element of the family is uniquely indexed by an element of the set I.

Basis and coordinates

A *basis* of $\mathcal{X}$ is a family of linearly independent vectors in $\mathcal{X}$ such that every vector in $\mathcal{X}$ is a linear combination of vectors of the family. In other words, a basis is a linearly independent spanning set of $\mathcal{X}$.

A vector space is *finite-dimensional* if it has a finite basis. All bases of a finite-dimensional vector space have the same number of vectors. The number is called the *dimension* of the vector space. If this number is I, the vector space is said to be *I-dimensional*.

Let $(\epsilon_i)_{i \in I}$ be a basis of the I-dimensional vector space $\mathcal{X}$, then any vector $x \in \mathcal{X}$ can be uniquely expressed as a linear combination of basis vectors, that is, $x = \sum x_i \epsilon_i$. The coefficients $(x_i)_{i \in I}$ are the *coordinates* (or components) of x over the basis $(\epsilon_i)_{i \in I}$.

> *Using coordinates in vector operations enables one to "think in terms of vectors" while "working with numbers".*

Numerical vector spaces

$\mathbb{R}^n$ as a vector space. If n is an integer with $n \geq 1$, the set $\mathbb{R}^n$ of the numerical n-tuples $(x_i)_{i=1,\ldots n}$ is naturally structured as an n-dimensional vector space with the canonical basis $(\epsilon_i)_{i=1,\ldots n}$ where $\epsilon_i = (0, \ldots, 1, \ldots 0)$ is the i-th canonical basis vector of $\mathbb{R}^n$ (with a 1 in the i-th slot).

$\mathbb{R}^I$ as a vector space. If I is a non-empty finite set, the set of maps $I \to \mathbb{R}$ is structured as an I-dimensional vector space, denoted $\mathbb{R}^I$, with the canonical basis $(\delta_i)_{i \in I}$; $\delta_i : I \to \mathbb{R}$ is such that $\delta_i(i) = 1$ and $\delta_i(i') = 0$ for all $i' \neq i$.

7.2.3 Linear Maps or Homomorphisms

A function between two vector spaces that preserves addition and scalar multiplication is called a homomorphism of vector spaces, or linear map.

Definition 7.2. *Let $\mathcal{X}$ and $\mathcal{Y}$ be two vector spaces over $\mathbb{R}$, a homomorphism between $\mathcal{X}$ and $\mathcal{Y}$ is a map $f : \mathcal{X} \to \mathcal{Y}$ satisfying the linearity property:*

$$\forall x, x' \in \mathcal{X}, \ \forall \lambda, \mu \in \mathbb{R}, \ f(\lambda x + \mu x') = \lambda f(x) + \mu f(x')$$

The *image of f* is the subspace $f(\mathcal{X})$ of $\mathcal{Y}$ defined as the set of vectors $y \in \mathcal{Y}$ such that $f(x) = y$, for some $x \in \mathcal{X}$. The nullspace or *kernel of f* is the set of $x \in \mathcal{X}$ such that $f(x) = 0$ (inverse image of the null vector). By definition, the *rank* of f is the dimension of its image $f(\mathcal{X})$.

Given linear maps $f : \mathcal{X} \to \mathcal{Y}$ and $g : \mathcal{Y} \to \mathcal{Z}$, the *composition* of f and g is the map $g \circ f : \mathcal{X} \to \mathcal{Z}$ such that $\forall x \in \mathcal{X} : g \circ f(x) = g(f(x)) \in \mathcal{Z}$.

A homomorphism $f : \mathcal{X} \to \mathcal{Y}$ is

1. *injective* if and only if $\forall x \in \mathcal{X} : f(x) = 0 \Rightarrow x = 0$;
2. *surjective* if and only if $f(\mathcal{X}) = \mathcal{Y}$;

3. an *isomorphism* if it is both injective and surjective.

An isomorphism f is *invertible*; its inverse $f^{-1} : \mathcal{Y} \to \mathcal{X}$ is such that

$$\forall x \in \mathcal{X}, f^{-1}(f(x)) = x \text{ and } \forall y \in \mathcal{Y}, f(f^{-1}(y)) = y.$$

A homomorphism from $\mathcal{X}$ to $\mathcal{X}$ is called an *endomorphism* of $\mathcal{X}$. An endomorphism f such that $f \circ f = f$ is said to be *idempotent*.

Projection

If $\mathcal{X}$ is the direct sum of two subspaces $\mathcal{L}$ and $\mathcal{L}'$, then for any $x = y + y'$ with $y \in \mathcal{L}$ and $y' \in \mathcal{L}'$, the *projection* of x onto $\mathcal{L}$ along $\mathcal{L}'$ is, by definition, the linear map p_L such that $p_L(x) = y$.

An endomorphism $p_L : \mathcal{X} \to \mathcal{X}$ is a projection if and only if it is idempotent, that is, $p_L \circ p_L = p_L$.

Linear form, bilinear form, quadratic form

A *linear form* on $\mathcal{X}$ is a linear map from $\mathcal{X}$ to $\mathbb{R}$. The set of all linear forms on $\mathcal{X}$ spans a vector space isomorphic to $\mathcal{X}$ called the *dual space* of $\mathcal{X}$.

A *bilinear form* on $\mathcal{X} \times \mathcal{Y}$ is a function $b : \mathcal{X} \times \mathcal{Y} \to \mathbb{R}$ that is linear in each argument separately. More precisely, $\forall x \in \mathcal{X}$ and $\forall y \in \mathcal{Y}$, the map $b_Y : \mathcal{X} \to \mathbb{R}$ defined by $b_Y(x) = b(x, y)$ and the map $b_X : \mathcal{Y} \to \mathbb{R}$ defined by $b_X(y) = b(x, y)$ are both linear forms.

A bilinear form s on $\mathcal{X}$ is *symmetric* if $\forall x, x' \in \mathcal{X}$, $s(x, x') = s(x', x)$, and it is *positive definite* if $s(x, x) > 0$ for every $x \neq 0$.

An application $q : \mathcal{X} \to \mathbb{R}$ is a *quadratic form* on $\mathcal{X}$ if it satisfies the following two properties: $\forall x, x' \in \mathcal{X}, \forall \lambda \in \mathbb{R}$,

1) $q(\lambda x) = \lambda^2 q(x)$;
2) Letting $b(x, x') = \frac{1}{2}(q(x+x') - q(x) - q(x'))$, b is bilinear and symmetric.

7.2.4 Matrix Formulas

Let $(\epsilon_i)_{i \in I}$ be a basis of the I-dimensional vector space $\mathcal{X}$ and let $(\delta_j)_{j \in J}$ be a basis of the J-dimensional vector space $\mathcal{Y}$. Let $\mathbf{x} = [x_i]$ be the I-column of the coordinates of any vector $x = \sum x_i \epsilon_i$ in $\mathcal{X}$ over the basis $(\epsilon_i)_{i \in I}$, and $\mathbf{y} = [y_j]$ be the J-column of the coordinates of any vector $y = \sum y_j \delta_j$ in $\mathcal{Y}$ over the basis $(\delta_j)_{j \in J}$.

Homomorphism

Every homomorphism $f : \mathcal{X} \to \mathcal{Y}$ is uniquely determined by the family of the images under f of the basis vectors $(\epsilon_i)_{i \in I}$ of $\mathcal{X}$. Letting $f(\epsilon_i) = \sum_j m_{ji} \delta_j$, the homomorphism f is represented by the $J \times I$-matrix $\mathbf{M} = [m_{ji}]$ whose entries in the i-th column are the $(m_{ji})_{j \in J}$ coordinates of the vector $f(\epsilon_i)$ over the basis $(\delta_j)_{j \in J}$ of $\mathcal{Y}$. Hence, $y = f(x)$ is written $\mathbf{y} = \mathbf{Mx}$.

The composition of homomorphisms corresponds to the *matrix multiplication*. Let $\mathbf{F}$, $\mathbf{G}$, $\mathbf{H}$ be the matrices associated with $f : \mathcal{X} \to \mathcal{Y}$, $g : \mathcal{Y} \to \mathcal{Z}$ and $h : \mathcal{X} \to \mathcal{Z}$, then $h = g \circ f$ is written $\mathbf{H} = \mathbf{GF}$.

Bilinear form

The matrix of the bilinear form $b : \mathcal{X} \times \mathcal{Y} \to \mathbb{R}$ is the $I \times J$-matrix $\mathbf{B} = [b_{ij}]$ with $b_{ij} = b(\epsilon_i, \delta_j)$; $\forall x, y \in \mathcal{X}$, $b(x, y)$ writes[4] $\mathbf{x}^\top \mathbf{By}$.

The matrix of a symmetric bilinear form $s : \mathcal{X} \times \mathcal{X} \to \mathbb{R}$ is the $I \times I$-matrix $\mathbf{S} = [s_{ii'}]$ with $s_{ii'} = s(\epsilon_i, \epsilon_{i'}) = s_{i'i}$.

Change of Basis

Let $(\epsilon_{i_1})_{i_1 \in I}$ and $(\epsilon'_{i_2})_{i_2 \in I}$ be two bases of $\mathcal{X}$, with $\epsilon'_{i_2} = \sum_{i_1 \in I} p_{i_1 i_2} \epsilon_{i_1}$ $(i_2 \in I)$.
$\mathbf{P} = [p_{i_1 i_2}]$ is the $I \times I$-matrix representation of the identity map $id_X : \mathcal{X} \to \mathcal{X}$ over the bases $(\epsilon'_{i_2})_{i_2 \in I}$ and $(\epsilon_{i_1})_{i_1 \in I}$, *in that order*. More precisely $\mathbf{P}$ is the matrix of identity $id_X : \mathcal{X}_2 \to \mathcal{X}_1$ where $\mathcal{X}_2$ is $\mathcal{X}$ equipped with the basis $(\epsilon'_{i_2})_{i_2 \in I}$ and $\mathcal{X}_1$ is $\mathcal{X}$ equipped with the basis $(\epsilon_{i_1})_{i_1 \in I}$.

The matrix $\mathbf{P}$ is called the *change of basis matrix from* $(\epsilon_{i_1})_{i_1 \in I}$ *to* $(\epsilon'_{i_2})_{i_2 \in I}$.

Vectors

Coordinates of vectors are *basis-bound*, that is, if the basis of a vector space changes, vectors remain the same, whereas their coordinates change.

We now show how to convert the matrix representation of a vector over one basis to the representation of the same vector over another basis.

Every vector $x \in \mathcal{X}$ can be referred either to $(\epsilon_{i_1})_{i_1 \in I}$ ("old basis") as $x = \sum_{i_1} x_{i_1} \epsilon_{i_1}$, or to $(\epsilon'_{i_2})_{i_2 \in I}$ ("new basis") as $x = \sum_{i_2} x'_{i_2} \epsilon'_{i_2}$. One therefore has: $x = \sum_{i_2} x'_{i_2} \left(\sum_{i_1} p_{i_1 i_2} \epsilon_{i_1} \right) = \sum_{i_1} \left(\sum_{i_2} p_{i_1 i_2} x'_{i_2} \right) \epsilon_{i_1}$. Hence:

$$\text{Passage formula:} \quad \forall i_1 \in I, \ x_{i_1} = \sum_{i_2 \in I} p_{i_1 i_2} x'_{i_2}$$

Let $\mathbf{x} = [x_{i_1}]$ be the I-column of the (old) coordinates of a vector over the basis $(\epsilon_{i_1})_{i_1 \in I}$ and $\mathbf{x}' = [x'_{i_2}]$ be the I-column of its (new) coordinates over the basis $(\epsilon'_{i_2})_{i_2 \in I}$:

$$\text{Passage formula:} \quad \mathbf{x} = \mathbf{Px}'$$

Remark. The new basis is given as a function of the old one, whereas the passage formulas give the old coordinates as a function of the new ones.

Homomorphisms

We now describe the effect of a change of bases on the matrix representation of a homomorphism.

---

[4]Writing $\mathbf{x}^\top \mathbf{By}$ for $b(x, y) = b(y, x)$ is masking the symmetry of the two arguments x and y of the bilinear form b, see *Critique du calcul matriciel*, Benzécri (1973, Vol. II, p. 58).

Let f be a homomorphism $\mathcal{X} \to \mathcal{Y}$. Let $(\delta_{j_1})_{j_1 \in J}$ and $(\delta'_{j_2})_{j_2 \in J}$ be two bases of $\mathcal{Y}$ and the $J \times J$-matrix $\mathbf{Q} = [q_{j_1 j_2}]$ be the change of basis matrix from $(\delta_{j_1})_{j_1 \in J}$ to $(\delta'_{j_2})_{j_2 \in J}$.

To convert the matrix representation $\mathbf{M}_1$ of homomorphism $f : \mathcal{X} \to \mathcal{Y}$ with respect to bases $(\epsilon_{i_1})_{i_1 \in I}$ and $(\delta_{j_1})_{j_1 \in T}$ into its representation $\mathbf{M}_2$ with respect to bases $(\epsilon'_{i_1})_{i_1 \in I}$ and $(\delta'_{j_1})_{j_1 \in T}$ we proceed as outlined in the arrow diagram on the right.

$$\mathbf{M}_2 = \mathbf{Q}^{-1}\mathbf{M}_1\mathbf{P}$$

$$
\begin{array}{ccc}
\mathcal{X} & \xrightarrow{\quad f \quad} & \mathcal{Y} \\
(\epsilon_{i_1}) & \mathbf{M_1} & (\delta_{j_1}) \\
id_X \big\uparrow \mathbf{P} & & \mathbf{Q}^{-1} \big\downarrow id_Y \\
\mathcal{X} & \xrightarrow[\quad f \quad]{\mathbf{M_2}} & \mathcal{Y} \\
(\epsilon'_{i_2}) & & (\delta'_{j_2})
\end{array}
$$

Bilinear forms

If $\mathbf{B_1}$ is the $I \times J$-matrix representation of the bilinear form $b : \mathcal{X} \times \mathcal{Y} \to \mathbb{R}$ with respect to bases $(\alpha_{i_1})_{i_1 \in I}$ and $(\delta_{j_1})_{j_1 \in J}$ and $\mathbf{B_2}$ is the one with respect to bases $(\alpha'_{i_2})_{i_2 \in I}$ and $(\delta'_{j_2})_{j_2 \in J}$, one has:

$$\mathbf{B}_2 = \mathbf{P}^{\top}\mathbf{B}_1\mathbf{Q}$$

7.3 Euclidean Vector Space

We will define now a Euclidean structure on a vector space which will allow us to deal with metric notions such as orthogonality, length (or distance) and angles. These notions are more intuitive in character than the purely vectorial notions, owing to the geometric representations they afford (see p. 230).

We begin by defining scalar (or inner) product and Euclidean space and by stating the Cauchy–Schwarz inequality (§7.3.1). We define orthogonality of vectors and of subspaces, orthogonal bases as well as orthonormal bases, and we show that every finite-dimensional Euclidean space has orthonormal bases using the Gram–Schmidt orthogonalization procedure (§7.3.2). Then, symmetric endomorphisms and orthogonal projections are defined and studied briefly (§7.3.3). We state the spectral theorem (§7.3.4), which is the basis of the search of principal directions of a cloud. Then (§7.3.5), we relate the inverse of a symmetric endomorphism after a perturbation of rank one to the inverse of the initial endomorphism, which is fundamental when we use the Mahalanobis distance in Chapters 4 and 5. We conclude with a section giving the numerical expressions and matrix formulas (§7.3.6).

7.3.1 Definitions and Basic Properties

Definition 7.3 (Euclidean space). *A finite-dimensional vector space $\mathcal{X}$ over $\mathbb{R}$ is a Euclidean space if and only if it is equipped with a symmetric bilinear form $b : \mathcal{X} \times \mathcal{X} \to \mathbb{R}$ that is positive definite (see p. 227).*

The number $b(x, x')$ is called *scalar product* (or *inner product*) of x and x' and denoted $\langle x|x'\rangle$.

Definition 7.4 (Euclidean norm). *The norm on $\mathcal{X}$ induced by the scalar product is called Euclidean norm; it is denoted $\|\cdot\|$ and defined by*

$$\forall x \in \mathcal{X}, \|x\| = \sqrt{\langle x|x\rangle}$$

The Euclidean norm satisfies the three properties of a norm:

1. $\forall x \in \mathcal{X}, \|x\| \geq 0; \|x\| = 0 \iff x = 0$ (positive definite)

2. $\forall x \in \mathcal{X}, \forall \lambda \in \mathbb{R} : \|\lambda x\| = |\lambda| \, \|x\|$ (here $|\lambda|$ is the absolute value of λ)

3. $\forall x, y \in \mathcal{X}, \|x + y\| \leq \|x\| + \|y\|$ (triangle inequality)

A *unit vector* is a vector x with $\|x\| = 1$.

Cauchy–Schwarz inequality: $\forall x, y \in \mathcal{X}, |\langle x|y\rangle| \leq \|x\| \, \|y\|$, the equality holding if and only if x and y are linearly dependent.

Geometric Representations

Whenever several vectors span a two-dimensional subspace of an n-dimensional vector space ($n \geq 2$), these vectors can be represented by segments in a plane geometric figure[5]. Thus, two linearly independent vectors x and x' can be represented by two sides of a triangle whose lengths are $\|x\|$ and $\|x'\|$ and whose angle φ is given by

$$\cos \varphi = \frac{\langle x|x'\rangle}{\|x\| \times \|x'\|} \qquad \text{with} \quad 0 \leq \varphi \leq \pi$$

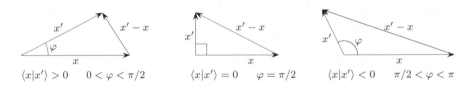

$\langle x|x'\rangle > 0 \quad 0 < \varphi < \pi/2$ $\langle x|x'\rangle = 0 \quad \varphi = \pi/2$ $\langle x|x'\rangle < 0 \quad \pi/2 < \varphi < \pi$

Figure 7.1
Geometric representation of two linearly independent vectors.

In a similar way, vectors spanning a three-dimensional subspace can be represented by a three-dimensional figure; thus three linearly independent vectors can be represented by three sides of a tetrahedron.

---

[5]Geometric representations follow from the canonical property of affine spaces stated below p. 238.

7.3.2 Orthogonality, Adjoint Homomorphisms

- Any two (non-null) vectors $x, x' \in \mathcal{X}$ are *orthogonal*, or perpendicular, if $\langle x|x' \rangle = 0$ (one writes $x \perp x'$).

- The family $(x_\ell)_{\ell \in L}$ of vectors in $\mathcal{X}$ is said to be *mutually orthogonal* when any two vectors of the family are orthogonal: if $\ell \neq \ell'$ then $\langle x_\ell | x_{\ell'} \rangle = 0$.

  An *orthogonal basis* of $\mathcal{X}$ is a basis of mutually orthogonal vectors.

- The family of vectors $(x_\ell)_{\ell \in L}$ is said to be *orthonormal* if vectors are unit vectors mutually orthogonal: $\langle x_\ell | x_{\ell'} \rangle = \begin{cases} 1 \text{ if } \ell' = \ell \\ 0 \text{ if } \ell' \neq \ell \end{cases}$.

  An *orthonormal basis* of $\mathcal{X}$ is an orthogonal basis of unit vectors.

- For any subset $\mathcal{L}$ of $\mathcal{X}$, the set
  $$\mathcal{L}^\perp = \{x' \in \mathcal{X} \,|\, \langle x|x' \rangle = 0, \text{ for all } x \in \mathcal{L}\},$$
  of all vectors orthogonal to all vectors in $\mathcal{L}$, is called the *orthogonal complement* of $\mathcal{L}$. One has $\mathcal{X} = \mathcal{L} \oplus \mathcal{L}^\perp$.

Theorem 7.1 (Adjoint homomorphisms). *Let $\mathcal{X}$ and $\mathcal{Y}$ be two Euclidean spaces where the scalar product on $\mathcal{X}$ is denoted as $\langle \cdot | \cdot \rangle_X$ and the scalar product on $\mathcal{Y}$ is denoted as $\langle \cdot | \cdot \rangle_Y$, and let $f : \mathcal{X} \to \mathcal{Y}$ be a homomorphism. There is a unique homomorphism $f^* : \mathcal{Y} \to \mathcal{X}$, such that*

$$\forall x \in \mathcal{X}, \forall y \in \mathcal{Y}, \ \langle x|f^*(y) \rangle_X = \langle f(x)|y \rangle_Y$$

The homomorphism f^* is called the adjoint of f.

Summarizing:

$$\mathcal{X} \underset{f^*}{\overset{f}{\rightleftarrows}} \mathcal{Y}$$

$$x \longmapsto f(x)$$
$$f^*(y) \longmapsfrom y$$
$$\langle x|f^*(y) \rangle_X = \langle f(x)|y \rangle_Y$$

Properties:
$f^{**} = f$ (symmetry of scalar product); $f(\mathcal{X})^\perp = \mathrm{Ker}f^*$; $f^*(\mathcal{Y})^\perp = \mathrm{Ker}f$.

7.3.3 Symmetric Endomorphisms, Orthogonal Projections

Definition 7.5 (Symmetric endomorphism). *An endomorphism $s : \mathcal{X} \to \mathcal{X}$ is said to be self-adjoint, or symmetric, if $s^* = s$.*

$$\forall x, x' \in \mathcal{X}, \ \langle s(x)|x' \rangle = \langle x|s(x') \rangle$$

- A symmetric endomorphism s is said to be *positive* if $\langle s(x)|x \rangle \geq 0$ for all x, *positive definite* if $\langle s(x)|x \rangle > 0$ for every $x \neq 0$.

- Given two adjoint homomorphisms $f : \mathcal{X} \to \mathcal{Y}$ and $f^* : \mathcal{Y} \to \mathcal{X}$, endomorphisms $f^* \circ f : \mathcal{X} \to \mathcal{X}$ and $f \circ f^* : \mathcal{Y} \to \mathcal{Y}$ are positive and symmetric.

Proposition 7.1. *Let s be a symmetric endomorphism of* $\mathcal{X}$.
(i) The function $\sigma : \mathcal{X} \times \mathcal{X} \to \mathbb{R}$ *defined such that*

$$\forall x, x' \in \mathcal{X},\ \sigma(x, x') = \langle x | s(x') \rangle_X = \langle s(x) | x' \rangle_X$$

is called symmetric bilinear form associated with s.
(ii) The function (also denoted σ) $\sigma : \mathcal{X} \to \mathbb{R}$ *such that*

$$\forall x \in \mathcal{X},\ \sigma(x) = \langle x | s(x) \rangle_X$$

is called quadratic form associated with s.

Definition 7.6 (Orthogonal projection). *Let* $\mathcal{L}$ *be an L-dimensional subspace of* $\mathcal{X}$ *(L $\geq$ 1), then the orthogonal projection of vector* x *onto* $\mathcal{L}$ *is the vector* $\widetilde{x} \in \mathcal{L}$ *such that the vector* $x - \widetilde{x}$ *is orthogonal to every vector in* $\mathcal{L}$.

If x is orthogonal to $\mathcal{L}$, $\widetilde{x}$ is the null vector; if $x \in \mathcal{L}$, $\widetilde{x} = x$.

dim. $\mathcal{L} = 1$ dim. $\mathcal{L} = 2$

Figure 7.2
Orthogonal projection of vector x onto a line, onto a plane.

The orthogonal *projection* p_α *onto a line* spanned by the non-null vector α is defined by

$$\forall x \in \mathcal{X} : p_\alpha(x) = \frac{\langle \alpha | x \rangle}{\|\alpha\|^2} \alpha$$

Theorem 7.2 (Gram–Schmidt orthogonalization). *Let* $(\epsilon_i)_{i=1,\ldots n}$ *be a basis of the n-dimensional vector space* $\mathcal{X}$, *then the vectors*

$$\alpha_1 = \epsilon_1 \ldots \alpha_i = \epsilon_i - \sum_{j=1}^{i-1} p_{\alpha_j}(\epsilon_i) \ldots \alpha_n = \epsilon_n - \sum_{j=1}^{n-1} p_{\alpha_j}(\epsilon_n)$$

form an orthogonal basis of $\mathcal{X}$.

Note that α_i is obtained by subtracting from ϵ_i the orthogonal projection of ϵ_i itself onto the orthogonal vectors α_1, ..., α_{i-1} that have already been computed. In addition, if we normalize each vector by dividing it by its length we end with an *orthonormal basis*.

For a proof of this classical property, see, for instance, Halmos (1958, pp. 127–128) or Gallier (2011, Proposition 10.8).

7.3.4 Spectral Theorem

The spectral theorem, alias *diagonalization theorem*, is the "mathematical heart" of the problem of principal directions of a cloud. It states that symmetric endomorphisms have real eigenvalues and that they can be "diagonalized" over an orthonormal basis.

Definition 7.7. *Given any endomorphism* $f : \mathcal{X} \to \mathcal{X}$, *a scalar* $\lambda \in \mathbb{R}$ *is called* eigenvalue *if there is some non-null vector* $x \in \mathcal{X}$ *such that* $f(x) = \lambda x$. *Then* x *is called an* eigenvector *associated with eigenvalue* λ.

If λ is an eigenvalue of f, the set of all vectors $x \in \mathcal{X}$ with $f(x) = \lambda x$ is a non-null subspace of $\mathcal{X}$ called *eigenspace* of λ; the dimension of this subspace is called the multiplicity of λ.

Theorem 7.3. *Let* $(\mathcal{X}, \langle \cdot | \cdot \rangle)$ *be a Euclidean space and let* $s : \mathcal{X} \to \mathcal{X}$ *be a symmetric endomorphism, then all eigenvalues of* s *are real and* $\mathcal{X}$ *has an orthonormal basis consisting of eigenvectors of* s.

This theorem is the classical version of the spectral theorem. Its proof can be found in all linear algebra textbooks (e.g., Gallier, 2011, Chapter 13).

Proposition 7.2. *Let* s *be a symmetric endomorphism on* $\mathcal{X}$, $(\alpha_\ell)_{\ell=1,\dots L}$ *be an orthonormal basis of* $s(\mathcal{X})$ *consisting of eigenvectors of* s *associated with non-null eigenvalues* $(\lambda_\ell)_{\ell=1,\dots L}$ *and* p_ℓ *be the orthogonal projection onto the line spanned by* α_ℓ; s *admits the* spectral decomposition:

$$s = \sum \lambda_\ell p_\ell \quad \text{with} \quad p_\ell : x \mapsto \langle x | \alpha_\ell \rangle \, \alpha_\ell$$

Definition 7.8. *Let* $\mathcal{X}'$ *be an invariant subspace of* $\mathcal{X}$ *under the endomorphism* $s : \mathcal{X} \to \mathcal{X}$ $(s(x) \in \mathcal{X}'$ *for all* $x \in \mathcal{X}')$, *the* restriction *of* s *to* $\mathcal{X}'$ *is an endomorphism of* $\mathcal{X}'$ *called "endomorphism induced by* s *onto* $\mathcal{X}'$".

The eigenvalues of a positive symmetric endomorphism are positive (or null). From the positive symmetric endomorphism s, we define the endomorphism of $\mathcal{X}$, denoted $s^{1/2}$, such as:

$$\forall \ell, \; s^{1/2}(\alpha_\ell) = \sqrt{\lambda_\ell} \, \alpha_\ell$$

7.3.5 Adding an Endomorphism of Rank 1

In this section, we state properties that are used in Chapters 4 and 5.

Proposition 7.3. *Given a positive definite symmetric endomorphism* $f : \mathcal{X} \to \mathcal{X}$, *a non-null vector* $u \in \mathcal{X}$, *and a positive number* a, *the inverse of endomorphism* $g \quad : \mathcal{X} \to \mathcal{X}$

$$x \mapsto g(x) = f(x) + a\langle u | x \rangle u$$

is such that

$$g^{-1} : \mathcal{X} \to \mathcal{X}$$
$$x \mapsto g^{-1}(x) = f^{-1}(x) - \frac{a[x|u]}{1 + a|u|^2} f^{-1}(u)$$

with $[x|u] = \langle f^{-1}(u) | x \rangle = \langle u | f^{-1}(x) \rangle$ *and* $|u|^2 = \langle f^{-1}(u) | u \rangle = [u|u]$,

Proof. g is symmetric since $\langle g(x)|y\rangle = \langle f(x)|y\rangle + a\langle u|x\rangle\langle u|y\rangle = \langle x|g(y)\rangle$;
g is positive since $\langle g(x)|x\rangle = \langle f(x)|x\rangle + a\langle u|x\rangle^2$, hence g is invertible.
One has $g\big(g^{-1}(x)\big) = f\big(g^{-1}(x)\big) + a\langle u|g^{-1}(x)\rangle u$ with:

- $f\big(g^{-1}(x)\big) = f\Big(f^{-1}(x) - \frac{a[x|u]}{1+a|u|^2}\, f^{-1}(u)\Big) = x - \frac{a[x|u]}{1+a|u|^2}\, u;$
- $\langle u|g^{-1}(x)\rangle = \langle u|f^{-1}(x) - \frac{a[x|u]}{1+a|u|^2}\, f^{-1}(u)\rangle = [x|u] - \frac{a[x|u]}{1+a|u|^2}|\vec{u}|^2 = \frac{[x|u]}{1+a|u|^2}.$

Hence $g\big(g^{-1}(x)\big) = f\big(g^{-1}(x)\big) + a\langle u|g^{-1}(x)\rangle u = x - a\frac{a[x|u]}{1+a|u|^2}\, u + \frac{a[x|u]}{1+a|u|^2}\, u = x.$
Similarly, one verifies $g^{-1}\big(g(x)\big) = x$ $\square$

Corollary 7.1. $\langle x|g^{-1}(x)\rangle = |x|^2 - \dfrac{a[x|u]^2}{1+a|u|^2}$ and $\langle u|g_u^{-1}(u)\rangle = \dfrac{|u|^2}{1+a|u|^2}$

Proof. $\langle x|g^{-1}(x)\rangle = \langle x|f^{-1}(x)\rangle - \frac{a[x|u]}{1+a|u|^2}\langle x|f^{-1}(u)\rangle = |x|^2 - \frac{a[x|u]^2}{1+a|u|^2}$. The second
equality follows. $\square$

Lemma 7.1. *Letting* $x' = f^{-1/2}(x)$, $u' = f^{-1/2}(u)$, $h_{u'} : x' \mapsto x' + \langle x'|u'\rangle\, u'$,
one has:

$$\langle x|g_u^{-1}(x)\rangle = \langle x'|h_{u'}^{-1}(x')\rangle$$

Proof. One has (Corollary 7.1) $\langle x|g_u^{-1}(x)\rangle = [x]^2 - \frac{a[x|u]^2}{1+a|u|^2}$. Letting $[x|u] =$
$\langle f^{-1}(x)|u\rangle = \langle f^{-1/2}(x)|f^{-1/2}(u)\rangle = \langle x'|u'\rangle$, and $|x|^2 = \|x'\|^2$ in the foregoing equa-
tion, we get $\langle x|g_u^{-1}(x)\rangle = \|x'\|^2 - \frac{a\langle x'|u'\rangle^2}{1+a\|u'\|^2} = \langle x' - \frac{a\langle x'|u'\rangle}{1+a\|u'\|^2}\, u'|x'\rangle$. By Proposition
7.3 (p. 233), $h_{u'}^{-1}(x') = x' - \frac{a\langle x'|u'\rangle}{1+a\|x'\|^2}\, u'$, hence the property. $\square$

Theorem 7.4. *Let* u *and* v *be two non-null vectors of* $\mathcal{X}$ *and let* g *be the
endomorphism of* $\mathcal{X}$ *defined by :*

$$\forall x \in \mathcal{X},\ g(x) = x + a\langle v|x\rangle u \quad (a \in \mathbb{R})$$

then u *is eigenvector of* g *associated with* $1 + a\langle v|u\rangle$ *and the orthogonal com-
plement of* v *is eigenspace of* f *associated with* 1.

Proof. $g(u) = \big(1 + a\langle v|u\rangle\big)u$, hence u is eigenvector of g associated with the eigen-
value $1 + a\langle v|u\rangle$. Let y be a vector orthogonal to v ($\langle v|y\rangle = 0$) hence $g(y) = y$: the
subspace $v^\perp$ is the eigenspace of g associated with eigenvalue 1. $\square$

Corollary 7.2. *If* $1 + a\langle v|u\rangle \neq 0$, *then* $g^{-1}(x) = x - \dfrac{a\langle v|x\rangle}{1 + a\langle v|u\rangle}u.$

Proof. If $1 + a\langle \vec{v}|\vec{u}\rangle \neq 0$, all eigenvalues of g are non-null, hence g is invertible
and its inverse has the same eigenvectors with eigenvalues equal to the inverse of
the eigenvalues of f. Hence $g^{-1}(u) = \frac{1}{1+a\langle v|u\rangle}\, u = (1 - \frac{a\langle v|u\rangle}{1+a\langle v|u\rangle})\, u$ et $g^{-1}(y) = y$
and $\forall x \in \mathcal{X}$ with $1 + a\langle v|x\rangle \neq 0$, $g^{-1}(x) = x - \frac{a\langle v|x\rangle}{1+a\langle v|u\rangle}\, u$ $\square$

7.3.6 Matrix Formulas

Let $\mathcal{X}$ be a Euclidean vector space where the scalar product is denoted $\langle\cdot|\cdot\rangle_X$. Let $(\epsilon_i)_{i\in I}$ be a basis of $\mathcal{X}$ with $q_{ii'} = \langle\epsilon_i|\epsilon_{i'}\rangle_X$ $(i, i' \in I)$ and let $\mathbf{Q} = [q_{ii'}]$ be the $I \times I$-matrix of scalar products between the basis vectors $(\epsilon_i)_{i\in I}$.

Scalar product

Let $\mathbf{x} = [x_i]$ and $\mathbf{y} = [y_i]$ be the I-columns of coordinates of vectors x and y over basis $(\epsilon_i)_{i\in I}$ of $\mathcal{X}$. The scalar product between x and y writes:

$$\langle x|y\rangle_X = \sum_{i\in I}\sum_{i'\in I} x_i q_{ii'} y_{i'} = \mathbf{x}^\top\mathbf{Q}\mathbf{y} = \mathbf{y}^\top\mathbf{Q}\mathbf{x}$$

The scalar product[6] is invariant, whereas its matrix representation is not. If $\mathcal{X}$ is referred to a basis $(\epsilon'_i)_{i\in I}$ with $q'_{ii'} = \langle\epsilon'_i|\epsilon'_{i'}\rangle_X$ and if $\mathbf{P}$ is the change of basis matrix from $(\epsilon_i)_{i\in I}$ to $(\epsilon'_i)_{i\in I}$, recall that (§7.2.4, p. 229) the matrix $\mathbf{Q}' = [q'_{ii'}])$ is such that

$$\mathbf{Q}' = \mathbf{P}^\top\mathbf{Q}\mathbf{P}$$

Adjoint homomorphisms

Let $\mathcal{Y}$ be a Euclidean vector space where the scalar product is denoted $\langle\cdot|\cdot\rangle_Y$, let $(\delta_j)_{j\in J}$ be a basis of $\mathcal{Y}$ and let $\mathbf{R} = [r_{jj'}]$ be the $J \times J$-matrix of scalar products between the basis vectors $(\delta_j)_{j\in J}$ of $\mathcal{Y}$: $r_{jj'} = \langle\delta_j|\delta_{j'}\rangle_Y$.

Let f be a homomorphism $\mathcal{X} \to \mathcal{Y}$. If $\mathbf{M} = [m_{ji}]$ is the matrix representation of homomorphism f with respect to bases $(\epsilon_i)_{i\in I}$ of $\mathcal{X}$ and $(\delta)_{j\in J}$ of $\mathcal{Y}$ (see p. 227), the matrix representation of the *adjoint homomorphism* $f^* : \mathcal{Y} \to \mathcal{X}$ (see definition p. 231) is $\mathbf{M}^* = \mathbf{Q}^{-1}\mathbf{M}^\top\mathbf{R}$, hence $\mathbf{Q}\,\mathbf{M}^* = \mathbf{M}^\top\mathbf{R}$.

Symmetric endomorphisms and bilinear forms

If $\mathbf{S}$ is the matrix representation of the symmetric endomorphism $s : \mathcal{X} \to \mathcal{X}$ with respect to the basis $(\epsilon_i)_{i\in I}$ of $\mathcal{X}$, one has $\mathbf{Q}^{-1}\mathbf{S}^\top\mathbf{Q} = \mathbf{S}$, i.e., $\mathbf{S}^\top\mathbf{Q} = \mathbf{Q}\mathbf{S}$. In general, the matrix of a symmetric endomorphism is not symmetric; it is symmetric when the basis of $\mathcal{X}$ is orthonormal, and then $\mathbf{Q} = \mathbf{I}$ and $\mathbf{S} = \mathbf{S}^\top$.

The matrix of the bilinear form $(x, x') \mapsto \langle x|s(x')\rangle_X$ is $\mathbf{Q}\mathbf{S} = \mathbf{M}^\top\mathbf{R}\mathbf{M}$ (symmetric matrix).

Diagonalization

Let $\mathcal{L} = s(\mathcal{X})$ be the eigenspace of s associated with non-null eigenvalues and let $(\alpha_\ell)_{\ell=1,\dots L}$ be an orthonormal basis of $\mathcal{L}$ consisting of eigenvectors of s. The matrix representation of the restriction of the endomorphism s to $\mathcal{L}$ is

---

[6]The matrix product $\mathbf{x}^\top\mathbf{y} = \mathbf{y}^\top\mathbf{x}$ is often called the "scalar product" of column-vectors $\mathbf{x}$ and $\mathbf{y}$: this terminology makes sense in terms of the elementary scalar product on $\mathbb{R}^n$. It becomes confusing whenever non-orthonormal bases are involved.

the $L{\times}L$ diagonal matrix $\mathbf{\Lambda} = [\lambda_\ell]$ (hence the name "diagonalization" given to the spectral theorem).

Let $\mathbf{A} = [a_{i\ell}]$ be the $I{\times}L$-matrix of the coordinates of the L orthonormal eigenvectors $(\alpha_\ell)_{\ell=1,\ldots L}$ of s over the basis $(\epsilon_i)_{i\in I}$ ($\alpha_\ell = \sum_i a_{i\ell}\epsilon_i$). Letting $\mathbf{S}$ be the $I{\times}I$-matrix representing endomorphism s with respect to the basis $(\epsilon_i)_{i\in I}$, the spectral decomposition of s leads to the following matrix expression:

$$\mathbf{SA} = \mathbf{A\Lambda} \quad \text{with} \quad \mathbf{A}^\top\mathbf{QA} = \mathbf{I}_L$$

or equivalently $\mathbf{A}^\top\mathbf{QSA} = \mathbf{\Lambda}$.

If the basis $(\epsilon_i)_{i\in I}$ is orthonormal, that is, $\mathbf{Q} = \mathbf{I}$, the matrix $\mathbf{S}$ is symmetric and the foregoing formulas reduce to the classical *diagonalization of a symmetric matrix*: $\mathbf{S} = \mathbf{A\Lambda A}^\top$ with $\mathbf{A}^\top\mathbf{A} = \mathbf{I}_L$.

Adding an endomorphism of rank one

If $\mathbf{F}$ denotes the matrix of $f : \mathcal{X} \to \mathcal{X}$ and $\mathbf{G_u}$ that of $g_u : x \to f(x) + \langle u|x\rangle u$ with respect to an orthonormal basis of $\mathcal{X}$, letting $\mathbf{u}^\top\mathbf{F}_u^{-1}\mathbf{u} = |u|^2$, one has:

$$\mathbf{G}_u = \mathbf{F} + \mathbf{uu}^\top; \quad \mathbf{G}_u^{-1} = \mathbf{F}^{-1} - \frac{\mathbf{F}^{-1}\mathbf{uu}^\top\mathbf{F}^{-1}}{1 + |u|^2}; \quad \mathbf{u}^\top\mathbf{G}_u^{-1}\mathbf{u} = \frac{|u|^2}{1 + |u|^2}$$

Here one finds the well-known *Sherman–Morrisson–Woodbury formulas* after a perturbation of rank one of matrix $\mathbf{F}$.

7.4   Multidimensional Geometry

Multidimensional geometry is the extension of plane and three-dimensional geometry (taught in high school) to higher finite-dimensional spaces. The primitive concepts of multidimensional geometry are basically those of elementary geometry, namely points, lines, planes, geometric vectors, etc. Multidimensional geometry enables us to carry over both the spatial intuition and the rigorous coordinate-free mode of reasoning proper to *pure geometry*.

Multidimensional geometry involves *affine properties*, dealing with lines, parallelism, etc., (roughly speaking, affine geometry is the study of properties invariant under affine bijections, the most basic result is the theorem of Thales) and *Euclidean properties*, dealing with Euclidean distances, orthogonality, angles (with the famous Pythagorean theorem), etc.

On the one hand, from a *mere mathematical standpoint*, going from vector spaces to geometry is a small step, because the underlying structures of multidimensional geometry are basically those of linear algebra. This presumably explains the limited coverage of multidimensional geometry in textbooks; see, however, Mac Lane and Birkhoff (1967, Chapter 12), Berger (2009, Chapters

2 & 3), Coxeter (1980), Suppes et al. (1989, Vol. 2, Chapter 12), Benzécri (1992, Chapters 9 and 10) and Gallier (2011, Chapter 20).

On the other hand, from a *conceptual standpoint*, going from vector spaces to geometry is a big step, because it establishes the specificity of geometric concepts with respect to vectorial ones, and first and foremost the notions of points and distances between points[7].

> **Notations**. The familiar *geometric notation* (see p. 10) can be preserved: capital letters (M, P, A, etc.) for points, and "arrowed" letters for *geometric vectors* ($\vec{v}$, $\overrightarrow{MN}$, etc.).

In this section, we first characterize the affine space (§7.4.1) and discuss barycentre in some detail (§7.4.2). Then we outline the fundamental notions of an affine subspace, namely Cartesian frame (§7.4.3), and affine map (§7.4.4). Finally, we review geometric properties (§7.4.5).

7.4.1 Affine Space

Definition 7.9. *An* affine space *is a nonempty set of points $\mathcal{U}$ for which there exists a finite-dimensional vector space $\mathcal{V}$ (over $\mathbb{R}$) and a map $\mathcal{V} \times \mathcal{U}$ into $\mathcal{U}$ with $(\vec{v}, M) \mapsto M + \vec{v}$ such that*

1. for all vectors $\vec{v_1}, \vec{v_2} \in \mathcal{V}$ and all points $M \in \mathcal{U}$,
$$(M + \vec{v_1}) + \vec{v_2} = M + (\vec{v_1} + \vec{v_2});$$

2. for all points $M \in \mathcal{U}$, $M + \vec{0} = M$;

3. for any two points (M, N), there is exactly one vector $\vec{v} \in \mathcal{V}$ with $N = M + \vec{v}$.

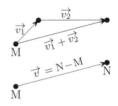

The elements of the affine space $\mathcal{U}$ are called *points*; the vector space $\mathcal{V}$ is called *vector space associated with $\mathcal{U}$*; the *dimension* of the affine space $\mathcal{U}$ is that of the vector space $\mathcal{V}$.

The vector associated with the bipoint (M,N) is denoted $\overrightarrow{MN}$ as the deviation (vectorial difference) from point M to point N, or (Grassmann's notation) $N - M$ ("terminal minus initial").

> **Remark.** One should be careful about the overloading of the addition symbol $+$. Addition is well defined on vectors, as in $\vec{v_1} + \vec{v_2}$; the translate $M + \vec{v}$ of a point $M \in \mathcal{U}$ by a vector $\vec{v} \in \mathcal{V}$ is also well defined, but the *addition of points M + P does not make sense.*
>
> In this respect, Grassmann's notation $N - M$ is somewhat confusing, since it suggests that points can be subtracted (but not added!).

---

[7]It is to emphasize the specificity of geometry that we have defined the Euclidean structure on a vector space not in terms of distance but merely in terms of norm.

Chasles's identity: $\forall A, B, C \in \mathcal{U}, \ \overrightarrow{AC} = \overrightarrow{AB} + \overrightarrow{BC}.$

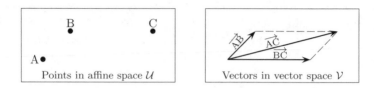

Figure 7.3
Illustration of the Chasles's identity.

The back and forth process between affine space and vector space is conveyed by the following two properties.

1. Any vector space $\mathcal{V}$ has an affine space structure specified by choosing $\mathcal{U} = \mathcal{V}$ with $\mathcal{V}$ as associated vector space, and by letting $+$ be addition in the vector space $\mathcal{V}$.

2. Any n-dimensional affine space $\mathcal{U}$ can be embedded in an $(n+1)$-dimensional vector space $\mathcal{V}^{\#}$.

These properties enable us, starting with a numerical vector space, to represent its vectors as points in an affine space $\mathcal{U}$, then as geometric vectors in the vector space $\mathcal{V}$ associated with $\mathcal{U}$, or alternatively in the extended vector space $\mathcal{V}^{\#}$ (as in projective geometry).

7.4.2 Barycentres

A fundamental concept for a vector space is that of linear combination of vectors; the corresponding one for affine space is that of barycentre (or affine combination) of points.

Proposition 7.4. *Given an affine space $\mathcal{U}$, let $(M^i)_{i \in I}$ be a family of points in $\mathcal{U}$, and let $(\lambda_i)_{i \in I}$ be a family of scalars. Given a point P in $\mathcal{U}$, the following properties hold:*

1. *If $\sum \lambda_i \neq 0$, then $\sum \lambda_i \overrightarrow{PM^i}$ defines a point G independent of P, called* barycentre, *such that $(\sum \lambda_i)\overrightarrow{PG} = \sum \lambda_i \overrightarrow{PM^i}$.*

2. *If $\sum \lambda_i = 0$, then $\sum \lambda_i \overrightarrow{PM^i}$ defines a vector independent of P.*

Proof. Given a point $P' \neq P$ in $\mathcal{U}$. By Chasles's identity, one has $\sum_i \lambda_i \overrightarrow{P'M^i} = (\sum \lambda_i)\overrightarrow{P'P} + \sum_i \lambda_i \overrightarrow{PM^i}$.

1) If $\sum \lambda_i \neq 0$, $\sum_i \lambda_i \overrightarrow{PM^i} = (\sum_i \lambda_i)\overrightarrow{PG}$, hence
$$\sum_i \lambda_i \overrightarrow{P'M^i} = (\sum \lambda_i)\overrightarrow{P'P} + (\sum_i \lambda_i)\overrightarrow{PG} = (\sum_i \lambda_i)(\overrightarrow{P'P} + \overrightarrow{PG}) = (\sum_i \lambda_i)\overrightarrow{P'G}.$$
2) If $\sum \lambda_i = 0$, one has $\sum_i \lambda_i \overrightarrow{P'M^i} = \sum_i \lambda_i \overrightarrow{PM^i}$. □

When $\sum \lambda_i \neq 0$, the barycentre G can be written $G = (\sum \lambda_i M^i)/(\sum \lambda_i)$, as the weighted average of points $(M^i)_{i \in I}$, or $G = P + (\sum \lambda_i \overrightarrow{PM^i})/(\sum \lambda_i)$.

When $\sum \lambda_i = 0$, $\vec{e} = \sum \lambda_i \overrightarrow{\text{PM}^i}$ is denoted $\sum \lambda_i \text{M}^i$, as the effect of contrast $\lambda_I = (\lambda_i)_{i \in I}$ among points $(\text{M}^i)_{i \in I}$; this notation extends Grassmann's notation for the difference between two points to a family of points.

Particular case: $I = \{i, i'\}$. For $\lambda_i = \lambda_{i'}$, G is the *midpoint* of $(\text{M}^i, \text{M}^{i'})$: $\text{G} = (\text{M}^i + \text{M}^{i'})/2$. For $\lambda_i = -1$ and $\lambda_{i'} = 1$, one gets the *deviation-vector* from point M^i to point $\text{M}^{i'}$: $\text{M}^{i'} - \text{M}^i = \overrightarrow{\text{M}^i\text{M}^{i'}}$.

The geometric construction of the barycentre of the two weighted points $(\text{M}^i, 3)$ and $(\text{M}^{i'}, 2)$ is exemplified on the figure on the right.

Corollary 7.3 (Barycentric property). *The weighted sum of the deviation-vectors from barycentre* G *to points* $(\text{M}^i)_{i \in I}$ *is the null vector:* $\sum \lambda_i \overrightarrow{\text{GM}^i} = \vec{0}$.

To sum up: An average of points is a point (barycentre); a contrast among points is a vector. These basic operations on points in an affine space are the extension of those on points onto a line or in a plane.

7.4.3 Affine Subspaces, Affine and Cartesian Frames

A vector subspace can be characterized as a non-empty subset of vectors that is stable with respect to linear combinations. In affine spaces, the corresponding notion is that of affine subspace that is stable with respect to barycentres.

Definition 7.10 (Affine subspace). *A subset* $\mathcal{M}$ *of* $\mathcal{U}$ *is an* affine subspace *if every barycentre of points in* $\mathcal{M}$ *is a point in* $\mathcal{M}$.

Given point $\text{P}_0 \in \mathcal{M}$, the vectors $\overrightarrow{\text{P}_0\text{P}}$, with $\text{P} \in \mathcal{M}$, generate a vector subspace independent of P_0, which is called the *direction* of the affine subspace.

The dimension of an affine subspace is that of its direction vector subspace. A one-dimensional affine subspace is called a *line*, a two-dimensional subspace is called a *plane*. An affine subspace whose dimension is equal to that of the whole space minus 1 (for instance a plane in a three-dimensional affine space) is called a *hyperplane*.

The direction of a line is a one-dimensional vector subspace: any vector $\vec{a}$ generating this vector subspace is called a *direction vector* of the line; the line is said to be of direction $\vec{a}$. A direction vector also provides an *orientation* of the line, that is, it defines an oriented line that is called *axis*.

An L-dimensional affine subspace $\mathcal{M}$ of $\mathcal{U}$ can be equipped with two sorts of frames: an *affine frame* or a *Cartesian frame*.

Definition 7.11 (Affine frame). *A family of points* $(\text{M}_\ell)_{\ell=1,...L+1}$ *is an* affine frame *of the affine subspace* $\mathcal{M}$ *if every point* P *in this subspace can be expressed in a unique way as a barycentre of points* $(\text{M}_\ell)_{\ell=1,...L+1}$.

$$\text{P} = \sum_{\ell=1}^{L+1} \varpi_\ell \text{M}_\ell \quad \text{with} \quad \sum_{\ell=1}^{L+1} \varpi_\ell = 1$$

The numbers ϖ_ℓ are called the *barycentric coordinates* of point P with respect to the affine frame $(M_\ell)_{\ell=1,\ldots L+1}$.

An affine frame consists in $L+1$ points for an L-multidimensional subspace, 3 points for a plane, 2 points for a line, etc.

Definition 7.12 (Cartesian frame). *A Cartesian frame of an L-dimensional subspace $\mathcal{M} \subseteq \mathcal{U}$ consists of an origin point $O \in \mathcal{M}$ together with a basis of L vectors[8] $(\vec{\alpha}_\ell)_{\ell \in L}$ of the direction of $\mathcal{M}$.*

By definition, the *Cartesian coordinates* of point $P \in \mathcal{M}$ with respect to the Cartesian frame $(O, (\vec{\alpha}_\ell)_{\ell \in L})$ are those of vector $\overrightarrow{OP}$ over the basis $(\vec{\alpha}_\ell)_{\ell \in L}$.

$$\overrightarrow{OP} = \sum x_\ell \vec{\alpha}_\ell$$

7.4.4   Affine Maps

Linear maps are characterized as maps preserving linear combinations. In affine spaces, the corresponding notion is that of affine maps that are defined as maps preserving barycentres (affine combinations).

Definition 7.13. *Given two affine spaces $\mathcal{U}$ and $\mathcal{U}'$, a map $f : \mathcal{U} \longrightarrow \mathcal{U}'$ is an affine map if and only if for every family $(M^i, \varpi_i)_{i \in I}$ of weighted points in $\mathcal{U}$ such that $\sum \varpi_i = 1$, we have*

$$f(\sum_{i \in I} \varpi_i M^i) = \sum_{i \in I} \varpi_i f(M^i)$$

There is a close connection between affine maps between $\mathcal{U}$ and $\mathcal{U}'$ and linear maps between their direction vector spaces $\mathcal{V}$ and $\mathcal{V}'$. More precisely, we have the following proposition.

Proposition 7.5. *Given an affine map $f : \mathcal{U} \to \mathcal{U}'$, there is a linear map $l : \mathcal{V} \to \mathcal{V}'$ such that*

$$\forall M \in \mathcal{U}, \forall \vec{v} \in \mathcal{V},\ f(M + \vec{v}) = f(M) + l(\vec{v})$$

An affine map is determined by the image of a point together with the associated linear map.

- *Homothety* (or central dilatation). Given any point $P \in \mathcal{U}$, and any scalar $\lambda \in \mathbb{R}$, a homothety with centre P and scale factor λ is the map $h_{P\lambda}$ defined by $h_{P\lambda} : M \mapsto P + \lambda \overrightarrow{PM}$ $(M \in \mathcal{U})$.

  One has the property $h_{P\lambda}(N) - h_{P\lambda}(M) = \lambda \overrightarrow{MN}$ (Thales' theorem).

- A *symmetry* with centre P, denoted s_P, is a homothety with scale factor $\lambda = -1$: $\forall M \in \mathcal{U}, s_P : M \mapsto P - \overrightarrow{PM}$.

  If $s_P(M) = N$, then $\overrightarrow{PM} = -\overrightarrow{PN}$.

---

[8]According to our general convention, L denotes both a finite set and its cardinality.

- *Translation.* Given a vector $\vec{v}$ of $\mathcal{V}$, a translation by vector $\vec{v}$ is the affine map, denoted $t_{\vec{v}}$, defined by $\forall M \in \mathcal{U}, t_{\vec{v}} : M \mapsto M + \vec{v}$. The associated linear map is the identity map of $\mathcal{V}$, since $\forall M, N \in \mathcal{U}, t_{\vec{v}}(M) = N + \vec{v} + \overrightarrow{MN} = t_{\vec{v}}(N) + \overrightarrow{MN}$.

  One has the property: if $M' = t_{\vec{v}}(M)$ and $N' = t_{\vec{v}}(N)$, then $\overrightarrow{M'N'} = \overrightarrow{MN}$.

7.4.5 Geometric Space

The Euclidean notions and properties of a multidimensional geometric space follow from those of a vector space (see p. 229).

Let $\mathcal{U}$ be a K-dimensional affine space and let $\mathcal{V}$ be the vector space over $\mathbb{R}$ associated with $\mathcal{U}$, now with the scalar product denoted $\langle \cdot | \cdot \rangle$, and the associated norm denoted $\| \cdot \|$. The space $\mathcal{U}$ becomes a *Euclidean space* if the *distance* between two points M and P in $\mathcal{U}$, denoted MP, is defined as the norm of geometric vector $\overrightarrow{MP}$: $MP = \|\overrightarrow{MP}\|$.

The *Euclidean distance* verifies the *properties of a distance* (metric space):

(1) *Positivity.* $\forall M, P \in \mathcal{U}$: $MP \geq 0$ *with* $MP = 0 \iff M = P$

(2) *Symmetry.* $\forall M, P \in \mathcal{U}$: $MP = PM$

(3) *Triangle inequality.* $\forall M, P, Q \in \mathcal{U}$: $MQ \leq MP + PQ$
with $MQ = MP + PQ \iff M, P, Q$ *are aligned with* P *between* M *and* Q.

For a one-dimensional space (geometric line), the Euclidean structure is canonically induced by the affine structure; for a multidimensional space, the *Euclidean distance must be specified*. A Euclidean distance induces the notion of *angle* between geometric vectors, as in elementary geometry.

Pythagorean theorem: $\forall M, P, Q \in \mathcal{U}$ one has:

$$\overrightarrow{MP} \perp \overrightarrow{PQ} \iff MP^2 + PQ^2 = MQ^2$$

7.5 Orthogonal Projection

In elementary geometry one defines the *orthogonal projection* of a point onto a line or a plane; the extension of this notion is the orthogonal projection of a point onto a subspace.

Definition 7.14. *Let* $\mathcal{M}$ *be a subspace of a geometric space; for any point* $P \notin \mathcal{M}$; *there exists a unique point* $P' \in \mathcal{M}$ *such that* PP' *is orthogonal to* $\mathcal{M}$. *Point* P' *is called the orthogonal projection (or projected point) of point* P *onto* $\mathcal{M}$.

By definition, if $P \in \mathcal{M}$, its projected point on $\mathcal{M}$ is P itself.

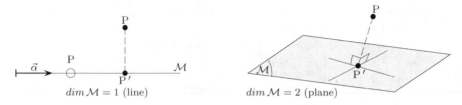

Figure 7.4
Orthogonal projection of point P onto a line, onto a plane.

Proposition 7.6 (Projection of a point onto a line). *Let $\mathcal{D}$ be the line passing through point O and with direction vector $\vec{\alpha}$, the orthogonal projection of point P onto $\mathcal{D}$ is the point P′ such that*

$$\mathrm{P}' = \mathrm{O} + \frac{\langle \overrightarrow{\mathrm{OP}} | \vec{\alpha} \rangle}{\|\vec{\alpha}\|^2}\, \vec{\alpha}$$

Definition 7.15. *Given a subspace $\mathcal{M}$ of $\mathcal{U}$, if P′ is the orthogonal projection of point P onto $\mathcal{M}$, an orthogonal affinity of basis $\mathcal{M}$ and ratio λ (with $\lambda \neq 0$) is the map such that $\mathrm{P}'' \mapsto \mathrm{P}' + \lambda\, \overrightarrow{\mathrm{P}'\mathrm{P}}$.*

7.5.1 Ellipsoid

The hyperellipsoid—for short, ellipsoid—is the multidimensional extension of the two-dimensional ellipse and the three-dimensional ellipsoid.

Ellipse

With respect to the orthonormal frame $(\mathrm{O}, \vec{\delta}_1, \vec{\delta}_2)$ of the plane, the Cartesian equation $x^2/a^2 + y^2/b^2 = 1$ (where $a \geq b > 0$) defines an ellipse with centre O, and whose lengths of principal semi-axes are equal to a (principal major axis) and b (principal minor axis).

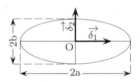

Figure 7.5
Ellipse in orthonormal frame (with $a = 1.5$, $b = 0.6$).

Definition 7.16. *Let $(\mathrm{O}, (\vec{\delta_k})_{k \in K})$ be an orthonormal Cartesian frame of $\mathcal{U}$ and $(a_k)_{k \in K}$ a family of strictly positive real numbers. The Cartesian equation $\sum x_k^2/a_k^2 = 1$ defines the K-dimensional ellipsoid $\mathcal{E}$ with centre O and principal axes $(\vec{\delta_k})_{k \in K}$ whose half-lengths are equal to $(a_k)_{k \in K}$.*

Consider the endomorphism f of $\mathcal{V}$ defined by $\forall k \in K, f(\vec{\delta_k}) = \vec{\delta_k}/a_k^2$, the ellipsoid $\mathcal{E}$ is the locus of points P such that $\langle f(\overrightarrow{\mathrm{OP}}) | \overrightarrow{\mathrm{OP}} \rangle = 1$.

The hyperplane with equation $\sum_{k \in K} u_k x_k = u_0$ is a *tangent hyperplane* to the ellipsoid $\mathcal{E}$ if and only if coefficients $(u_k)_{k \in K}$ satisfy the following equation: $\sum_{k \in K} a_k^2 u_k^2 = u_0^2$ (tangential equation of the ellipsoid).

Dimensional considerations and homogeneity.

In classical geometry, the distance between points is a *length* whose numerical expression depends on the length taken as a unit. Thus $3\,\text{cm} = 30\,\text{mm}$: with a unit ten times smaller, the numerical expression of length is multiplied by ten, then the length is a magnitude of *dimension 1*. The product of two lengths has numerical expression multiplied by 100, and is as such of *dimension 2* (area). The ratio of two lengths is invariant, that is, it is "without dimension".

Pure geometric properties do not depend on unit length; the terms involved in such properties are of identical dimensions and are said to be *dimensionally homogeneous*. For instance, in Pythagorean theorem, both sides of the "=" sign are of dimension 2. In pure geometry, non-homogeneous relations (such as $AB^2 = AC$) are meaningless.

The pure geometry concept extends to multidimensional geometry, assimilating the Euclidean distance with a length. A geometric vector will be called *unit vector* if its length is equal to the distance unit. A Cartesian frame of a geometric space will be called *orthonormal* if its basis vectors are orthogonal and their norms are equal to the distance unit. Dividing a vector by its norm enables one to state pure geometric properties.

In all cases, translating the geometric construction into matrix notation is an easy task, whereas the converse "translation"—that is, starting from matrix formulas, deciphering the rationale of the procedure—is a real brainteaser. There is often a haze of vagueness and arbitrariness around the choice of coordinates in graphical displays stemming from GDA and related methods; this haze is fostered by the matrix approach to statistics, which is simply not powerful enough to cope with geometric structures.

Bibliography

Anderson, M. and Ter Braak, C. (2003). Permutation tests for multi-factorial analysis of variance. *Journal of Statistical Computation and Simulation*, 73(2):85–113.

Anderson, T. (2003). *An Introduction to Multivariate Statistical Analysis*. New York: Wiley, 3rd edition.

Beh, E. J. and Lombardo, R. (2014). *Correspondence Analysis: Theory, Practice and New Strategies*. John Wiley & Sons.

Benzécri, F. & J.-P. (1984). *Pratique de L'Analyse des Données. 1 analyse des correspondances, exposé élémentaire*. Paris: Dunod.

Benzécri, J.-P. (1992). *Correspondence Analysis Handbook*. Dekker: New York. (adapted from J.-P. & F. Benzécri, Paris: Dunod, 1984).

Benzécri, J.-P. & *coll*. (1973). *L'Analyse des Données. 1 Taxinomie, 2 L'analyse des correspondances*. Paris: Dunod.

Berger, M. (2009). *Geometry I*. Berlin: Springer-Verlag.

Berry, K. J., Johnston, J. E., and Mielke Jr, P. W. (2014). *A Chronicle of Permutation Statistical Methods: 1920–2000, and Beyond*. New York: Springer.

Bienaise, S. (2013). *Méthodes d'inférence combinatoire sur un nuage euclidien/Etude statistique de la cohorte EPIEG*. PhD thesis, Université Paris Dauphine, CEREMADE.

Bienaise, S. and Le Roux, B. (2017). Combinatorial typicality test in geometric data analysis. *Statistica Applicata – Italian Journal of Applied Statistics*, 29(2-3):331–348.

Bourdieu, P. (2000). *Les structures sociales de l'économie*. Paris: Le Seuil.

Cohen, J. (1977). *Statistical Power Analysis for the Behavioral Sciences* (revised ed.). New York: Academic Press.

Cox, D. and Hinkley, D. (1974). *Theoretical Statistics*. London: Chapman & Hall/CRC.

Coxeter, H. (1961/1980). *Introduction to Geometry*. New York: Wiley.

Cramér, H. (1946). *Mathematical Methods of Statistics*. Princeton: Princeton University Press.

Daudin, J., Duby, C., and Trecourt, P. (1988). Stability of principal component analysis studied by the bootstrap method. *Statistics: A journal of Theoretical and Applied Statistics*, 19(2):241–258.

Davison, A. C. and Hinkley, D. V. (1997). *Bootstrap Methods and Their Application*, volume 1. Cambridge University Press.

de Leeuw, J. and van der Burg, E. (1986). The permutational limit distribution of generalized canonical correlations. In Diday, E., editor, *Data Analysis and Informatics*, volume IV, pages 509–521. North-Holland.

Dwass, M. (1957). Modified randomization tests for nonparametric hypotheses. *The Annals of Mathematical Statistics*, 28(1):181–187.

Edgington, E. (2007). *Randomization Tests*. London: Chapman & Hall/CRC, fourth edition.

Efron, B. and Tibshirani, R. J. (1993). *An Introduction to the Bootstrap*. London: Chapman & Hall/CRC.

Ferrandez, A.-M. and Blin, O. (1991). A comparison between the effect of intentional modulations and the action of L–Dopa on gait in Parkinson's disease. *Behavioural Brain Research*, 45:177–183.

Fisher, R. A. (1925). *Statistical Methods for Research Workers*. Edinburg: Oliver & Boyd.

Fisher, R. A. (1935). *The Design of Experiments*. Edinburg: Oliver & Boyd.

Freedman, D. and Lane, D. (1983). A nonstochastic interpretation of reported significance levels. *Journal of Business & Economic Statistics*, 1(4):292–298.

Freedman, D., Pisani, R., Purves, R., and Adhikari, A. (1991). *Statistics*. New York: W.W. Norton & Co, 2 edition.

Gabriel, K. and Hall, W. (1983). Rerandomization inference on regression and shift effects: Computationally feasible methods. *Journal of the American Statistical Association*, 78(384):827–836.

Gallier, J. (2011). *Geometric Methods and Applications: For Computer Science and Engineering*. Berlin: Springer.

Gifi, A. (1990). *Nonlinear Multivariate Analysis*. London: Wiley.

Gilula, Z. and Haberman, S. J. (1986). Canonical analysis of contingency tables by maximum likelihood. *Journal of the American Statistical Association*, 81(395):780–788.

Godement, R. (1966). *Algebra*. Boston: Houghton Mifflin. (translated from *Cours d'algèbre*, Paris: Hermann).

Good, P. (2011). *A Practitioner's Guide to Resampling for Data Analysis, Data Mining, and Modeling*. New York: Chapman & Hall/CRC.

Gower, J. and Hand, D. (1996). *Biplots*. London: Chapman & Hall/CRC.

Greenacre, M. (1984). *Theory and Applications of Correspondence Analysis*. London: Academic Press.

Greenacre, M. (2016). *Correspondence Analysis in Practice*. London: Chapman and Hall/CRC.

Halmos, P. (1958). *Finite Dimensional Vector Spaces*. New York: Van Nostrand.

Hotelling, H. (1951). A generalized T test and measure of multivariate dispersion. In Neyman, J., editor, *Proceedings of the Second Berkeley Symposium on Mathematical Statistics and Probability*, pages 23–41. University of California Press.

Johnston, J. E., Berry, K. J., and Mielke, P. W. (2007). Permutation tests: Precision in estimating probability values. *Perceptual and Motor Skills*, 105(3):915–920.

Lafon, P. (1984). *Dépouillements et Statistiques en Lexicométrie*. Paris: Slatkine–Champion.

Le Roux, B. (1998). Inférence combinatoire en analyse géométrique des données. *Mathématiques et sciences humaines*, 144:5–14.

Le Roux, B. (1999). Analyse spécifique d'un nuage euclidien: application à l'étude des questionnaires. *Mathématiques et sciences humaines*, 146:6–83.

Le Roux, B. (2014a). *Analyse géométrique des données multidimensionnelles*. Paris: Dunod.

Le Roux, B. (2014b). Structured data analysis. In Blasius, J. and Greenacre, M., editors, *Visualization and Verbalization of Data*, chapter 12. London: Chapman & Hall/CRC.

Le Roux, B. and Rouanet, H. (1984). L'analyse multidimensionnelle des données structurées. *Mathématiques et sciences humaines*, 85:5–18.

Le Roux, B. and Rouanet, H. (2004). *Geometric Data Analysis: From Correspondence Analysis to Structured Data Analysis (Foreword by P. Suppes)*. Dordrecht: Kluwer.

Le Roux, B. and Rouanet, H. (2010). *Multiple Correspondence Analysis, 163*. QASS. Thousand Oaks (CA): Sage Publications.

Lebaron, F. and Dogan, A. (2016). Do central bankers' biographies matter? *Sociologica*, 10(2):1–37.

Lebart, L. (1976). The significancy of eigenvalues issued from correspondence analysis of contingency tables. *COMPSTAT 1976-Proceedings in computational statistics*.

Lebart, L. (2007). Which bootstrap for principal axes methods? In *Selected Contributions in Data Analysis and Classification*, pages 581–588. Springer.

Lebart, L., Morineau, A., and Piron, M. (1995/2006). *Statistique exploratoire multidimensionnelle*. Paris: Dunod.

Lunneborg, C. E. (2000). *Data Analysis by Resampling: Concepts and Applications*. Brooks/Cole.

Mac Lane, S. and Birkhoff, G. (1967). *Algebra*. New York: Macmillan.

Malinvaud, E. (1981). *Méthodes statistiques de l'économétrie*. Paris: Dunod.

Murtagh, F. (2005). *Correspondence Analysis and Data Coding with Java and R*. London: Chapman & Hall/CRC.

Pagès, J. (2014). *Multiple Factor Analysis by Example Using R*. London: Chapman and Hall/CRC.

Pearson, K. (1901). On lines and planes of closest fit to systems of points in space. *The London, Edinburgh, and Dublin Philosophical Magazine and Journal of Science*, 2:559–572.

Pesarin, F. (2001). *Multivariate Permutation Tests: With Applications in Biostatistics*. Chichester: Wiley.

Pesarin, F. and Salmaso, L. (2010). *Permutation Tests for Complex Data: Theory, Applications and Software*. Chichester: Wiley.

Pitman, E. J. (1937). Significance tests which may be applied to samples from any populations. *Journal of the Royal Statistical Society*, 4:119–130.

Rao, C. (1973). *Linear Statistical Inference and its Applications*. New York: Wiley.

Richard, J.-F. (2004). *Cours de psychologie: processus et applications*, volume 6, chapter 3. Paris: Dunod.

Romano, J. P. (1989). Bootstrap and randomization tests of some nonparametric hypotheses. *The Annals of Statistics*, 17:141–159.

Rouanet, H. (1996). Bayesian methods for assessing importance of effects. *Psychological Bulletin*, 119(1):149.

Rouanet, H., Bernard, J.-M., Bert, M.-C., Lecoutre, B., Lecoutre, M.-P., and Le Roux, B. (1998). *New Ways in Statistical Inference: From Significance Tests to Bayesian Methods (Foreword by P. Suppes)*. Bern: Peter Lang.

Rouanet, H., Bernard, J.-M., and Le Roux, B. (1990). *Analyse inductive des données*. Paris: Dunod.

Rouanet, H., Bernard, J.-M., and Lecoutre, B. (1986). Nonprobabilistic statistical inference: A set-theoretic approach. *The American Statistician*, 49:60–65.

Rouanet, H., Lebaron, F., Le Hay, V., Ackermann, W., and Le Roux, B. (2002). Régression et Analyse Géométrique des Données. *Mathématiques et sciences humaines*, 160:13–45.

Rouanet, H. and Lecoutre, B. (1983). Specific inference in anova: From significance tests to bayesian procedures. *British Journal of Mathematical and Statistical Psychology*, 36(2):252–268.

Rouanet, H. and Lépine, D. (1970). Comparison between treatments in a repeated–measurement design: Anova and multivariate methods. *British Journal of Mathematical and Statistical Psychology*, 23:147–163.

Rozencwajg, P. (1994). *Stratégies de résolution de problèmes scolaires et différences individuelles*. PhD thesis, Université Paris Descartes.

Saporta, G. and Hatabian, G. (1986). Régions de confiance en analyse factorielle. *Data analysis and informatics*, IV:499–508.

Student (1908). The probable error of a mean. *Biometrika*, 6(1):1–25.

Suppes, P., Krantz, D., Luce, R., and Tversky, A. (1989). *Foundations of Measurement*, Volume II: *Geometrical, Threshold and Probabilistic Representations*. San Diego: Academic Press.

Tukey, J. W. (1977). *Exploratory Data Analysis*. Reading (Mass.): Addison–Wesley.

Weinberg, S. L., Carroll, J. D., and Cohen, H. S. (1984). Confidence regions for INDSCAL using the jackknife and bootstrap techniques. *Psychometrika*, 49(4):475–491.

Author Index

Subject Index

(Page numbers in slanted characters refer to places where notions are applied.)